IP over WDM

IP over WDM

Kevin H. Liu

QOptics Inc, Oregon, USA

JOHN WILEY & SONS, LTD

Other Wiley Editorial Offices

John Wiley & Sons Inc., 111 River Street, Hoboken, NJ 07030, USA

Jossey-Bass, 989 Market Street, San Francisco, CA 94103–1741, USA

Wiley-VCH Verlag GmbH, Boschstr. 12, D–69469 Weinheim, Germany

John Wiley & Sons Australia Ltd, 33 Park Road, Milton, Queensland 4064, Australia

John Wiley & Sons (Asia) Pte Ltd, 2 Clementi Loop 02–01, Jin Xing Distripark, Singapore 129809

John Wiley & Sons Canada Ltd, 22 Worcester Road, Etobicoke, Ontario, Canada M9W 1L1

British Library Cataloguing in Publication Data

A catalogue record for this book is available from the British Library

ISBN 0–470–84417–5

Typeset in 10.5/12pt Optima by Deerpark Publishing Services Ltd, Shannon, Ireland.
Printed and bound in Great Britain by TJ International, Padstow, Cornwall, UK.
This book is printed on acid-free paper responsibly manufactured from sustainable forestry in which at least two
trees are planted for each one used for paper production.

Contents

List of Figures

List of Tables

About the Author

Kevin H. Liu, Ph.D., is a lead software architect at QOptics, currently working on scalable common control plane for optical networks. He was a research scientist at Internet Architecture Research Lab, Telcordia Technologies (formerly Bell Communications Research or Bellcore). Before joining Bellcore, he held research and teaching positions at Rutgers University, New Brunswick, New Jersey, and Victoria University of Technology, Melbourne, Victoria, Australia. He has worked on several U.S. government DARPA-funded IP/WDM research projects such as the MONET (Multiwavelength Optical Networking) NC&M (Network Control and Management) project, the NGI (Next Generation Internet) SuperNet NC&M project, and the NGI Optical Label Switching project.

He has published over thirty journal and conference papers in the areas of network control and management, WDM optical networks, and distributed systems, and applied patent in IP/WDM networks. Among the publications, his articles have appeared in *IEEE Transactions on Communications, IEEE Journal on Selected Areas in Communications, IEEE Journal of Lightwave Technology, IEEE Network Magazine,* and *IEICE Transactions on Communications.* His biography has been included in the "Who's Who in the World" and the "International Who's Who of Information Technology".

The author can be reached by email at kliu@ieee.org.

Preface

It is widely believed that Internet Protocol (IP) provides the only convergence layer in the global and ubiquitous Internet. Above the IP layer, there are a great variety of IP-based services and appliances that are still evolving from its infancy. The inevitable dominance of IP traffic makes it apparent that the engineering practices of the network infrastructure should be optimised for IP. On the other hand, fibre optics as a dispersive technology revolutionises the telecom and networking industry by offering enormous network capacity to sustain next-generation Internet growth. WDM (Wavelength Division Multiplexing) as a fibre bandwidth exploring technology is state-of-the-art. Using WDM over existing fibre networks can increase network bandwidth significantly as well as maintain the same network operational footprint. It has been proved as a cost-efficient solution for long-haul networks.

As worldwide deployment of optical fibres and WDM technologies such as WDM components and control systems mature, WDM-based optical networks have been deployed not only in backbones but also in metro, regional, and access networks. WDM optical networks are no longer just point-to-point pipes providing physical transport services, but blend with a new level of network flexibility. Integrating IP and WDM to transport IP traffic over WDM-enabled optical networks efficiently and effectively becomes an urgent yet important task. As the WDM and IP over WDM industries emerge and grow, related industry forums have been established, such as OIF and ODSI, and a sub-IP Area including seven working groups has been formed in IETF, focusing on standardisation and IP/WDM inter-networking.

To our knowledge, this book is the first to focus on IP over WDM optical networks with in-depth discussion on IP/WDM network control and IP/WDM traffic engineering. It not only summarises the fundamental mechanisms and the recent development and deployment of WDM optical networks but also details both the network and the software architecture to implement WDM-enabled optical networks designed to transport IP traffic.

The book is targeted for the following audiences:

- Technical engineers and network practitioners, network designers and analysts, network managers and technical management personnel interested in the recent development, implementation, and deployment of IP/WDM networks.
- First-year graduate students or senior undergraduate students majoring in networking and/or network control and management.

A modest background is required to understand the material presented in this book. The reader is expected to have a basic understanding of computer networks and computer systems. Example introductory materials can be found in *Computer Network* [Tane96], *Computer Architecture* [Tane99], and *Operating Systems* [Nutt02].

IP/WDM networking is designed for transporting IP traffic in a WDM-enabled optical network, to leverage IP universal connectivity and the WDM massive bandwidth capacity. This book answers the following questions:

- What is a WDM network?
- What is IP over WDM?
- Why IP over WDM?
- What is the recent (research and commercial) development on WDM optical networks?
- How to interconnect and/or inter-network IP and WDM?
- How to control and manage IP/WDM networks?
- What is IP/WDM traffic engineering?
- What are the benefits of virtual topology reconfiguration?
- What are the IP/WDM network applications?
- What are the other specific issues related to IP/WDM networks?

As communication and computer networks are quickly converging, the conventional Telecom and IP network equipment, their control and management software, and their services and applications will inevitably interact and eventually integrate to support both data and voice traffic. On the one hand, the conventional telecom networks developed three switching modes: circuit switching such as the T1 transport network, packet switching such as X.25 and frame relay, and cell switching such as ATM, and developed and deployed the control software such as SS7 for physical circuits and UNI, NNI, and PNNI for ATM virtual circuits. On the other hand, IP networks become widespread and captured nearly all the data traffic market due to the popularity and the emergence of the Internet. IP follows a packet switching paradigm, and with MPLS, IP can support different qualities of service. IP control protocols are widely accepted, deployed, and proved to be scalable.

In a previous computer and communication networking integration attempt, IP and ATM were connected. A classical IP over ATM approach forms a static client server networking system, where IP plays the role as the client network and ATM works as the server transport network. This approach is complex since both IP and ATM have complete network control and management systems such as routing and signalling protocols. The IP network and the ATM network use different addressing so they cannot interact with each other directly. As a result, address resolution between IP and ATM has to be provided. The classical IP over ATM approach is static and does not scale.

To some extent, WDM networks can be treated as ATM networks with parallel transmission technology. An ATM switch has fibre ports whereas a WDM switch has wavelength ports (multiple wavelength channels per fibre). In addition, a WDM switch can be implemented in pure optical domain so that the latency is kept very low. IP over WDM is the topic of this book. We present several IP/WDM networking models such as overlay, augmented, or peer-to-peer. The selection for implementation also depends on network ownership and administration authority. We will present IP/WDM traffic engineering for optimal resource allocation in IP/WDM networks.

The book reviews the IP/WDM history as well as the international effort on the next-generation Internet. It also surveys the current IP/WDM standardisation. A detailed review on IP/WDM background information is provided. To assist the reader to understand the challenge posed in transporting IP traffic over WDM, we will present a chapter on the characteristics of the Internet and the IP routing (Chapter 3), and a chapter on WDM optical networks (Chapter 4). Topics on IP/WDM are organised into four chapters: IP over WDM internetworking models (Chapter 5), IP/WDM network control (Chapter 6), IP/WDM traffic engineering (Chapter 7), and IP/WDM specific issues (Chapter 8).

Kevin Liu
New Jersey

Acknowledgements

Writing this book has been an interesting experience in my life. I would like to thank all the people around me. In particular, I would like to thank my colleagues at Bellcore/Telcordia. The information in this book originates from several projects: the MONET NC&M project, the NGI SuperNet NC&M project, and the NGI OLS project.

Some of the material in the book has appeared in my early publications [Liu00a], [Liu00b], and [Liu02]. I would like to thank the coauthors of these papers in particular Dr John Wei, Brian Wilson, and Dr Changdong Liu. I also would like to thank IEEE for generously assigning me the right and the publisher to use these publications in this book.

I would like to thank the members of the MONET NC&M team in particular Dr John Wei, Brian Wilson, Jorge Pastor, Dr Ned Stoffel, Dr Mike Post, Dr Tsanchi Li, and Kenneth Walsh for many stimulating discussions. I also would like to thank colleagues at the government agencies (DARPA, DIA, DISA, NASA, NRL, and NSA) for their help, support, and feedback during the course of the MONET NC&M work. I would like to thank DARPA MONET Program Manager, Dr Burt Hui, for his guidance and support.

I would like to thank the members of the NGI SuperNet NC&M team in particular Dr Yukun Tsai, Dr Narayanan Natarajan, Dr Tsong-Ho Wu, Dr Changdong Liu, Dr Ramu S. Ramamurthy, and Arunendu Roy. I would like to thank the members of the NGI OLS team in particular Dr Gee-Kung Chang, Dr Sung-Yong Park, Brian Meagher, Jeffery Young, and Dr George Ellinas. I would like to thank DARPA NGI Program Manager, Dr Mari Maeda, for her guidance and support during the course of the NGI projects.

I would like to thank the editors at John Wiley & Sons, Ltd for their help and support, and the reviewers of my book draft and the book proposal for valuable suggestions.

Finally, I would like to thank my family especially my wife, Guohua, for her love, encouragement, and support throughout the writing of this book.

1

Introduction

- What is a WDM-enabled optical network?
- Why IP over WDM?
- What is IP over WDM?
- Next-generation Internet
- IP/WDM standardisation
- Summary and subject overview

1.1 What is a WDM-enabled Optical Network?

Conventional copper cables can only provide a bandwidth of 100 Mbps (10^6) over a 1 Km distance before signal regeneration is required. In contrast, an optical fibre using wavelength division multiplexing (WDM) technology can support a number of wavelength channels, each of which can support a connection rate of 10 Gbps (10^9). Long-reach WDM transmitters and receivers can deliver good quality optical signals without regeneration over a distance of several tens of kilometres. Hence, optical fibre can easily offer bandwidths of tens of Tbps (10^{12}). (Note throughout the book we use 'b' for bits and 'B' for bytes.) In addition to high bandwidth, fibre, made of glass (which is in turn made mainly from silica sand), is cheaper than other conventional transmission mediums such as coaxial cables.

Glass fibre transmission has low attenuation. Fibre also has the advantage of not being affected by electromagnetic interference and power surges or failures. In terms of installation, fibre is thin and lightweight, so it is easy to operate. An existing copper-based transmission infrastructure can be (and has been) replaced with fibre cables. In the fibre infrastructure, WDM is considered as a parallel transmission technology to exploit the fibre bandwidth using non-overlapping wavelength channels.

An individual optical transmission system consists of three components:

- the optical transmitter
- the transmission medium
- the optical receiver.

The transmitter uses a pulse of light to indicate the '1' bit and the absence of light to represent the '0' bit. The receiver can generate an electrical pulse once light is detected. A single-mode fibre transmission requires the light to propagate in a straight line along the centre of the fibre. The use of single-mode fibre results in a good quality signal, so it is used for long-distance transmission.

A light ray may enter the fibre at a particular angle and go through the fibre through internal reflections. A fibre with this property is known as multimode fibre. The basic optical transmission system is used in an optical network, which can be a local access network (LAN), a metropolitan local exchange network (MAN), or a long-haul inter-exchange network (also known as Wide Area Network, WAN).

1.1.1   TDM vs. WDM

There is a continuous demand for bandwidth in the construction of the Internet. It is also relatively expensive to lay new fibres and furthermore to maintain them. To explore the existing fibre bandwidth, two multiplexing techniques have been developed as shown in Figure 1.1.

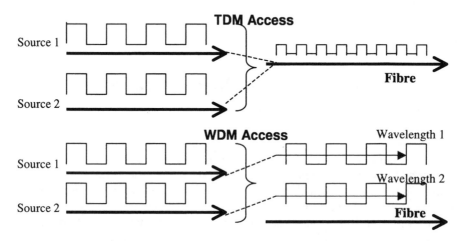

Figure 1.1   TDM vs. WDM.

Time Division Multiplexing (TDM) is achieved through multiplexing many lower speed data streams into a higher speed stream at a higher bit rate by means of non-overlapping time slots allocated to the original data streams.

Wavelength Division Multiplexing (WDM) is used to transmit data simultaneously at multiple carrier wavelengths through a single fibre, which is analogous to using Frequency Division Multiplexing (FDM) to carry multiple radio and TV channels over air or cable. We focus our discussion on WDM optical networks since TDM technology requires extremely high-speed electronics for high-speed data transmission and commercial high-speed TDM development lags far behind WDM. However, TDM and WDM can be used together in such a way that TDM provides time-sharing of a wavelength channel, for example, through aggregating access network traffic for backbone network transport.

WDM requires parallel transmission channels at different wavelengths within a single physical fibre. Splitting the useable wavelength bandwidth into a number of slots (wavelength channels) not only demands sophisticated equipment but also increases the likelihood of inter-channel interference. As such, WDM equipment cost may dominate the total cost in a LAN and MAN environment. As of November 2001, commercial WDM optical switches are able to support 256 wavelength channels, each of which can support a data rate of OC-192 (10 Gbps).

A point-to-point WDM-enabled optical transmission system consists of Wavelength Add/Drop Multiplexer (WADM) and Wavelength Amplifier (WAMP) (see Figure 1.2(a)). A WDM-enabled optical network employs point-to-point WDM transmission systems and requires Wavelength Selective Crossconnect (WSXC) that is able to switch the incoming signal unto a different fibre possibly a different optical frequency. An example of a four-wavelength double-ring WDM network is shown in Figure 1.2(b).

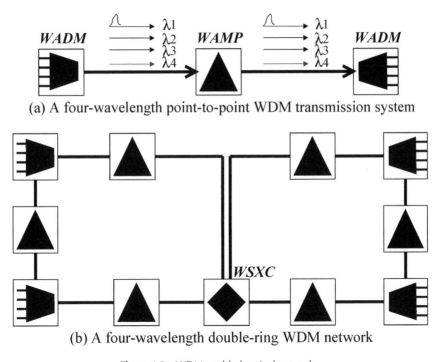

(a) A four-wavelength point-to-point WDM transmission system

(b) A four-wavelength double-ring WDM network

Figure 1.2   WDM-enabled optical networks.

In the optical network industry, WADM, WSXC, and WAMP are also known as Optical Add/Drop Multiplexer (OADM), Optical Cross Connect (OXC), and Optical Amplifier (OAMP). We use the terms of WADM, WSXC, and WAMP to emphasise the network elements in a WDM-enabled optical network. Details of the WDM network elements are presented in Chapter 4.

1.1.2   WDM Optical Network Evolution

The first generation of WDM provides only point-to-point physical links that are confined to WAN trunks. WAN WDM network configurations are either static or use manual configurations. The WDM link itself only supports relatively low-speed end-to-end connectivity. The technical issues of the first-generation WDM include design and development of WDM lasers and amplifiers, and static wavelength routing and medium access protocols. The WADM can also be deployed in MANs, for example, using a ring topology. To interconnect WADM rings, Digital Cross Connects (DCX) are introduced to provide narrowband and broadband connections. Generally these systems are used to manage voice switching trunks and T1 links.

The second generation of WDM is capable of establishing connection-orientated end-to-end lightpaths in the optical layer by introducing WSXC. The lightpaths form a virtual topology over the physical fibre topology. The virtual wavelength topology can be reconfigured dynamically in response to traffic changes and/or network planning.

The technical issues of the second-generation WDM include the introduction of wavelength add/drop and cross-connect devices, wavelength conversion capability at cross-connects, and dynamic routing and wavelength assignment. Also in second-generation WDM networks, network architecture begins to receive attention, in particular the interface for interconnection with other networks. Both first-generation and second-generation WDM networks have been deployed in carrier operational networks. Their cost efficiency in long-haul networks has been widely accepted.

The third-generation of WDM offers a connectionless packet-switched optical network, in which optical headers or labels are attached to the data, transmitted with the payload, and processed at each WDM optical switch. Based on the ratio of packet header processing time to packet transmission cost, switched WDM can be efficiently implemented using label switching or burst switching. Pure photonic packet switching in all optical networks is still under research.

The bufferless, all-optical packet router brings a new set of technical issues for network planning:

- contention resolution;
- traffic engineering;
- over-provisioning;
- over-subscription;
- interoperability with conventional IP (Internet Protocol) routers (destination-based routing).

Examples of third-generation WDM devices are:

- optical label switch routers
- optical Gigabit routers
- fast optical switches.

Interoperability between WDM networks and IP networks becomes a major concern in third-generation WDM networks. Integrated routing and wavelength assignment based on MultiProtocol Label Switching (MPLS), also known as Generalised MPLs (GMPLS), starts to emerge. Other key software technical issues include bandwidth management, path reconfiguration and restoration, and Quality of Service (QoS) support.

Figure 1.3 describes WDM network evolution. Traffic granularity refers to both the volume of the traffic and the size of each traffic unit. Traffic in access networks is aggregated/multiplexed before riding over backbone networks. A review of WDM research activities and optical network testbeds is provided in Chapter 2.

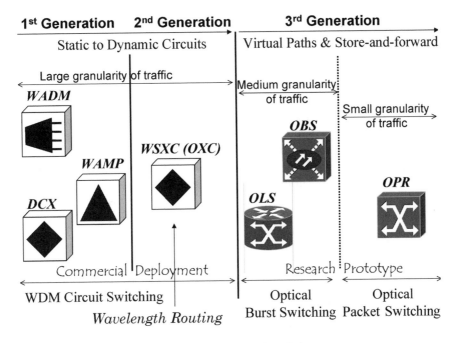

Figure 1.3   WDM network evolution.

Optical circuit switching is used for large-sized aggregated traffic (such as data trunks), so that once circuits are set up, the formed topology does not change often. This provides cost-efficiency in the long haul network because a few add/drop points are needed by the traffic and only physical transport link services are required. Optical packet switching is used for small-sized data packets. It offers efficient, flexible resource sharing by introducing complexity to the control system.

A compromise between packet and circuit switching is optical burst switching, which switches traffic bursts over a packet-orientated network. Packet switching offers cut-through paths, known as 'layer 2 switching'. The cut-through paths lower the network latency by avoiding intermediate node layer 3 functions. A packet routed network employs a store-and-forward paradigm, where each node maintains a routing table and a forwarding table. Once a packet arrives at the node, by comparing the packet header with the local forwarding table, the packet is sent to the next hop on the routing path.

1.2   Why IP over WDM?

It is widely believed that IP provides the only convergence layer in the global and ubiquitous Internet. IP, a layer 3 protocol, is designed to address network level interoperability and routing over different subnets with different layer 2 technologies. Above the IP layer, there are a great variety of IP-based services and appliances that are still evolving from its infancy. An example is IP-based home networking inter-connecting a wide range of electronic devices. Hence, the inevitable dominance of IP traffic makes apparent the engineering practices that the network infrastructure should be optimised for IP. Below the IP layer, optical fibre using WDM is the most promising wireline technology, offering an enormous network capacity required to sustain the continuous Internet growth.

WDM technology will become more attractive as the price of WDM systems lowers. With the continuous worldwide deployment in optical fibre and the maturity of WDM, WDM-based optical networks have been deployed not only in the back-bone but also in metro, regional, and access networks. In addition, WDM optical networks are no longer just point-to-point pipes providing physical link services, but blend well with any new level of network flexibility requirements.

The control plane is responsible for transporting control messages to exchange reachability and availability information and computing and setting up the data-forwarding paths. The data plane is responsible for the transmission of user and application traffic. An example function of the data plane is packet buffering and forwarding. IP does not separate the data plane from the control plane, and this in turn requires QoS mechanisms at routers to differentiate control messages from data packets.

A conventional WDM network control system uses a separate control channel, also known as a data communication network (DCN), for transporting control messages. A conventional WDM network control and management system, e.g. according to the TMN framework, is implemented in a centralised fashion. To address scalability, these systems employ a management hierarchy. Combining IP and WDM means, in the data plane, one can assign WDM optical network resources to forward IP traffic efficiently, and in the control plane, one can construct a unified control plane, presumably IP-centric, across IP and WDM networks. IP over WDM will also address all levels of interoperability issues on intra- and inter-WDM optical networks and IP networks.

The motivation behind IP over WDM can be summarised as follows:

- WDM optical networks can address the continuous growth of the Internet traffic by exploiting the existing fibre infrastructure. The use of WDM technology can significantly increase the use of the fibre bandwidth.
- Most of the data traffic across networks is IP. Nearly all the end-user data application uses IP. Conventional voice traffic can also be packetised with voice-over-IP techniques.
- IP/WDM inherits the flexibility and the adaptability offered in the IP control protocols.

- IP/WDM can achieve or aims to achieve dynamic on-demand bandwidth allocation (or real-time provisioning) in optical networks.

  - By developing the conventional, centralised controlled optical networks into a distributed, self-controlled network, the integrated IP/WDM network can not only reduce the network operation cost, but can also provide dynamic resource allocation and on-demand service provisioning.

- IP/WDM hopes to address WDM or optical Network Element (NE) vendor interoperability and service interoperability with the help of IP protocols.

  - WDM optical networks require a scalable and unified control plane across subnets supplied by various WDM vendors. IP control protocols are widely deployed and proved to be scalable. The emergence of MPLS not only complements conventional IP with traffic engineering and a different quality of service capability, but also proposes a unified IP-centric control plane across networks.
  - The existence of a variety of WDM network equipment demands vendor interoperability. For example, opaque WADMs require specific signal formats such as SONET/SDH (Synchronous Optical Network/Synchronous Digital Hierarchy) signals on their add/drop client interfaces. WDM interoperability requires the presence of a network layer, which is IP.

- IP/WDM can achieve dynamic restoration by leveraging the distributed control mechanisms implemented in the network.
- From a service point of view, IP/WDM networks can take advantage of the QoS frameworks, models, policies, and mechanisms proposed for and developed in the IP network.
- Given the lessons learned from IP and ATM integration, IP and WDM need a closer integration for efficiency and flexibility. For example, classical IP over ATM is static and complex, and IP to ATM address resolution is mandatory to translate between IP addresses and ATM addresses.

IP/WDM integration will eventually translate into an efficient optical network transport reducing the cost of IP traffic (i.e. cost per bit/mile) and increasing the utilisation of the optical network.

1.3 What is IP over WDM?

IP/WDM network is designated to transmit IP traffic in a WDM-enabled optical network to leverage both IP universal connectivity and massive WDM bandwidth capacity. Figure 1.4 shows transporting IP packets or SONET/SDH signals over WDM networks. A software controller controls the switching fabric. IP, as a network layer technology, relies on a data link layer to provide:

- framing (such as in SONET or Ethernet);
- error detection (such as cyclic redundancy check, CRC);
- error recovery (such as automatic repeat request, ARQ).

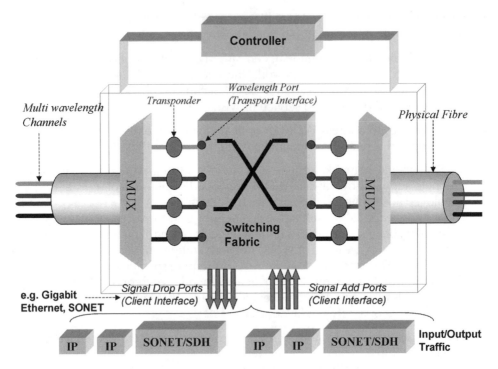

Figure 1.4   Transporting IP packets over wavelengths.

Some of the link layer functionality is implemented in the interface, for example, the add/drop client interfaces or the pass-by physical transport interfaces.

An objective of optical networking is to provide optical transparency from end to end so that the network latency is minimised. This requires all-optical interfaces and all-optical switching fabric for the edge and intermediate network elements. Transponders are used to strengthen the optical signal. There are all-optical transponders (tunable lasers) and Optical-Electrical-Optical (O-E-O) transponders. The figure shows two types of traffic, IP (e.g. Gigabit Ethernet) and SONET/SDH, which in turn requires Gigabit Ethernet and SONET/SDH interfaces. In the case of multiple-access links, a sublayer of the data link layer is the Media Access Protocol (MAC) that mediates access to a shared link so that all nodes eventually have a chance to transmit their data. The definition of a protocol model to efficiently and effectively implement an IP/WDM network is still an active research area.

Figure 1.5 shows three possible approaches for IP over WDM. The first approach transports IP over ATM (Asynchronous Transfer Mode), then over SONET/SDH and WDM fibre. WDM is employed as a physical layer parallel transmission technology. The main advantage of this approach by using ATM is to be able to carry different types of traffic onto the same pipe with different QoS requirements.

Another advantage introduced by ATM is its traffic engineering capability and the flexibility in network provisioning, which complements the conventional IP best effort traffic routing. However, this approach is offset by complexity, as IP over ATM is more complex to manage and control than an IP-leased line network.

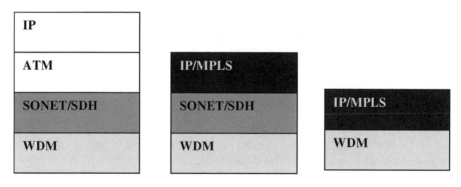

Figure 1.5 Three approaches for IP over WDM (data plane).

ATM uses a cell switching technology. Each ATM cell has a fixed 53-byte (5-byte header and 48-byte user data) length, so application traffic has to be packetised into cells for transport and reassembled at destination. ATM cell packetisation is the responsibility of the ATM SAR (Segmentation and Reassembly) sublayer. SAR becomes technically difficult above OC-48. Having an ATM circuit layer between IP packet and the WDM circuit seems superfluous. The statement is strengthened by the emergence of the MPLS technique of the IP layer. The key features of MPLS include:

- Use of a simple, fixed-length label to identify flows/paths.
- Separating control from data forwarding, control is used to set up the initial path, but packets are shipped to next hop according to the label in the forwarding table.
- A simplified and unified forwarding paradigm, IP headers are processed and examined only at the edge of MPLS networks and then MPLS packets are forwarded according to the 'label' (instead of analysing the encapsulated IP packet header).
- MPLS provides multiservice. For example, a Virtual Private Network (VPN) set up by MPLS has a specific level of priority indicated by the Forwarding Equivalence Class (FEC).
- Classification of packets is policy-based, with packets being aggregated into FEC by the use of a label. The packet-to-FEC mapping is conducted at the edge, for example, based on the class of service or the destination address in the packet header.
- Providing enabling mechanisms for traffic engineering, which can be employed to balance the link load by monitoring traffic and making flow adjustments actively or proactively. In the current IP network, traffic engineering is difficult if not impossible because traffic redirection is not effective by indirect routing adjustment and it may cause more congestion elsewhere in the network. MPLS provides *explicit path routing* so it is highly focused and offers class-based forwarding. In addition to explicit path routing, MPLS offers tools of *tunneling, loop prevention and avoidance, and streams merging* for traffic control.

IP/MPLS over SONET/SDH and WDM is the second approach in the figure. SONET/SDH provides several attractive features to this approach:

- First, SONET provides a standard optical signal multiplexing hierarchy by which low-speed signals can be multiplexed into high-speed signals.
- Second, SONET provides a transmission frame standard.
- Third is the SONET network protection/restoration capability, which is completely transparent to upper layers such as the IP layer.

SONET networks usually employ a ring topology. SONET protection scheme can be provided:

- as *1 + 1* meaning data are transferred in two paths in the opposite direction and the better signal is selected at the destination;
- as *1:1* indicating there is a separate signalled protection path for the primary path;
- or as *n:1* representing where primary paths share the same protection path.

The design of SONET also enhances OAM&P (Operations, Administration, Maintenance, and Provisioning) to communicate alarms, controls, and performance information at both system and network levels. However, SONET carries substantial overhead information, which is encoded in several levels. Path overhead (POH) is carried from end-to-end. Line overhead (LOH) is used for the signal between the line terminating equipment, such as OC-n multiplexers. Section overhead (SOH) is used for communication between adjacent network elements, such as regenerators. For an OC-1 pipe with 51.84 Mbps transmission rate, its payload has the capacity to transport a DS-3 with 44.736 Mbps digital bit rate.

The third approach for IP/WDM employs IP/MPLS directly over WDM, which is the most efficient solution among the possible approaches. However, it requires that the IP layer looks after path protection and restoration. It also needs a simplified framing format for transmission error handling. There are a few alternatives providing IP over WDM framing format. Several companies are developing a new framing standard known as Slim SONET/SDH, which provides similar functionality as in SONET/SDH but with modern techniques for header placement and matching frame size to packet size.

Another example is to adopt the Gigabit Ethernet framing format. The new 10-Gigabit Ethernet is especially designed for dense WDM systems. Using the Ethernet frame format, hosts (Ethernet) on either side of the connection do not need to map to another protocol format (e.g. ATM) for transmission.

Conventional IP networks use in-band signalling so data and control traffic is transported together over the same link and path. A WDM optical network has a separate data communication network for control messages. Hence, it uses out-of-band signalling as shown in Figure 1.6.

In the control plane, IP over WDM can support several networking architectures, but the architecture selection is subject to constraints on existing network environments, administrative authority, and network ownership.

The details of IP over WDM network control architecture and internetworking models are presented in Chapter 5. A unified IP-centric control plane promises a peer-to-peer networking system whereas non-IP controlled networks and/or any proprietary solutions are likely to form overlay networks (when connecting to IP networks).

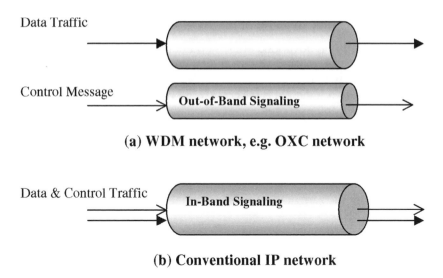

(a) WDM network, e.g. OXC network

(b) Conventional IP network

Figure 1.6 Data and control traffic in IP and WDM networks.

1.4   Next-generation Internet

The goal of developing a truly ubiquitous and peer-to-peer network as the next-generation Internet will succeed only through the same spirit of co-operative partnership that developed the first global Internet. In fact, it requires more worldwide commitment than the first Internet because there are already a large number of deployed networks, among them non-IP or proprietary networks and different regions and countries which would like different features supported by the Internet. A range of next-generation Internet projects is described below.

US Internet-related research and development partnerships include not only entities that are directly focused on Internet development such as IETF but also general standard organisations such as IEEE and ANSI and federal government agencies such as DARPA (Defense Advanced Research Projects Agency) (www.darpa.mil) and NSF (National Science Foundation) (www.nsf.gov). The Next Generation Internet (NGI) initiative (www.ngi.gov) was established in 1998 for the period of 5 years, through which government agencies will cooperate to create next-generation Internet capabilities to allow for enhanced support for their core missions, as well as to advance the state-of-the-art in advanced networking. The NGI initiative will:

- develop new and more capable networking technologies to support Federal agency missions;
- create a foundation for more powerful and versatile networks in the 21st century;
- form partnerships with academia and industry that will keep the US at the cutting edge of information and communication technologies;
- enable the introduction of new networking services that will benefit businesses, schools, and homes.

The primary goals of the initiative are:

- conducting research in advanced end-to-end networking technologies, including differentiated services, particularly for digital media, network management, reliability, robustness, and security;
- prototyping and deploying national-scale testbeds that are able to provide 100 to 1000 times current Internet performance;
- developing revolutionary new applications requiring high performance networks.

To achieve the second goal, the SuperNet testbed has been set up, where two classes of networks have been deployed to satisfy the 100 × and 1000 × requirements. The SuperNet network is referred to as the NGI 1000 × network, to which approximately thirty sites are now connected, with access rates of several gigabits per second.

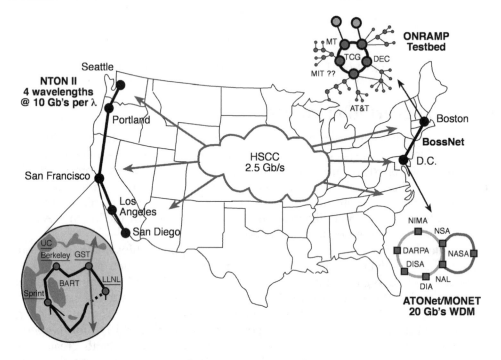

Figure 1.7   NGI SuperNet testbed (source: SuperNet www.ngi-super.org).

The SuperNet testbed, shown in Figure 1.7, is composed of several interconnected and interoperating 'regional' testbeds including:

- NTON II network connecting metropolitan areas on the West Coast over four wavelengths, each with 10 Gbps capacity.
- HSCC (High Speed Connectivity Consortium) network connecting cross-country sites and regional networks at 2.5 Gbps end-to-end.
- ONRAMP network fielding advanced metropolitan area and regional access technologies.

- BOSSNET network enabling physical layer networking and communication experiments over multiple dark fibres.
- ATDNET/MONET (Advanced Technology Demonstration Network) testbed based on multi-wavelength, reconfigurable optical networking technology.

SuperNet aims at the development of a foundation for terabits per second, which requires research and development related to the design and creation of ultra-high-speed, multiplexing, switching, and transmission technologies. SuperNet also intends to streamline networking protocol stacks as displayed in Figure 1.8.

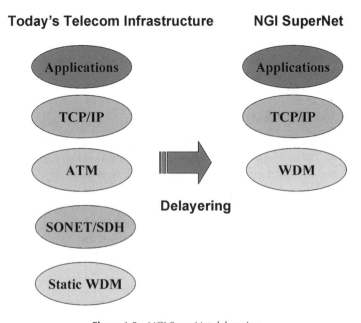

Figure 1.8   NGI SuperNet delayering.

CANARIE, the Canadian Network for the Advancement of Research, Industry, and Education, is a non-profit corporation supported by its members, project partners and the Canadian government to accelerate Canada's advanced Internet development and facilitate the widespread adoption of faster, more efficient networks and enable the next generation of advanced products, applications and services.

In February 1998, the Canadian government provided CANARIE with a $55 million grant towards the $120 million project to develop a national optical R&D Internet, known as CA*net III. Industry members provided the remaining part of the funding. Using new fibre optic-based technology and Dense Wavelength Division Multiplexing (DWDM), CA*net 3 intends to deliver unrivalled network capability with a potential for OC768 (40 Gbps) to Canadian research institutions and universities. Phase I was completed in October 1998, where an optical Internet backbone was set up between Toronto, Ottawa and Montreal. Currently Phase II is in progress, through which the optical Internet Backbone will be extended west from Toronto to Vancouver and east from Montreal to Atlantic Canada.

In Europe, an extensive testbed infrastructure dedicated to advanced research including projects related to communications and networking was funded by the European ACTS (Advanced Communications Technologies and Services) program. In addition, many European NRNs (National Research Networks) have established national high-performance advanced network infrastructures. To address Internet working and provide transport services between countries, a series of R&D projects have been completed, e.g. DANTE, TEN-34, TEN155, and EuropaNet.

In Asia Pacific, a number of countries have participated in the APAN (Asia Pacific Advanced Networking) initiative, which aims at conducting research and development for advanced applications and services and providing a network environment for local research communities and for international collaboration. The APAN was established in June 1997 by its founding members including Australia, Japan, Korea, and Singapore. Eastern Europe, Russia, and the Middle East are also developing a variety of advanced networking initiatives.

1.5 IP/WDM Standardisation

Community interests on IP/WDM come from two main groups, the Internet Engineering Task Force (IETF) (www.ietf.org) and the International Telecommunication Union, Telecommunication Standardisation Sector (ITU-T) (www.itu.org) respectively. IETF is an open international community of network designers, operators, vendors, and researchers concerned with the evolution of the Internet architecture and the smooth operation of the Internet, established in 1986. The actual technical work of the IETF is conducted in its working groups, which are organised by topic into several areas (e.g., routing, transport, security, etc.). The newly formed IETF *sub-IP* Area has seven workgroups listed in Table 1.1.

Table 1.1   IETF sub-IP workgroups

ccamp	Common Control and Measurement Plane
gsmp	General Switch Management Protocol
ipo	IP over Optical
iporpr	IP over Resilient Packet Rings
mpls	Multiprotocol Label Switching
ppvpn	Provider Provisioned Virtual Private Networks
tewg	Internet Traffic Engineering

In particular, IETF has been focusing on these IP/WDM-related issues:

- MPLS/MPλS (Multiprotocol Lambda Switching)/GMPLS (Generalized MPLS).
- Layer 2 and layer 3 functionalities within optical networks.
- NNI (Network to Network Interface) standard for optical network.

ITU-T was created on March 1, 1993, within the framework of the 'new' ITU, replacing the former International Telegraph and Telephone Consultative Committee (CCITT) whose origins go back to 1865. The ITU-T mission is to ensure an efficient and on-time production of high quality standards covering all fields of telecommunications except radio aspects. Standardisation work is carried out by 14 study groups in which representatives of the ITU-T membership develop Recommendations for the various fields of international telecommunications on the basis of the study of questions.

At present, there are more than 2600 Recommendations (Standards) enforced by ITU-T. ITU-T has two related Study Groups (SG), SG 15 and SG 13. SG 15 is responsible for studies on optical and other transport networks, systems and equipment, which encompasses the development of transmission layer standards for the access, metropolitan and long haul sections of communication networks. SG 13 is responsible for studies related to Internet working of heterogeneous networks encompassing multiple domains, multiple protocols and innovative technologies with a goal to deliver high-quality, reliable networking. Specific aspects are architecture, interworking and adaptation, end-to-end considerations, routing and requirements for physical layer transport. In particular, ITU-T has been focusing on these IP/WDM-related issues:

- Layer 1 features of the OSI model.
- Architectures and protocols for next-generation optical networks, also known as optical transport network (OTN), defined in G.872.
- Architecture for the automatic switched optical network, defined in G.ason.

With the advance of the optical networking, on April 20, 1998, Cisco Systems and Ciena Corporation announced an industry-wide initiative to create the Optical Internetworking Forum (OIF) (www.oiforum.org) an open forum focused on accelerating the deployment of optical Internet works. The founding members of the forum were AT&T, Bellcore (now Telcordia Technologies), Ciena Corporation, Cisco Systems, Hewlett-Packard, Qwest, Sprint and WorldCom (now MCI WorldCom). The mission of the OIF is to provide a venue for equipment manufacturers, users and service providers to work together to resolve issues and develop key specifications to ensure the interoperability of optical networks.

Currently, there are five working groups within OIF: Architecture, Carrier, OAM&P, Physical and Link Layer, and Signalling. In particular, OIF has been working on the following issues:

- Optical UNI (User-to-Network Interface)
- Optical NNI.

The concept of UNI and NNI is used in client server networking systems. UNI is the protocol between the user (i.e. client) and the network (i.e. server) whereas NNI is the protocol between networks. NNI can be either INSI (Intra Network Switching Interface) or ICI (Inter Carrier Interface). UNI and NNI have been used in ATM and Frame Relay networks. For example, ATM networks defined two types of ATM cells: UNI and NNI cells. Both types of cell header have the length of five bytes but they differ in the last four bits of the first byte of the header. NNI headers use the entire first byte to represent the virtual path identifier (VPI); UNI headers use the first four bits for VPI and the remaining four bits of the first byte for generic flow control (GFC) directing multiple traffic streams over the interface.

1.6   Summary and Subject Overview

This chapter introduced the topic of this book, IP over WDM. To clarify the concepts and motivate the readers, we answered the questions: 'What is a WDM-enabled optical network?' 'Why IP over WDM?' 'What is IP over WDM?' We reviewed the worldwide effort on next-generation Internet applications. We also listed the current IP/WDM standardisation organisations and their technical focus.

Figure 1.9 provides a conceptual view of this book. A detailed review of IP/WDM background information is provided in Chapter 2. Example topics include optical communications, WDM network testbeds and product comparisons, communication protocols, Internet architecture, IPv4 addressing, Gigabit Ethernet, MPLS, and distributed systems. To assist the reader to understand the challenges posed in transporting IP traffic over WDM, we present a chapter on the characteristics of the Internet and IP routing, and a chapter on WDM optical networks. IP traffic characteristics and models are presented in Chapter 3. We also explain the existing TCP (Transmission Control Protocol) traffic control policies. Internet routing is explained. In particular, we will focus on traffic engineering, OSPF (Open Shortest Path First protocol), and BGP (Border Gateway Protocol). We will also present a section on IPv6. For WDM optical networks, in Chapter 4, we present discussions on the enabling technologies, the optical components, the optical network control and management, the software architecture and design, an information model, WDM signalling, and the WDM network visualisation. Chapters 3 and 4 are independent, so they can be studied in parallel.

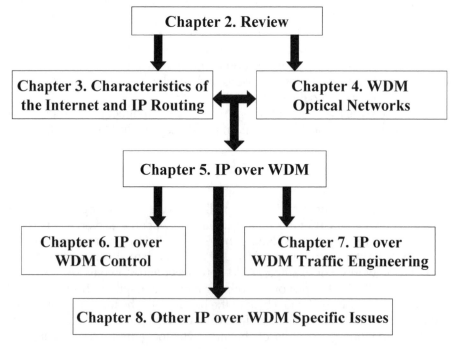

Figure 1.9   Conceptual view of this book.

Topics on IP/WDM are organised into four chapters, IP over WDM, IP/WDM network control, IP/WDM traffic engineering, IP/WDM-specific issues, respectively. Chapter 5 presents IP over WDM architectural models, IP/WDM internetworking models, and IP/WDM software architectural models. Chapter 6 presents IP/WDM network control. Our discussion follows these key issues: IP/WDM network addressing, topology discovery, IP/WDM routing, signalling, WDM access control, IP/WDM restoration, GMPLS, inter-domain network control, and WDM network element control and management protocol.

Chapter 7 introduces IP/WDM traffic engineering. A framework is presented and the related functional components within the framework are detailed. MPLS traffic engineering is reviewed. The chapter emphasises IP over WDM traffic engineering. In particular, we present lightpath virtual topology design and topology migration for OXC networks and reconfiguration for packet switched WDM networks. We also present a simulation study on IP over WDM reconfiguration.

IP/WDM network-specific issues are discussed in Chapter 8. The chapter covers topics on group communication in IP/WDM networks, IP/WDM network and service management, and TCP over optical networks. Finally, the book is concluded in Chapter 9, where we provide a summary, and discuss IP/WDM network applications and future research issues.

2

Review

- Telecommunication networks
- Optical communications
- WDM network testbed and product comparison
- Communication protocols
- Internet architecture
- IPv4 addressing
- Gigabit Ethernet
- MPLS
- Distributed systems

2.1 Telecommunication Networks

A telecom network consists of nodes that are either terminals or network nodes and links that interconnect the nodes (see Figure 2.1). An access network connects terminals to the network nodes; a trunk network connects the network nodes to each other. An access network can be a point-to-point network or a shared medium network. Figure 2.1 shows a shared bus access network in which terminals have to compete for resources of the shared medium and consequently multiple access techniques are needed.

As in any natural hierarchy between an access network and a trunk network, a telecom network is typically constructed level-by-level, i.e. hierarchically. Traditionally, AT&T networks have used a 5-level hierarchy, but the current telecom networks uses only three levels:

- dynamic routing tandem/international switching centre
- toll centre
- end-office.

There are three switching modes in telecom networks:

- circuit switching
- packet switching
- cell switching.

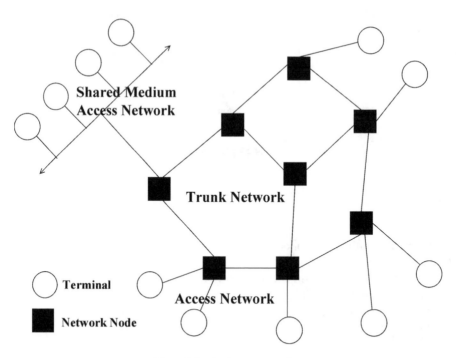

Figure 2.1   A telecom network.

Circuit-switched networks offer connection-orientated services, where connections are set up end-to-end before information transfer and resources are reserved for the whole duration of connection. In circuit-switched networks, delays occur before and after information transfer, but there is no extra delay and no overhead during information transfer. Packet-switched networks offer connectionless services in which there is no connection setup and no resource reservation. It also offers connection orientated services in which virtual connections are set up end-to-end before information transfer but there is no resource reservation. Information transfers as discrete packets each of which has a varying length. The connectionless packet switching requires that each packet carries its global address of destination; the connection-orientated packet switching uses only a local address, i.e. the logical channel index. In connectionless packet switched networks, there is no delay before information transfer, but during information transfer, packets have to carry their header (overhead), expect packet-processing delays, and may suffer queuing delays (when packets compete for joint resources).

Cell-switching networks offer connection orientated services, where virtual connections are set up end-to-end before information transfer, and where resource reservation is possible but not mandatory. Information transfers as cells, each of which has a fixed length and uses a local address. In cell-switched networks, there is no delay before information transfer, but during information transfer, each cell carries a header (overhead), expects packet-processing delays, and may suffer from queuing delays (if resources are not reserved beforehand).

Table 2.1 presents a summary of circuit switching, connectionless packet switching, connection-orientated packet switching, and cell switching. The different switching modes are compared with respect to the use of network resources (utilisation), the traffic sequence integrity, the real-time guarantee, the flexibility, the different traffic types (with QoS), and the in-use examples. The original telecom networks such as PSTN were purely analogue networks but ISDN and GSM are completely digital networks including terminal and the access parts. Packet- and cell-switched networks are completely digital networks.

Table 2.1 Circuit, packet, and cell switching

	Circuit switching	Connectionless packet switching	Connection orientated packet switching	Cell switching
Use of network resources	Not efficient	Very efficient	Efficient	Efficient
Traffic sequence integrity	Guaranteed	Not guaranteed	Guaranteed	Possible
Real-time guarantee	Yes	No	No	Possible
Flexibility	Poor	Very good	Good	Good
Different traffic types	Yes	No	Yes	Yes
In-use examples	PSTN	IP, SS7	X.25, Frame Relay	ATM

2.2   Optical Communications

We consider data transmission over optical fibres (guided media) in this book. During optical fibre transmission, a light emitted from a source enters the cylindrical glass or plastic core of a fibre, passes along the optical core, and hits the receiver circuit at the other end.

There are two types of light sources, the light-emitting diode (LED) and the injection laser diode (ILD). LED and ILD are semiconductor devices that can generate beams of lights when a voltage is applied. The LED is comparatively cheaper and has a longer operational life. LEDs can operate in a greater range of temperatures. ILDs operate on the laser principle and potentially produce higher data rates due to smaller range of light frequencies generated and less dispersion. A cornerstone of optical networking is the creation of a tuneable laser diode operating around 1.55 μm. 'Tuneable' means the same physical device (i.e. the laser diode) can be used to generate different wavelengths of light.

An optical receiver is a semiconductor device that detects the light and then generates a flow of electricity. So it works as the opposite of an LED. Optical receivers can be described in terms of these parameters: receiver efficiency (i.e. the ratio of output current power to input optical power), the range of optical wavelengths (over which the receiver operates), response time (how quickly the receiver can react to changes in the input optical power), and the noise level of the receiver.

Three optical bandwidth regions with low attenuation have been selected in fibre transmission. They are centred on 0.85, 1.3 and 1.55 μm. The first wavelength band is relatively inexpensive to use, but it can only support lower data rates for limited distances. The other two wavelength bands can be used in high-speed WDM transmission. Figure 2.2 shows the two WDM-usable bandwidth regions with very low attenuation, which, in total, can provide a bandwidth of 50 TeraHz.

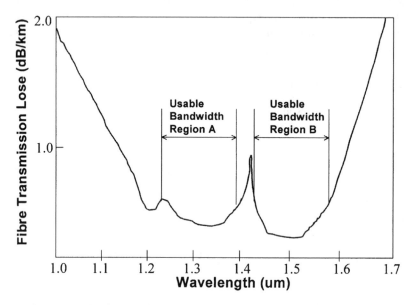

Figure 2.2   Fibre bandwidth regions with low attenuation for WDM transmission.

In a communication system, analogue/digital data is transmitted by means of signals. Figure 2.3 shows an example of the fundamental continuous (analogue) signal, the sine wave, which can be represented by three parameters: amplitude (A), frequency (f), and phase (φ). The amplitude is the peak value or strength of signal over time. The frequency is the rate at which the signal repeats. A related parameter is the period of a signal (T), which is the amount of time of one cycle. The phase is a measure of the relative position in time from a reference point within one single period of the signal.

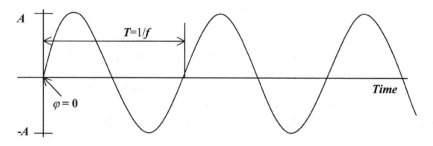

Figure 2.3 $A\sin(2\pi ft + \varphi)$.

The wavelength of a signal represents the length of one cycle, i.e. the distance of the two points of corresponding phase of two consecutive cycles. Therefore, wavelength (λ) can be defined as $\lambda = vT = v/f$, where v represents the velocity the signal is travelling.

On the other hand, digital signals are discrete and represent a stream of 1s and 0s. Digital signals not only carry digital data but also can encode analogue data. A periodic digital signal is the *square wave*, which can also represented by A, f, and φ (see Figure 2.4).

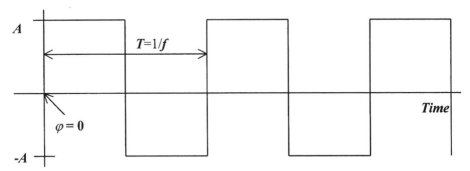

Figure 2.4 A Square Wave (A, f, φ).

2.2.1 Optical Communication Impairments

This book focuses on communication protocols, architecture, and software aspects of IP/WDM networks. In this section, we review some topics related to optical communication transmission and physical layer impairment concepts that will be used frequently.

Attenuation

The phenomenon that signal power gradually reduces over distance as the signal propagates is known as attenuation. To allow for attenuation, the signal must have sufficient strength for the receiver to detect that signal and therefore we must maintain a signal level sufficiently higher than any noise in the signal channel.

The signal power can be improved using amplifiers, for example, in the case of long distance transmission. In addition, attenuation is an increasing function of frequency. As indicated in Figure 2.2, the lowest attenuation in fibre, 0.2 dB/km, occurs at 1.55 μm. Beyond that point, as wavelengths increase, the attenuation starts to pick up and then becomes high quickly. Attenuation in fibre is logarithmic as expressed below:

$$N_{dB} = -10\log_{10}\frac{P'}{P}$$

where N_{dB}, P' and P represent the fibre attenuation in units of dB/km, the signal power measured at the output of 1 km fibre, and the input signal power respectively.

Dispersion

Dispersion is the effect that different frequency components of the transmitted signal travel at different velocities in the fibre, arriving at different times at the receiver. There are different types of dispersion in optical communication. Examples include multimode distortion, polarisation dispersion, and chromatic dispersion. Multimode distortion occurs in multimode fibres, in which the signal is spread in time because the velocity of propagation of the optical signal is not the same for all modes.

Polarisation describes the light wave orientation. The different polarisations of light travelling at different speeds through optical fibre can cause polarisation dispersion. Chromatic dispersion specifies the wavelength dependence of the velocity of propagation (of the optical signal) on the bulk material of which the fibre is made. The amount of dispersion is wavelength dependent. Dispersion is a problem because it results in inter-signal interference if fibre lengths (before amplifiers) are too long. One way to reduce dispersion is to increase the distance between the light pulses, but this lowers the signalling rate and so the overall data rate.

Nonlinear effects

A linear system describes the output as a linear function of the input. When the optical power within an optical fibre is small such as in low bit-rate systems, fibre can be regarded as a linear medium, i.e. the loss and Refractive Index (RI) of the fibre are independent of the signal power. RI is a property of the fibre core and determines how fast the light travels in the fibre. As the advent of WDM systems and the increasing demand of higher bit rates, the amount of optical power within the fibre increases. When power levels reach a fairly high level in the system, the impact of nonlinear effects arises since both the loss and RI depend on the optical power of the signal in the fibre.

Nonlinear effects include Kerr and Scattering effects. Kerr effects refer to the relationship between the refractive index and the light intensity of the signal. This can result in:

- **self-phase modulation**, where a wavelength can spread out into adjacent wavelengths;
- **cross-phase modulation**, where different wavelengths spread out into each others; or
- **four-wave mixing**, where several wavelengths interact to create a new wavelength.

Scattering effects refer to the signal loss and stimulation due to the contact between light and fibre. Scattering effects include Raman Scattering and Brillouin Scattering. The former re-emits a longer wavelength due to the loss of energy, while the latter occurs because of the generated acoustic waves. Nonlinear effects generally are non-desirable since they make it hard to read signals, and it is possible to misinterpret signals at the receiver. However, Raman Scattering boosts signal power so it is useful for amplification.

Crosstalk

Crosstalk represents the undesired coupling of a signal to another optical signal. Crosstalk in fibre transmission is also known as optical coupling. Interchannel crosstalk occurs when the two interfering optical signals have different wavelengths. Intrachannel crosstalk can take place when two light sources are transmitting using the same wavelength (or very close wavelengths) and a small amount of light of the first signal 'leaks' to the second's receiver.

Crosstalk can also happen when there are multiple paths for an optical signal. In that case, light is leaked into alternate path(s). The branching signal also reaches the receiver and therefore causes receiver confusion.

2.2.2   Optical Switching

We describe optical switching and networking-related concepts in this section. These concepts have significant influence on route computation, path setup, wavelength access, and WDM network and equipment modeling.

Switching domains

There are several types of optical switching technology classified by switching domains. The classification is also related to the switching traffic granularity (i.e. the size of the signal that a carrier needs to switch):

- **Fibre switching**: switches an incoming fibre including all the wavelength channels on it to an outgoing fibre.
- **Wavelength band switching**: switches a set of wavelengths on an incoming fibre to an outgoing fibre.
- **Wavelength switching**: switches an individual wavelength to a wavelength on an outgoing fibre.
- **Subwavelength switching**: in the case of aggregated traffic, switches subwavelength payloads onto the outgoing fibre, e.g. TDM slots.
- **Space switching**: switches a signal from one input port to several different (possibly all) output ports.
- **Time switching**: each input port is given a time slot for admitting a signal. Time switching is used in conjunction with other switching techniques.

Client interface vs. physical transport interface

Client interfaces represent the boundary between the WDM network and external networks. A client interface provides an interface from the WDM provider network to a client network. A client interface may request specific client signal formats, e.g. SONET/SDH signals.

Client interfaces that are located at the WDM network edge-switch are directly connected to a switching fabric. An add port can insert a client signal into the WDM network through the switching fabric. The client signal is switched to a corresponding multiplexer for wavelength multiplexing and then amplified before being sent through the fibre to the receiver. A drop port can 'drop' a wavelength channel from the WDM network. The incoming wavelengths go through de-multiplexing, fabric switching, and then the drop port. Depending on the interface card supported at the drop port, the wavelength is converted into a specific client signal format.

Within a WDM network, WDM switches are connected using physical transport interfaces. Being WDM-capable, a physical transport interface has a number of ports, each of which supports one wavelength channel. When a WDM signal in a fibre comes into the physical transport interface, the signal goes through a demultiplexer, takes an input port on the switching fabric, switches onto a fabric output port, multiplexes, and transmits. If the fabric does not allow optical frequency interchange, the input and output port must have the same wavelength.

Wavelength continuity vs. wavelength interchange

In WDM optical networks, optical cross-connect (OXC) requires wavelength switching, which connects a specific wavelength on an incoming fibre to the same wavelength on one or more ongoing fibres. Wavelength switching for an incoming signal depends not only on the availability of the specific optical frequency on the outgoing fibres but also on the capability of the fabric supporting wavelength interchange. Therefore, wavelength continuity and interchange are the two conditions to describe the wavelength at the fabric physical transport interface. Wavelength continuity means lack of wavelength interchange, i.e. the same wavelength or frequency is required in the signal for end-to-end transmission. Supporting wavelength interchange not only can reduce the network bandwidth waste but also helps with contention for specific wavelengths.

1R, 2R, and 3R regeneration

The degree of transparency depends on the type of signal regeneration at the fabric. 1R simply just relays the signal by amplification without retiming and reshaping. 2R amplifies and then reshapes the signal to remove noise and dispersion without retiming during signal regeneration. 2R offers transparency to bit rate but does not support different modulation formats. 3R regenerates signals through amplification and reshaping and then retiming by synchronising with the network clock. 3R eliminates transparency to bit rates and framing formats completely since the signal is re-clocked. By comparison, 3R produces a cleaner signal at each regeneration node so 3R signals can travel a relatively long distance 'safely'. However, complex regeneration is expensive and time-consuming.

Arbitrary concatenation

Concatenation and grooming are used for improving bandwidth utilisation. A grooming switch has the ability to divide the signal into smaller payload granularities and directs the payloads to different ports. Figure 2.5 shows wavelength and subwavelength concatenation. Arbitrary concatenation refers to the capability of precise band-

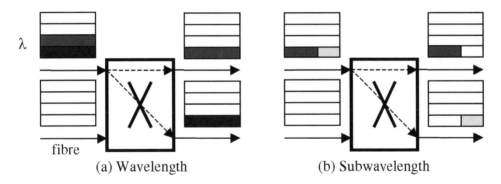

Figure 2.5 Grooming and concatenation.

width provisioning that is not confined to the standard SONET line rates. For example, a conventional OC-48 OADM only handles OC-3 and OC-12, but with arbitrary concatenation, a switch can combine, for example, exactly 17 STS-1s to create an OC-17. Arbitrary concatenation is an important feature for a WDM switch because it introduces flexibility and quality of service to traffic with finer granularities.

A WDM system

ITU-T has an optical transport network layer model that clearly describes a WDM system. Figure 2.6 shows three major layers in a WDM system, defined by ITU-T. The bottom layer is the optical transmission section layer that represents the physical fibre path between network elements within a WDM system. The middle layer is the

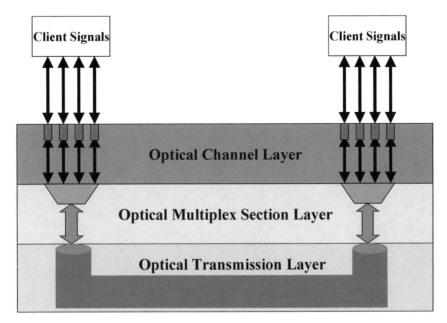

Figure 2.6 WDM system as defined in ITU optical transport network layer model.

optical multiplex section layer that represents the path of the aggregate multiplexed signal through a WDM system. The top layer is the optical channel layer that represents the path of any individual wavelength through the WDM system. Client signals such as Gigabit Ethernet, SONET, ATM, IP, etc. are assigned to wavelengths at the optical channel layer and then transported through the network in the optical transmission layer.

2.2.3   Opaque vs. Transparent Switching

WDM switching functionality can be implemented in either the electrical or the optical domain. In the case of electrical switching, the optical signal is terminated at the entry of the switching fabric, converted to electrical signals, and regenerated as an optical signal at the output of the fabric. By comparison, in an all-optical switching, the original optical signal passes through the fabric without the electrical and optical conversion. Electrically switched fabric is also known as opaque fabric whereas the all-optical switched fabric is referred to as transparent fabric.

In a circuit-switched network, switching control is provisioned through a shadowed data communication network. A transparent optical network refers to the capability of supporting an end-to-end optical channel, upon which there is no O-E-O conversion in the intermediate hops.

Opaque systems are attractive for applications where subwavelength grooming and signal processing are required. The majority of available opaque switches support or will support grooming at subwavelength granularity and provide the ability to arbitrarily concatenate lower-speed signals.

Transparent systems are suitable for switching the entire wavelength and groups of wavelengths at the network core. Table 2.2 presents a feature comparison between opaque and transparent switching.

Table 2.2 Opaque vs. transparent switching

Features	Opaque	Transparent
Wavelength conversion	Intrinsic	May or may not
Optical impairments	Less as signal regenerates	Need special mechanisms
Interoperability	Yes, multi-vendor	Not yet
Traffic grooming	Can be supported	Not supported
Ability to manipulate data	Yes	No
Scalability to high speed	Less scalable	Scalable
Scalability in ports	Less scalable	Scalable
Efficiency for higher speed traffic	Low	High
Technology maturity	Mature	Less mature
Transmission cost per bit	High	Low
Bit rate and protocol transparency	No	Yes
Bandwidth utilisation	High	Low
Flexibility	High	Low
Network management information	Easy to obtain	Difficult to obtain

2.3 WDM Network Testbed and Product Comparison

There are a number of WDM network testbeds which were developed in the 1990s. Commercial WDM network deployment started in 1994 with point-to-point networks. Currently, optical networking is one of the hottest sectors in the telecommunication industry and has developed into several industrial fields:

- optical communication related semiconductors
- optical components
- optical network equipment
- optical network systems.

This section describes early WDM testbeds used for research and development, and then compares commercial WDM network products.

2.3.1 WDM Network Testbeds

Bellcore's LAMBDANET and IBM's Rainbow are the two early WDM local area network testbeds. Both TDM and WDM technology are used in the Bellcore's LAMBDANET system. In the LAMBDANET, each node has a fixed wavelength for each transmitter (using a Distributed FeedBack laser, DFB) and an array of receivers (the size of the array is equal to the number of nodes in the network). The incoming wavelengths are separated using a grating demultiplexer. The design of LAMBDANET architecture aims at simplicity and also supports multicasting. Each transmitter, in time slots, multiplexes the traffic destined to all other nodes into a single wavelength. Each receiving node simultaneously receives all the traffic, buffers it, and selects the traffic destined for it in the electronic domain. Two sets of experiments were performed, with 16 and 18 wavelengths, running at 1.5 and 2.0 Gbps per wavelength, respectively.

IBM's Rainbow network was originally designed to interconnect 32 IBM PS/2 workstations using 32×200 Mbps channels.

The Rainbow testbed was the first to demonstrate tunable components. The network uses fixed transmitters and tunable receivers. Each node has a DFB laser transmitter that is associated with a specific wavelength channel. Before the transmitter sends data, the receiver needs to tune to the transmitter wavelength. The synchronisation is completed in the connection setup process using out-of-band signalling. If a receiver is idle, it will check all the wavelength channels for a connection setup request. Once the receiver finds the channel, it sends an acknowledgement to the corresponding transmitter. Thereafter, the channel is set up and ready for transmission.

The All-Optical Networking (AON) Consortium, formed by AT&T, DEC, and MIT Lincoln Laboratory also under DARPA sponsorship, tried to develop architectures and technologies to exploit the unique properties of fibre optics for advanced broadband networking including both WDM and TDM. AON deployed a static wavelength routing testbed in the Boston metropolitan area to demonstrate the feasibility and interaction of architectures, optical technologies and applications. The testbed uses space switches to implement wavelength routers and converters. It allows the establishment of semi-permanent physical circuits for teleconferencing or other scheduled services. This is supported by DFB lasers with tuning times of tens of seconds.

Dynamic virtual circuits are implemented by distributed Bragg-reflector (DBR) lasers with tuning ranges of 10 nm and tuning times from 2 to 8 ns. This circuit-switched service is implemented as a scheduled time division multiplexing system for delivering data from a source to a large number of destinations. A best-effort datagram service is implemented by providing optical Ethernet channels linked by conventional Internet routers. The testbed network supports 20 wavelengths. User nodes are connected to the testbed by 100 Mbps FDDI links.

Long-distance WDM transmission was started at AT&T Bell-Labs and British Telecom. Field trials were performed on the AT&T testbed in 1992 with four wavelengths, running at 1.7 Gbps each. The wavelengths are sliced from the lowest attenuation region 1.55 μm, and WDM transmission is tested to a distance of 840 km.

The European RACE (Research and development in Advanced Communications technologies in Europe) MWTN (Multi Wavelength Transport Network) program involves a consortium of European companies and universities. In one of the BT RACE field trials, the demonstration network spans 500 km and uses four wavelengths (in the 1.3 μm band), running at 2.5 Gbps each. The RACE MWTN program has developed two basic optical networking elements: the optical cross-connect (OXC) and the optical add/drop multiplexer (OADM).

Multiwavelength Optical Networking (MONET), a five-year program funded by the DARPA and MONET Consortium (including AT&T, Bell Atlantic, BellSouth, Lucent, Pacific Telesis, SBC/TRI, and Telcordia), aimed at developing the technologies needed to build a flexible, reliable, high-capacity, high-performance, and cost-effective national scale optical WDM network. The first phase of the MONET program was completed in 1997, where the MONET New Jersey Network was constructed and deployed. This network consisted of three interconnected testbeds: the LEC testbed at Red Bank, NJ, the Long Distance Testbed at Holmdel, NJ, and the Cross-connect testbed at Holmdel, NJ. The network demonstrated for the first time managed a multiwavelength network elements and connection setup across two networks and control administrative domains with transmission at 2.5 Gbps on eight wavelengths and over 2000 km.

The second phase of the MONET program was completed in 1999 with the deployment of the MONET Washington DC Area Network. The DC network consists of several types of WDM network equipment: Wavelength Selective Cross-connect (WSXC), Wavelength Amplifier (WAMP), and Wavelength Add/Drop Multiplexer (WADM). The field trial network consists of two separate ring networks. The west ring (O-E-O network) consists of WADMs and WAMPs supplied by Tellium, and the east ring (all-optical network) consists of WSXCs and WAMPs from Lucent. Both rings are managed by a single network management system. This system also integrates management of the optical, SONET, and ATM layers. The control messages are transported over a shadowed Data Communications Network (DCN). The MONET DC network supports eight wavelength channels, each of which operates at 10 Gbps. For the first time, WDM reconfigurability and end-user applications were demonstrated over a WDM network.

European's ACTS (Advanced Communications Technology and Services) program aims at accelerating deployment of advanced communications infrastructures and services. ACTS has a program of funding from the European Union, and has five partners: BT, Ericsson, Telia, Italtel, and University of Essex and with the following focus: design and management of optical network, customer access networks, multiplexing and transmission, key optical components, and switching and routing.

In 1998, DARPA established several key NGI projects. Among them, there is the SuperNet project, which aims to advance networking technologies and applications through deployment of national-scale testbeds vastly superior to today's Internet.

On the US West Coast, the National Transparent Optical Network (NTON) deployed over a distance of 2500 km linking Seattle and San Diego, with a bandwidth of 10–20 Gbps. NTON is a WDM network employing in-place commercial fibres. NTON links government, research and private sector labs and provides the ability to interface with most of the broadband research networks in the US.

On the US East Coast, NGI Optical Network for Regional Access using Multiwavelength Protocols (ONRAMP) deployed in the Boston area to investigate the cost-effective distribution of network bandwidth for regional access, and the Boston to Washington DC fibre optic network (BoSSNET) constructed using a dark fibre connection and a reliable OC-3 link. The link will utilise an embedded TrueWave fibre plant and approximately 15 optical amplifier huts spaced by 60~90 miles.

Figure 2.7 shows the list of WDM network testbeds. Table 2.3 shows five WDM network testbed comparison.

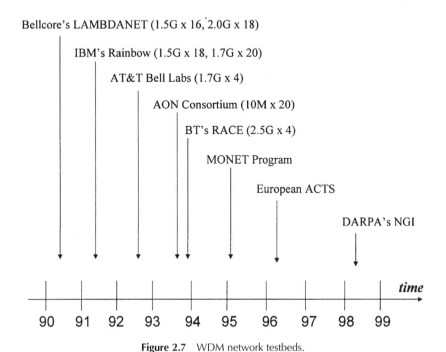

Bellcore's LAMBDANET (1.5G x 16, 2.0G x 18)

IBM's Rainbow (1.5G x 18, 1.7G x 20)

AT&T Bell Labs (1.7G x 4)

AON Consortium (10M x 20)

BT's RACE (2.5G x 4)

MONET Program

European ACTS

DARPA's NGI

time

90   91   92   93   94   95   96   97   98   99

Figure 2.7 WDM network testbeds.

Table 2.3 WDM network testbed comparison

Field trail	AON	Photo	OPEN	Meton	MONET
Region/area	Boston	German	France	Sweden	NJ/DC
Distance	LAN	500km	380km	MAN	WAN
Bits/sec per wavelength	10M/10G	2.5/10G	2.5G	2.5G	2.5G/10G
Number of wavelengths	20	4/8	4	4	8/16
Device	Giga-switch	OXC	OXC	OXC/ OADM	WSXC/ WADM

2.3.2 Product Comparison

Table 2.4 shows a comparison of products from Lucent, Ciena, Corvis, Tellium, Nortel, Alcatel, Calient, and Village Networks. The products are compared in terms of these features:

- switching technology
- fabric type
- switch capacity
- client interface signal format/line rate
- number of wavelengths
- granularity
- total ports.

Switching technology embraces one of these values: O-E-O circuit switching, all optical circuit switching, optical packet switching, and optical burst switching. Switch capacity specifies the maximum capacity provided by the switch.

2.4 Communication Protocols

A protocol is a set of rules and formats that govern the communication between peers. The key elements of a protocol include message syntax, message semantics, and timing. Message syntax specifies the fields a message contains and the format the message is encoded; message semantics interpret the meaning of the message; timing includes speed matching and sequencing between the sender and the receiver.

Example functions of protocols are:

- fragmentation and reassembly
- encapsulation
- connection control
- ordered delivery
- flow control
- error control
- addressing
- multiplexing
- transmission services.

Table 2.4 WDM product comparison[a]

Manufacturer	Product	Switching technology	Fabric type	Switch capacity	CI signal format/line rates	number of wavelengths	Granularity	Total ports
LUCENT	Lambda Router	All-optical circuit switching	256x256 3D MEMS	2.6 Tbps	ANY	NA	Lambda	256 OC-192
	Lambda Router2	All-optical circuit switching	3D MEMS	48 Tbps	ANY	NA	Lambda	1200 OC-768
CIENA	CoreDirector	OEO circuit switching	Electrical CLOS	7.68Tbps	OC-3- OC-192	16	STS-1	3072 OC-48 or 718 OC-192
CORVIS	Optical Switch	All-optical circuit switching	NA	2.4 Tb/s	OC-48, OC-192	NA	Lambda/Fibre	6 fibre ports or 960 OC-48
Tellium	Aurora	OEO circuit switching	Electrical 256x256 MEMS	1.28 Tb/s	OC-3- OC-192	128	OC-48	512 OC-48
	Aurora Full Spectrum	All-optical circuit switching		10.2 Tb/s	SONET	NA	Lambda	256 OC-192
Nortel	OPTera PX	All-optical circuit switching	3D MEMS	14 Tb/s	OC-3 - OC-192	32	Lambda	256/1024 OC-192/OC-768
Alcatel	OPTera HDX Crosslight	OEO circuit switching / All-optical circuit switching	Electrical CLOS Agilent 32x32 Bubble	1.33 Tb/s	NA	NA	STS-1	800 OC-192
	Optinex 1680 OGM	OEO circuit Switching	I-Cube Electrical	813 Gb/s	NA	NA	STS-1	192 16000 DS-3 or 325 OC-48
Calient	Diamond Wave4K	All-optical circuit Switching	4096x4096 3D MEMS	164 Tb/s	NA	NA	Lambda	4096 ports
Village Networks	Optical Packet Node	Optical Packet switching	NA	640 Gpbs	Mutli-rate/multi-protocol packet interfaces (GbE, OC-3/STM-1 to OC-192/ STM-64 / POS ATM, Ethernet, MPLS, MPLambdaS)	128	Flow switching	256 OC-192

[a]Information in this table is collected from manufacturers' web sites and advertising brochures.

A network protocol basically provides a service to network applications. Protocols can be structured into layers with dependencies, where each layer offers certain functionality and lower layer protocols provide services to higher layer protocols. A set of such protocols forms a protocol stack. The interface between an upper and a lower layer protocol in the protocol stack is a Service Access Point (SAP).

Protocols can be implemented either in user space or in kernel. User space protocols are easier to maintain and customise but have poorer performance since each data access involves two context switches. A context switch is needed whenever there is a mode switch, for example, user mode switching into kernel mode. A kernel-located protocol is efficient but hard to customise and maintain.

A protocol can be symmetric or asymmetric. Symmetric protocol is used between peer entities. Most of the network layer (layer 3) protocols, such as OSPF and BGP, are peer-to-peer protocols. Asymmetry in protocols can be dedicated by the logic of a message exchange or the desire to keep one of the entities as simple as possible.

An example of the logic for message passing in an asymmetric protocol is client/server relationship. Master/slave protocol is another example of asymmetric protocol. In addition to interfaces, protocol dynamics and behaviours can be described using a finite state machine (FSM). A FSM represents an abstract machine that at any time is in exactly one of its finite sets of state. An input to a FSM triggers the transition to another state. The state after the transition depends only on the current state and the input. Figure 2.8 shows the FSM for TCP congestion control.

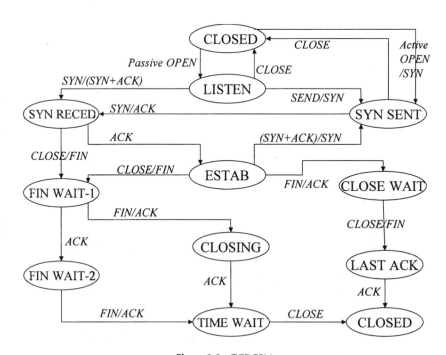

Figure 2.8   TCP FSM.

2.5 Internet Architecture

The continuous growth of the Internet is one of the most exciting and interesting phenomena in networking. Various kinds of applications utilising the Internet have appeared, such as e-commerce and collaborative groupware, but Internet-based applications and services are still evolving and maturing. Computer networking is a complex subject because many technologies exist and many corporations have created networking protocols independently, which are not always all compatible. A global Internet requires conformance to certain standards in order to interwork with each other. The current Internet standardisation is referred to as the TCP/IP model. Table 2.5 describes the TCP/IP model with reference to the OSI communication model.

Table 2.5 TCP/IP vs. OSI model

Layer No.	OSI model	TCP/IP model	Functionality
7	Application	Application	Provides application level protocols, e.g. TELNET for virtual
6	Presentation		terminal, FTP for file transfer, SMTP for electronic mail, SNMP
5	Session		for network management, HTTP for fetching web pages.
4	Transport	Transport	Provides end-to-end data transport management from the source machine to the destination machine, e.g. TCP and UDP.
3	Network	Network	Provides a (virtual) network layer that route packets or make connections across the network, e.g. network addressing, packet routing protocols, data transfer functions, ICMP, ARP, and RARP.
2	Data link	Data link	Provides a reliable transfer of data across the physical link, e.g. framing and clock synchronisation. Also regulates access to the network, e.g. MAC protocols.
1	Physical	Physical	Transmits the unstructured data over the physical medium

Physical layer is the actual layer that transmits and carries data across a physical link. So it is concerned with the transmission media and its properties and signalling agreements. Example physical layer devices include amplifiers, repeaters, and concentrators. The data link layer encapsulates data according to framing formats. Framed data are transmitted over the physical layer connection. Data frames can use error detection and error correction. In addition to transmission error control, data link layers can provide flow control that prevents a sender from overrunning a slow receiver.

Example data link layer protocols provide error and flow control including:

- SDLC (Synchronous Data Link Control)
- HDLC (High level Data Link Control)
- LAPB (Link Access Procedure, Balanced)
- SLIP (Serial Line IP)
- PPP (Point to Point Protocol).

In the case of multiple-access links, a sublayer of the data link layer is the MAC (Media Access Control) sublayer that mediates access to a shared link so that all nodes eventually have a chance to transmit their data. A well-known example of MAC layer is the CSMA/CD (Carrier Sense Multiple Access/Collision Detection) protocol.

The network layer is aimed at network-level issues such as routing and interoperability. IP as a network layer protocol has been widely accepted and deployed. IP control protocol suite includes:

- IGP (Interior Gateway Protocol) routing protocol
- EGP (Exterior Gateway Protocol) routing protocol
- ICMP (Internet Control Message Protocol)
- RSVP (Resource Reservation Protocol).

A network layer can be viewed as a virtual layer across different types of subnetworks. To enable universal routing and control message protocols, physical interfaces are given virtual addresses, i.e. the IP addresses. ARP provides the address translation from an IP address to the physical address in a LAN.

The transport layer sitting on top of the network layer is responsible for end-to-end delivery management. Two well-known services provided in the transport layer are TCP and UDP (User Datagram Protocol). TCP offers connection-orientated services, which provide reliable transport across the network. UDP offers a connectionless service, which provides the IP best-effort service to applications.

The application layer is comprised of application protocols, which can be either a network utility protocol or an application-specific protocol. The former includes:

- FTP (File Transfer Protocol)
- TELNET (remote terminal protocol)
- SMTP (Simple Mail Transfer Protocol)
- HTTP (Hypertext Transfer Protocol)
- SNMP (Simple Network Management Protocol).

The latter includes various kinds of application specific and proprietary protocols.

IP, a layer 3 protocol, is the foundation of the Internet architecture. IP not only offers a network-level interconnection for building diverse distributed applications, but also hides the details of the underlying networking technology from applications by providing a simple packet delivery service.

The basic Internet architecture interconnects networks using a special network device, known as a router. In this way, the Internet can be seen as routers interconnecting networks as shown in Figure 2.9. An IP router employs a store-and-forward packet switching paradigm, where a routing table is constructed with the destination network addresses and the next hop addresses, and an incoming packet is examined and forwarded to the next hop according to its destination address. The routing table is stored and maintained by the routing protocols such as OSPF.

2.6 IPv4 Addressing

An important function of IP is IP addressing, which divides the IP address into host portion and network portion. Within a single IP network, IP packets can be directly

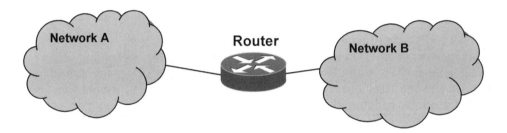

Figure 2.9 Internet architecture.

forwarded to the destination host. If the IP packet is destined for a different network after examining the packet header, the packet is forwarded. If the forwarding system is a host, the packet will be forwarded to the default router. The IP address is designed as a global identifier.

An IPv4 address is a 32-bit value that uniquely identifies an interface, not a host or router. Hence, an IP address describes a connection to a network not an individual computer. Having a network and a host portion in the address introduces a hierarchical structure to the addressing space.

IPv4 has five classes of address which are shown in Figure 2.10. However, global addresses are now allocated using the CIDR scheme.

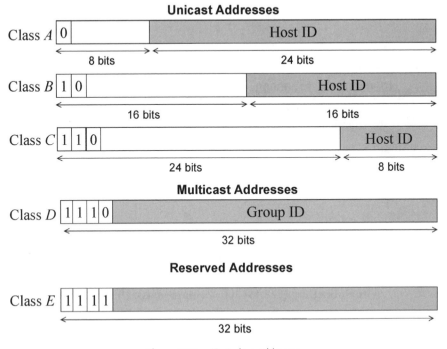

Figure 2.10 IPv4 class addresses

2.6.1   Subnetting

The IPv4 addressing structure is static. With the continuous growth of the Internet and the increase in Internet users and devices, IPv4 addresses are becoming scarce. At the same time, Class A and B addresses can accommodate a large number of hosts within a network. To efficiently use the existing unicast addresses, subnet addressing is introduced, by which multiple physical networks can share a common network prefix.

The original host portion of the address space can be further divided into physical network portion and host identifier. Subnet addressing also allows a finer granularity of IP routing since there will be more layers in the routing hierarchy. For instance, Bellcore/Telcordia has a Class B address 128.196.0.0. Within Telcordia, one octet can be used for representing physical networks and the last octet is used for the host identification.

Subnet addressing schemes are flexible since each site or company decides its own scheme. To avoid ambiguity and further enable subnet routing, a subnet mask must be provided for each network implementing a subnet. In the Telcordia example given, the subnet mask has the value of 255.255.255.0 (see Figure 2.11). Furthermore, a network can even be configured with different masks, also known as variable length subnet masks (VLSM). Without VLSM, only one subnet mask can be applied to a network. This limits the flexibility on the number of hosts given the number of subnets. In other words, VLSM allows further partition of the remaining host space in addition to the first subnet boundary. For example, VLSM can divide a network with 256 addresses into three subnets: subnet 1 with 128 addresses, subnet 2 with 64 addresses, and subnet 3 with 64 addresses.

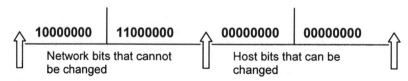

Without Subnetting (128.196.0.0 & default subnet mask 255.255.0.0)

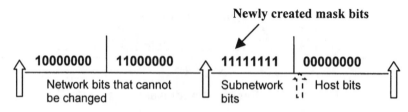

With Subnetting (128.196.0.0 & subnet mask 255.255.255.0)

Figure 2.11   An example (128.196.0.0) of IP subnetting.

2.6.2   Unnumbered Addresses

Sometimes, there are a number of point-to-point networks within a company's network. To save network addresses, the point-to-point network can be configured as an 'anonymous' network with unnumbered links. In that case, the two physical interfaces connecting each link are not assigned IP addresses. Instead, they are explicitly declared as 'unnumbered' and given aligned addresses. The aligned address can be other physical interface's IP address at the same router. In the routing table, the destination network connected using the unnumbered links appears as the default route and the next hop address is the remote site's aligned IP address.

Figure 2.12 gives an example of an IP unnumbered link in an anonymous point-to-point network. There is an unnumbered link between routers A and B, each of which has two physical interfaces, le0 and le1. The unnumbered link uses router A's le0 interface and router B's le1 interface. The interfaces are configured as described in the figure. Router A and B's routing tables are also shown in the figure.

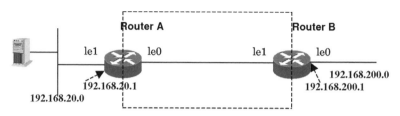

Interface Configuration:

Router A
interface le1 192.168.20.1
interface le0 unnumbered 192.168.20.1

Router B
interface le0 192.168.200.1
interface le1 unnumbered 192.168.200.1

Router A Routing Table:

Dest. Network	Next Hop	Physical Interface
default	192.168.200.1	le0
192.168.20.0	host addresses	le1

Router B Routing Table:

Dest. Network	Next Hop	Physical Interface
default	192.168.20.1	le1
192.168.200.0	other addresses	le0

Figure 2.12   IP unnumbered link.

2.6.3   Secondary Addresses

IP also allows a physical interface to have more than one IP address. The first address is the primary address while other addresses are known as secondary addresses. A secondary address is usually considered more dynamic or temporary than the primary address. Secondary addresses can be used for the following purposes:

- In a mobile environment such as in mobile IP, an interface can have multiple addresses. The primary address is the home network address, while secondary addresses can be assigned dynamically to facilitate roaming.
- Secondary addresses can be used to create a single network from subnets that are physically separated by another network. An example of this usage is the Virtual LAN.

- Sometimes there may not be enough host addresses for a particular network segment. To accommodate the additional hosts, the default router can be given secondary addresses to emulate two logical subnets using one physical subnet.

2.6.4   Classless Inter-Domain Routing (CIDR)

In 1993, the restriction on class-based address and routing in IPv4 was lifted to tackle the worldwide shortage of IPv4 addresses, the growing popularity of the Internet, inefficient class-based address allocation, and the growing size of the Internet core routers' routing table. CIDR allows an IP network to be represented with a variable prefix length and not just the 8, 16, 24 bit boundaries of class A, B, and C respectively.

The CIDR variation has the format of $< prefix/length >$, in which the length specifies the number of the leftmost contiguous significant bits representing the network portion of the address. For example, you can use 200.10.0.0/16 to represent a network, where the left-hand 16 bits provide the network part of the address.

2.7   Gigabit Ethernet

The first Ethernet was conceived and implemented at Xerox PARC (Palo Alto Research Centre) in 1973, where the lab prototype operated at 2.94 Mb/s. In 1980, the IEEE organised its first meeting on LAN technology. In the meeting, several working groups were created. IEEE 802.3 is the standard for LANs based on CSMA/CD technology, and IEEE 802.4 and 802.5 are the standards for LANs using Token Bus and Token Ring.

In the early 1980s, 10 Mb/s Ethernets over twisted pair cables started to appear, and in 1990, the IEEE approved the well-known 10Base-T standard. However, Ethernet remains simple as a transmission technology and Ethernet devices include only hubs and repeaters. To interconnect Ethernet LANs, Bridges are introduced and full-duplex twisted-pair links between the host and the hub can be used. The introduction of Bridges/Switches allowed LAN bandwidth to be dedicated to a single computer, rather than its being shared among multiple devices. Therefore, Ethernet is no longer just a shared access medium, which has to avoid collisions, for example, by using CSMA/CD (Carrier Sense Multiple Access with Collision Detection) protocol. CSMA/CD provides network access in a shared access Ethernet.

In CSMA/CD, to send data, a node first listens to the network to see if it is busy. When the network is not busy, the node will send data and then sense collisions. If a collision occurs, the data is re-transmitted. Due to Ethernet's popularity and end applications requirement, Ethernet technology is demanded with higher and higher data rate. Fast Ethernet was introduced in 1992 with an operating data rate of 100 Mbps. In 1996, an IEEE 802.3 Task Force was formed to develop a standard allowing Ethernet operating at 1000 Mbps, also known as Gigabit Ethernet. The Gigabit Ethernet evolution is shown in Figure 2.13.

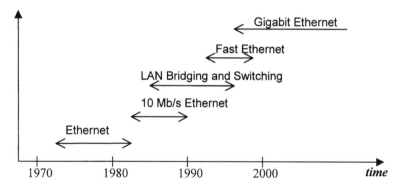

Figure 2.13   Gigabit Ethernet evolution.

IEEE 802.3 also developed a shorthand notation that refers to any particular Ethernet or Gigabit Ethernet implementation. The notation uses the template *rate-signal-phy*, where *rate* represents the data rate in megabits per second, *signal* indicates the signalling on the channel, being either baseband (BASE) or broadband (BROAD), and *phy* refers to the nature of the physical medium. For example, 10BASE5 represents the original Ethernet with 10 Mbps bandwidth, baseband channel (i.e. the channel is dedicated to the Ethernet), and a maximum length of 500 m.

1000BASE-T refers to the Gigabit Ethernet standard for copper UTP. The standard allows up to 100 m of four pairs of Category 5 UTP. 1000BASE-X represents the Gigabit Ethernet standard for Fibre Channel Physical Layer. Gigabit Ethernet reuses the existing Fibre Channel technology, which is used to interconnect supercomputers, storage devices, and workstations. Three types of medium are supported in 1000BASE-X:

- **1000BASE-CX**: two pairs of Shielded Twisted-Pair (STP), 25 meters maximum.
- **1000BASE-SX**: two multimode or single-mode optical fibres using shortwave laser optics (e.g. 850 nm laser)
- **1000BASE-LX**: two multimode or single-mode optical fibres using longwave laser optics (e.g. 1300 nm laser).

Figure 2.14 shows the Ethernet frame format. Destination and source addresses identify the receiving and sending station for the frame. The length/type field represents the number of valid data octets contained within the data field. The data field

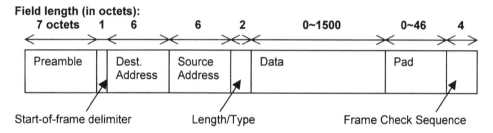

Figure 2.14   Ethernet frame format.

carries the payload information. The pad field is used to fill out the frame to the minimal size, for example, for collision detection. The last four octets encodes a checksum based on the frame contents excluding the first eight octets. To ensure that the frame is long enough to be detected in case of collisions, Ethernet requires a minimum frame size of 64 bytes.

Recently, Gigabit Ethernet vendors have began to support Jumbo Frame that extends the frame size up to 10,240 octets. As a result, the data carried per Gigabit Ethernet frame has been considerably increased.

2.7.1 Gigabit Ethernet Architecture

Gigabit Ethernet is a data link layer and physical layer technology, so it is transparent to network and other higher layers. Figure 2.15 shows the Gigabit Ethernet protocol architecture. GMII and XGMII are the interfaces between the MAC layer and the physical layer, and use the same management interface as MII (Media Independent Interface) defined in Fast Ethernet. The Gigabit MII provides separate 8-bit wide receive and transmit data paths so it supports both half-duplex and full-duplex operations. It offers two media status signals: one represents the presence of the carrier and another represents the absence of collision. The reconciliation sublayer maps these signals to physical signalling primitives defined in the MAC layer. Therefore, using Gigabit MII allows the same MAC layer to connect various media types.

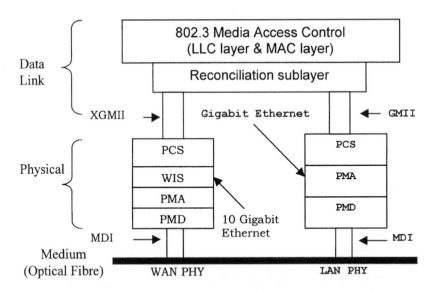

MDI = Medium Independent Interface   PMD = Physical Medium Dependent
PCS = Physical Coding Sublayer   WIS = WAN Interface Sublayer
PHY = Physical Layer Device   GMII = Gigabit Media Independent Interface
PMA = Physical Medium Attachment   XGMII = 10 Gb Media Independent Interface

Figure 2.15   Gigabit Ethernet protocol architecture.

The physical layer has three sublayers for Gigabit Ethernet and four sublayers in 10-Gigabit Ethernet. PCS provides a uniform interface to the Reconciliation layer for all physical media. Carrier sense and collision detect notifications are generated in this sublayer. PMA offers a medium independent means for the PCS to support various serial bit-orientated physical media. PMD maps the physical medium to the PCS so this sublayer defines the physical layer signalling used for various media.

10-Gigabit Ethernet has a WIS used for WAN environment. In the physical layer (PHY), two standards are defined. LAN PHY operates at a data rate of 10 Gb/s to support data transmission over dark fibre and dark wavelengths. WAN PHY operates at a data rate compatible with the payload rate of OC-192c (i.e. 9.58 Gb/s) to provide support for transmission of Ethernet on installations based on SONET. In WAN PHY, the SONET frame format is adopted but only limited overhead is implemented.

A Gigabit Ethernet Network Interface Card (NIC) comprises an Ethernet controller (i.e. a MAC entity), an Encoder/Decoder (ENDEC), and a set of drivers and receivers for the particular physical medium in use.

2.7.2 Gigabit Ethernet Applications

One of the key applications of Gigabit Ethernet is inter-campus networking. Each campus has a number of LANs. To interconnect campus networks, Gigabit Ethernet offers an attractive cost-efficient solution. Most of the complexity of an electronic router comes from the layer-3 functionality, i.e. network layer routing. However, when LANs are connected, they use the same layer 2 technology so layer 3 can be omitted. A layer 2 Gigabit Ethernet switch is simpler and has higher performance than layer 3 routers. Routing in layer 2 switched networks uses the spanning tree algorithm specified in IEEE 802.1D. The spanning tree routing is simple but static and has scaling problems (comparing to layer 3 routing such as OSPF).

Figure 2.16 shows typical applications for Gigabit Ethernet. Multi-tenant buildings use LAN technology. An access Gigabit Ethernet layer 2 switch is installed to connect the building LANs and bridge them to a core Gigabit Ethernet switch. The core switch has layer 3 (IP) functionality since it is used to interface with IP routers and inter-connect different technology networks. As shown in the figure, Gigabit Ethernet networks can be deployed in a metropolitan area to interconnect servers, super-computers, databases, ISPs, and user traffic.

Figure 2.17 shows 10 Gigabit Ethernet applications. 10 Gigabit Ethernet links can be used to connect MAN traffic to WAN. A typical example of WAN is long haul WDM network that supports several wavelength channels, operating at OC-192 each. End user or application traffic can be aggregated through Gigabit Ethernet and then 10-Gigabit Ethernet and connected to WAN for long distance transmission.

Gigabit Ethernet (or 10-Gigabit Ethernet) is not new. The existing solutions in the MAN include ATM and SONET rings. However, Gigabit Ethernet offers a cost-efficient alternative. It may not catch up with SONET rings in terms of 'carrier grade' reliability, which requires heavy performance monitoring and high levels of protection/restoration. But this also further increases its cost advantage. Gigabit Ethernet not only has a lower per port cost but also offers even lower cost on network operations

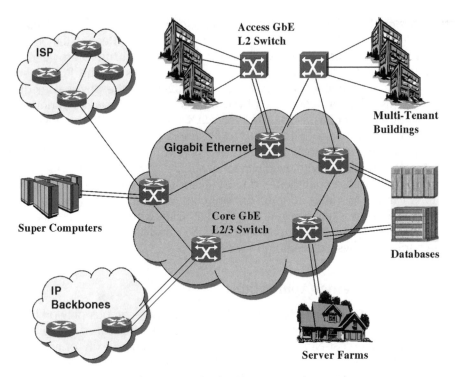

Figure 2.16 Gigabit Ethernet networks.

and provisioning. Currently, there are strong interests in traffic management and QoS in the Gigabit Ethernet industry.

2.8 Multiprotocol Label Switching (MPLS)

Conventional IP networks offer flexibility and scalability, but they provide only best-effort service. However, as the deployment of IP networks continues to grow and IP networks compete to become a major standard in not only data networking but also in telecommunication networks, certain applications, especially voice-based traffic, have strong demands for quality of service.

MPLS is a technology integrating the label-switch forwarding paradigm and network layer routing. In a MPLS network, packets are encapsulated with labels at the ingress edge node, and the labels are then used to forward the packets along label switched paths (LSP) before being removed at the egress edge node. MPLS is a value-added service to IP since the LSPs can be explicitly configured along specific paths. Without explicit routing entities, MPLS supports only the IP best-effort forwarding.

A MPLS packet has a 32-bit 'Shim Header' as shown in Figure 2.18. Within the header, the Label field specifies the MPLS label and uses 20 bits. The EXP field denotes 'experimental' although its proposed use is to indicate Per Hop Behaviour of labelled packets travelling between LSRs. The 1-bit S field indicates the presence of a label stack. The 1-octet TTL (Time To Live) field is decremented at each LSR hop

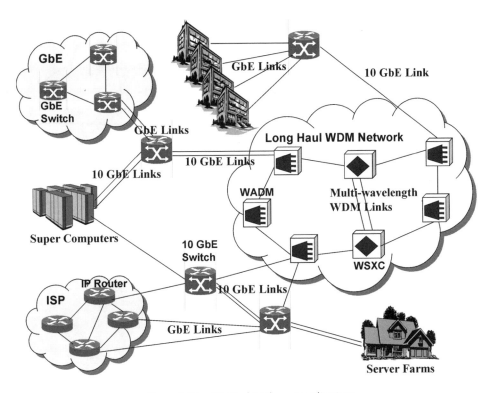

Figure 2.17   10 Gigabit Ethernet applications.

and is used to throw away looping packets. The value of MPLS TTL is copied from IP header at ingress and copied back to IP header at egress.

Figure 2.19 illustrates MPLS with its key concepts. A MPLS network consists of Label Switch Routers (LSR), which are able to switch packets according to labels embedded in the packet header. There are two types of LSR in MPLS networks: edge and core LSR. Each core LSR has a label forwarding table, in which there are at least two columns, the in-label and out-label. Once a packet arrives, the label-forwarding table decides its next hop based on the label carried in the MPLS packet header. The

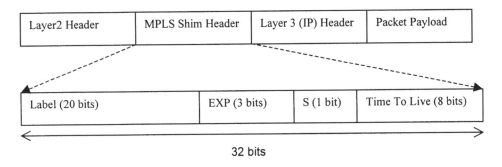

Figure 2.18   MPLS header.

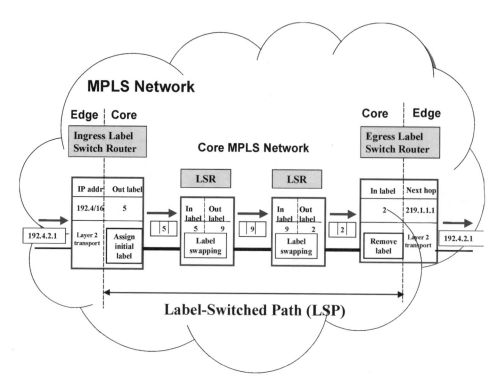

Figure 2.19   MPLS concepts.

outgoing packet is transmitted with the new label. This process is called label swapping. The in-label and out-label are negotiated during the process of LSP setup.

The edge LSR also supports traffic mapping, which defines a Forward Equivalent Class (FEC), and maps IP traffic to FEC and then to a LSP. A FEC is a grouping function, according to which, a set of packets (however different they may be) share the same forwarding function. An example of FEC is the destination address in the IP packet header. In an ingress LSR forwarding table, instead of in-label, an IP address is used; in an egress LSP forwarding table, instead of out-label, the next hop address is specified. At the ingress, non-MPLS packets are encapsulated with an initial label; at the egress, the labels are removed from the MPLS packets before sending to other networks. A LSP is an end-to-end data-forwarding path across the MPLS network, and it consists of an ingress LSR, a series of label switching hops, and an egress LSR.

2.8.1   Label Distribution

A LSP can be set up using a distributed signalling protocol, where a major task is label distribution. Two well-known signalling protocols used in MPLS are the Label Distribution Protocol (LDP) and the Resource Reservation Protocol (RSVP). Figure 2.20 shows an example of LSP setup using RSVP. Once a LSP is computed by the ingress LSR, the ingress LSR RSVP daemon is informed to send a PATH message,

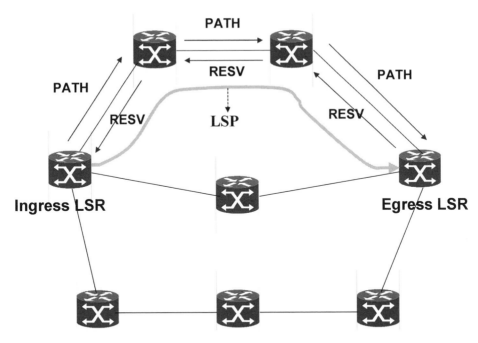

Figure 2.20   Signalling with RSVP.

which is used to request labels for the new LSP. A PATH message contains the following objects:

- **Session Object**: uniquely identifies the LSP tunnel.
- **Session Attribute**: controls LSP priority, pre-emption, reservation scheme and fast reroute.
- **Sender_TSPEC**: transmission specification to indicate desired bandwidth reservation.

There are two 'flavours' on routing PATH message. One is to use source routing, in which the PATH route is embedded into the PATH message. Source routing is suitable in the case of explicit routed LSPs since an IP routing protocol provides only a best-effort routing path. Another is to address the PATH messages to the IP address of the egress LSR. This has the assumption of using the default routing path provided by IP. In the RSVP implementation, there is an Explicit Route Object (ERO) that can specify the PATH message route. ERO can be specified as either 'loose', which relies on the routing table to find the destination, or 'strict', which details the directly connected next hop. ERO allows a route having both loose and strict components.

Once the egress LSR receives the PATH message, the egress LSR allocates a label and starts to send RESV messages to its LSR upstream. Upon receipt of the RESV message, each LSR on the upstream LSP determines if it can satisfy the requests for bandwidth, QoS, and other aspects of the LSP policy. If the LSR can comply with the request, it distributes a label to its upstream LSR along the signalling path. In such a way, labels are distributed from downstream to upstream LSR until the ingress LSR receives the final label. This is known as downstream initiated label distribution. After receiving the RESV message, the ingress LSR formally announces the new LSP by updating its traffic engineering database and routing table. The traffic engineering database is located at each LSR and synchronised by the IP routing protocol (i.e. OSPF).

In addition to the functionality provided in RSVP, LDP provides an LSR neighbour-hood discovery mechanism to enable LSR peers to locate each other and establish communication automatically. The LDP discovery protocol works by sending 'Hello' messages using UDP to a multicast address ('all the routers in the current network' group). Hello messages can also be sent through tunnels, which represent the existing LSPs. Once it discovers a neighbour, a LDP session can be formed between two LSRs using a TCP connection, for example, to exchange labels.

2.8.2   Traffic Engineering

The most important application of MPLS initially is traffic engineering. MPLS traffic engineering aims to precisely control traffic flows in the network in order to optimise network usage. It does this by shifting traffic from an IGP-defined path to a less congested path.

There are two approaches for MPLS traffic engineering: offline and online traffic engineering. Offline traffic engineering works like a simulation model. It aims at global resource optimisation by taking consideration of all link and node resource constraints and all LSPs in the network. Due to the complexity involved in this approach, the solution may no longer be optimal once it is generated. This is because the routing topology and traffic patterns themselves are dynamic in nature.

The online traffic engineering uses constraint-based routing, where operators can configure LSP constraints at the ingress LSR. Examples of the constraints are band-width reservation, including or excluding a specific link, and including specific node traversals. Figure 2.21 shows the entities and flows in the constraint-based routing.

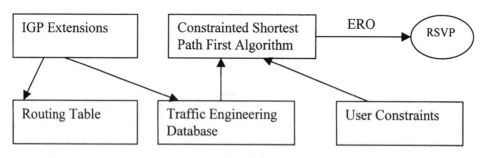

Figure 2.21   Constraint-based routing.

An example of an IGP is the OSPF protocol. OSPF extensions for traffic engineering refer to the design and implementation of opaque traffic engineering LSA (Link State Advertisement) with type 10. The opaque LSA does not have any effects on the standard LSA (types 1 to 5). Using this routing protocol extension, Traffic Engineering Database (TED) is populated with detailed network link attributes and topology information. TED has a similar format to the standard IGP link-state database but it is maintained independently of the IGP database.

Based on TED, the Constrained Shortest Path First (CSPF) algorithm computes LSPs across the network topology (i.e. IP topology). An explicit route calculated by CSPF is handed to RSVP as an ERO.

2.8.3   Quality of Service (QoS)

IP has two QoS models, integrated services (*intserv*) and differentiated services (*diffserv*). Intserv is designed to provide end-to-end QoS guarantees for individual applications, i.e. fine-grained QoS (or QoS per connection/flow). Diffserv is designed to provide a small number of classes of services across the IP networks, i.e. coarse-grained QoS (or QoS for large aggregated traffic streams). Although a MPLS network is equipped with virtual connection (LSP) setup and teardown mechanisms including RSVP, it matches diffserv better (than intserv).

Intserv isolates the controlled service traffic from the guaranteed service traffic, which has a bound on the maximum delay. It requires that the networks implement these mechanisms:

- admission control
- flow specification
- flow reservation protocol RSVP
- packet scheduling for QoS guarantees.

The flow specification has two parts: request specification for requested service types and traffic specification for bandwidth requirements. To support intserv, each router has to implement connection admission control to decide whether the connection can be admitted without degrading the QoS of other already admitted connections, and packet scheduling to associate each packet with the appropriate reservation, for example, based on source/destination address, protocol number, source/destination port.

Intserv uses RSVP for bandwidth reservation. The soft-state maintained reservation paths generate a large amount of control traffic (possibly for many hours) so the approach is not scalable with the number of flows. Intserv also introduces router complexity in the implementation of admission control and packet classification and scheduling. Hence, it is not suitable for core networks.

Diffserv has better scalability because the complex classification and conditioning functions are only performed at the network boundary, while within the network, traffic aggregates are treated according to the per-hop behaviour (PHB). There is no signalling required; instead service agreement. As a result, QoS is not guaranteed but differential, i.e. higher priority traffic gets better QoS than low priority traffic.

Flow-level traffic management is only performed at the edge router. Packets are marked with a Differentiated Services Class Point (DSCP) at trust boundary. The DSCP is carried in the 6-bit differentiated service field of the IP packet header, which was formerly known as the Type of Service field.

Class-level traffic management is performed at the core routers, which implement PHBs. The DSCP within the packet header indicates the packet's PHB. Standard PHB include Best Effort (BE), Expedited Forwarding (EF), and Assured Forwarding (AF). Diffserv is aimed at aggregated traffic not individual flows. The Service Level Agreement (SLA) between networks sharing a border with a diffserv network is manually configured and static. There is no signalling for dynamic SLA in diffserv.

MPLS is a natural match to the IP diffserv architecture since both MPLS and diffserv scalability comes from traffic aggregation at the edge and aggregate processing in the core. In order to support diffserv using MPLS, the DSCP needs to be encoded into the MPLS header.

There are two methods for MPLS support of diffserv. The first method uses the EXP bits in the MPLS header to represent the packet DSCP, which is known as E-LSP. The second method uses the label identifier in the MPLS header to represent class of service, which is known as L-LSP. Figure 2.22 shows a comparison between E-LSP and L-LSP. Both of them need to set up LSPs using label-binding protocols such as RSVP and LDP. However, L-LSP also requires PHB or PHB scheduling group signalling to bind a packet/flow queue to a label. Both of them implement queues but the E-LSP queue is inferred from the label and the EXP field, whereas the L-LSP queue is inferred exclusively from the label. In the figure, E-LSP supports EF and AF1 services on the single LSP, whereas EF and AF1 packets are routed into different queues.

L-LSP supports EF and AF1 services on separate LSPs. E-LSP is simple and carries

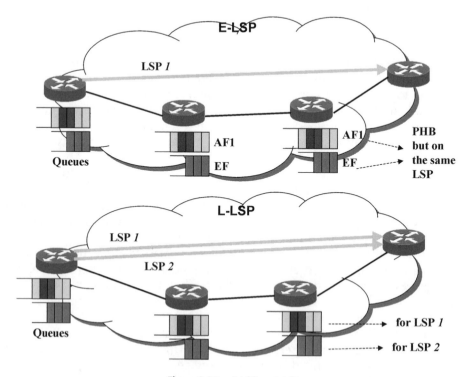

Figure 2.22   E-LSP vs. L-LSP.

up to eight behaviour aggregates on an LSP for a given FEC. When more traffic needs to be accommodated, however, load sharing is needed. L-LSP uses separate labels for the support of each PHB scheduling class, so it is able to support an arbitrarily large number of PHBs. However, this also introduces more signalling operations and label management complexity.

2.8.4 Virtual Private Network (VPN)

VPN technologies allow a company to build a 'virtual' and 'private' network over open or public networks to interconnect its hosts and routers at different sites. VPN is an encrypted or encapsulated communication that transfers data from one point to another point securely. In fact, security is one of the most important issues in VPN. In this section, we cover only network architecture issues. VPN security discussion can be found elsewhere.

The user/company that uses the VPN is the VPN customer. The VPN service providers provide the facilities that construct and maintain the VPN. A VPN is virtual and private because the VPN customers are likely to share a segment of or the entire open networks without each other traffic being affected. Within each customer network, hosts and routers at different sites can communicate as if they are directly connected.

VPN can be implemented either with an overlay approach or with a peer approach. In the overlay approach, the VPN service provider provides a point-to-point connection to the VPN customer. The connection can be based on any existing layer 2 technologies, e.g. ATM and frame relay. The VPN service provider can use MPLS to set up the circuit. Once the virtual link is up, the customer access points (i.e. the IP routers) immediately become IP next hop neighbours.

Running standard IP routing protocols over virtual links will discover the entire customer network topology and enable network layer routing. However, overlay VPNs cannot scale to large VPN service deployment. The reasons are three-fold. First, an overlay network has little network resource information shared between customer and provider but a VPN customer has the desire to design and operate their own virtual backbone. Second, an overlay model does not allow a service provider serving a large number of customers simultaneously. Third, configuration changes require a number of site updates, so it involves considerable complexity. For example, a full meshed VPN needs *n* virtual connections when a new site is added.

The peer approach presents VPN provider networks as peers to the VPN customer networks. MPLS has a more interesting role in the peer approach. Figure 2.23 shows a VPN peer approach implemented using MPLS and BGP. The details of IP routing protocols are discussed in the next chapter.

Within the provider network, LDP is used to distribute IGP labels. In the data plane, the traffic is IP packets between CE (Customer Edge) and PE (Provider Edge) and MPLS packets with label identifiers between PE and P (Provider core) and within P.

In the control plane, CE and PE are EBGP (Exterior BGP) peers, and within the provider network, LSRs including PEs are IBGP (Interior BGP) peers. However, to scale up the peer VPN solution, several technologies are proposed. First, PE maintains multiple VPN routing and forwarding tables, also known as VRF. CE in a VPN is

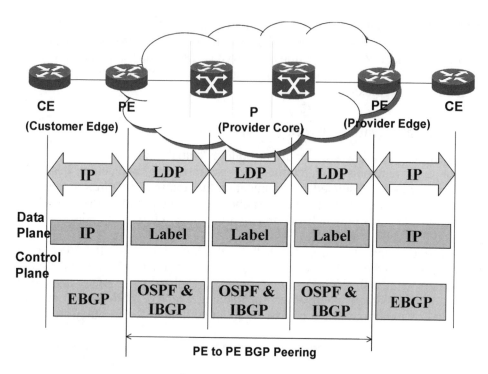

Figure 2.23   MPLS and BGP.

associated with a VRF. A particular packet's IP destination address is looked up in a particular VRF only if that packet has arrived directly from a site associated with that table. This allows per-VPN segregation of routing information and therefore has control over inter-VPN-site connectivity on a per-VPN basis. Second, each VPN has its own address space. The VPN IP address has an eight-byte Router Distinguisher in addition to the IP address. The Router Distinguisher contains a 2-octet type field, a 2-octet AS number field, and a 4-octet assigned number field. The use of Router Distinguisher does not require any modification to the BGP protocol. Third, routing information is not freely distributed among sites within a VPN. This is achieved by route filtering on BGP community attribute. The BGP community acts as an identifier, which can be attached to an advertised route.

2.9   Distributed Systems

A network by itself has only routers and switches interconnected by physical links. However, end users prefer to interface with a system that has control and management software in addition to the hardware. Usually, the software can be written for two purposes: automation and optimisation.

A key objective of software is its usability, i.e. software as an industry is designed to improve user productivity. A major task of a distributed system is resource management for sharing. From a software point of view, a distributed system is one in which components located at networked computers communicate and co-ordinate their actions by passing messages. A distributed system implementation offers increased reliability and addresses incremental growth and scalability compared to a centralised system. The Internet is an example of a distributed system, where a large number of networks using different networking technologies are interconnected. On the Internet, a large number of services are created to facilitate users. Examples of Internet services include E-mail and World Wide Web (WWW).

The popularity of distributed systems also applies to the strong demand for applications. Conventional applications such as banking systems require that customer databases are distributed to branches, and Automatic Telling Machines and computers are installed in a distributed fashion and networked with a single system image. A new group of applications has evolved as networks along with the Internet become more prevalent and network bandwidth becomes more plentiful. As a result, users that are geographically dispersed can now work and play together.

This application group is known as computer-supported co-operative work (CSCW). Representative applications include video conferencing, mission planning, distance education, electronic whiteboard, and Internet games. Remote services have also emerged to help users to access or share information held by others in their network or system. For example, Napster and Gnutella are popular Internet file sharing programs.

However, distributed systems also face challenges. First, they are complex. Distributed systems require distributed software. Designing, implementing, testing, and maintaining distributed software may not be an easy task. Second, certain resources in a distributed system may become overloaded. How to monitor the system and avoid congestion and message loss is another challenge. Third, security has become a bigger concern in distributed systems. TCP/IP is a widely known public protocol suite, and easy and convenient data access from anywhere creates additional security problems.

2.9.1 Design Objectives

A simple network system can be relatively static and consists of only long-lived entities. A complex distributed system, on the other hand, comprises numerous components. Any of these components can start and stop without bringing down the whole system, and the system as a whole can respond to changes in its environment and its constitution in a reliable and predictable way. There are several design issues that must be addressed in a distributed system [Tane95]:

- **Transparency**: is used to provide a single system view to users. From a user perspective, transparency hides application distribution; from an application program perspective, transparency hides system distribution. There are several kinds of transparency in a distributed system, i.e. location, migration, replication, concurrency, and parallelism forms of transparency.
- **Flexibility**: is an objective for any system or network. For example, to replace the monolithic operating system, a microkernel is created. A microkernel supports only the very basic operating system functions: IPC, some memory management, a

small amount of process management, and low-level I/O. Other functions can be designed and performed by user-level programs or by additional dynamically loaded modules. The microkernel design is more flexible than that of the monolithic O/S.

- **Reliability**: is another common objective for any 'operational' system or network. Reliability can be provided in several aspects: data must not get lost, the system must be secure, and the system should be fault tolerant. A measurable criterion of reliability is availability that refers to the time that the system is usable.
- **Performance**: an acceptable system always has performance requirements. There are different performance metrics in a system. Applications are interested in response time and end-to-end delay. Systems can be measured in terms of utilisation and throughput. Networks can be characterised by jitter and packet drop rate.
- **Scalability**: one of the key advantages of distributed systems over centralised systems is scalability in number of nodes. So a distributed system should scale in size.

2.9.2 Architectural Models

Distributed systems encompass different types of resources: CPUs, memory, disk, and physical links. The architectural model of a distributed system here refers to the service model implemented over distributed systems. Figure 2.24 shows a taxonomy of service models for distributed systems. A centralised model is one in which users are directly connected to the machine and applications are hosted on the machine. A centralised model does not need networking. A processor pool model is a parallel processing model, where a collection of CPUs is maintained in a pool and can be

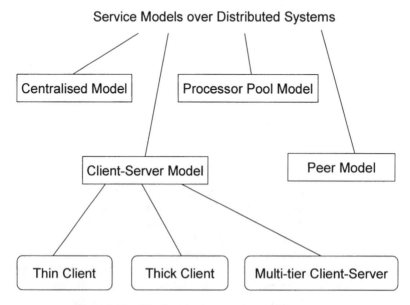

Figure 2.24   Distributed system service model taxonomy.

dynamically assigned to processes. An early version of this model is also known as a network of workstations that is designed to utilise the computation resources of idle workstations.

The most common service model is client/server, where client and server can be located on different machines. There are three components to a client/server model: client, server, and service. The client site is the machine requesting a service; the server site is the machine performing the service; the service is the task or function performed by the interaction of server and client.

Client/server can be implemented in different 'flavours'. A 'thin' client is designed to keep the amount of client software minimal and perform complex tasks and bulk processing on the server. An example of thin client is a Java application written as an Applet. A 'thick' client conducts the bulk of data processing. The server is rather simple and responsible for only storing and retrieving data. A Microsoft Windows PC is an example of a thick client, which requires the PC having certain capacity. A complex client/server application is usually implemented in a multi-tier structure since clients may require services from different servers and the services also need inter-service communication. An example of a multi-tier client-server application is the Java Enterprise Application Framework in which there are four tiers: client, web server, application server, and data storage.

Peer models can be treated as client/server located on every machine. IP routers are an example of the peer model. A peer model assumes each machine has similar capabilities so no machine is dedicated as a server.

2.9.3 Clustering

A cluster is a collection of autonomous machines interconnected by a communication network and has a single system view to external applications and users. A cluster API provides a collection of system interfaces to perform operations, for example, querying the set of nodes within a cluster, monitoring all states, and launching applications.

Clustering can be used for scalability. The performance of the nodes within a cluster is not subject to external factors. Clustering makes load balancing easier. The interconnect network among the clustering elements is a separate network that could be isolated from the external network, making the network load determinable only by the applications running on the cluster.

Clustering can also be used for reliability. Three design options on cluster application failover are cold, warm, and hot failover. A cold failover restarts an application once a cluster detects that it has stopped (or failed). A warm failover restarts an application in its last checkpoint image. Checkpoints are saved automatically and logged into an external file. A hot failover requires a backup system and the backup takes over when an application dies. The backup system is clockstep synchronised to the application states.

2.9.4 API for Distributed Applications

To facilitate distributed application development, two levels of APIs (Application Programming Interfaces) have been defined. The low-level API in IP networks is also known as the socket API; the high-level API includes RMI (Remote Method Invocation) and RPC (Remote Procedure Call) library calls and CORBA (Common Object Request Broker Architecture) IDL (Interface Definition Language) [Stev98].

Figure 2.25 shows the layout of low level and high level APIs. The socket APIs are developed originally in the early 1980s for BSD UNIX, and have been re-implemented many times for other platforms, for example, Winsock on Windows NT. Sockets provide an abstraction of the communication endpoints. Once a socket is established, it is perceived as an endpoint address (i.e. a file handler) to which processes can read or write. The endpoint address is represented by an IP address combined with a port number, e.g. 192.168.10.69:80. In the kernel, sockets are the implementation of protocol stacks. IP offers connection-orientated TCP and connectionless UDP communication services.

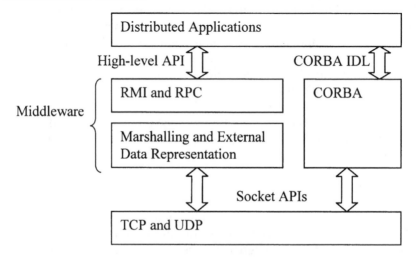

Figure 2.25   Socket APIs and high-level APIs.

Socket API can be grouped into three sets:

- **setup API such as** socket(), bind(), connect(), listen(), accept()
- **shutdown API such as** close(), shutdown()
- **communication API such as** read(), write(), recvfrom(), sendto().

Low-level APIs are efficient but applications have to cope with external data representation and marshalling especially in an heterogeneous environment. Marshalling is the process of assembling a collection of data items into a form suitable for transmission in a message. To further simplify application development in a distributed environment, high level APIs are introduced to automate many common network programming tasks such as parameter marshalling and unmarshalling, object registration, location, and activation.

3

Characteristics of the Internet and IP Routing

- IP router overview
- Internet traffic engineering
- TCP traffic policing
- Internet traffic characteristics and models
- Internet routing
- Open Shortest Path First Protocol (OSPF)
- Border Gateway Protocol (BGP)
- IPv6

3.1 IP Router Overview

As described in Chapter 2, the basic building block of the Internet is an IP router. Figure 3.1 shows the four main functional groups of an IP router:

- routing functions
- packet forwarding functions
- packet processing and classification functions
- encapsulation and conversion functions.

An IP router uses packet switching technology of a store-and-forward paradigm. Routing functions discover and maintain the network topology and compute routing paths across the network to destinations. They provide network-wide control in the IP network. For scalability, routing functions are implemented in a distributed fashion and the routing information base is synchronised using soft state protocols with timers and periodical refreshing and update. Routing functions are responsible for updating the routing table. We will cover IP routing protocols later in the chapter.

Once a packet arrives at a router, the packet header needs to be examined, which is the function of packet processing and classification. Packet classification can be based upon the addresses (source or destination) and the 'type of service fields' in the packet header. Packet processing also includes queuing different packet flows for

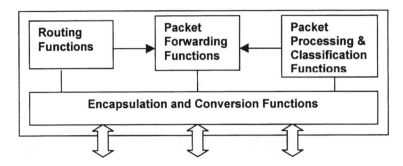

Figure 3.1   Router functionality.

prioritised services and data link layer conversions. Packet forwarding uses a table lookup process, which compares the incoming packet with the routing table and decides on the outgoing interface for the packet. Encapsulation and conversion is used to interface with a variety of network media so that data are encapsulated and/ or converted between link layer protocols as necessary. Hence, routers can interconnect networks with different layer 2 technologies, for example, ATM, frame relay, Ethernet, Token ring.

3.1.1 IPv4 Datagram

IP packets are transferred in a standard unit, known as an IP datagram. The IPv4 datagram format is shown in Figure 3.2. In the figure, all the fields except the data field belong to the IP datagram header.

A common header without options uses 20 octets. The 'options' field does not have a fixed length. For example, source routing lists all intermediate-hop IP

0 1 2 3 4 5 6 7   0 1 2 3 4 5 6 7   0 1 2 3 4 5 6 7   0 1 2 3 4 5 6 7

Ver	Header Length	Type of Service	Total Length	
Identification			Flags	Fragment Offset
TTL		Protocol	Header Checksum	
Source IP Address				
Destination IP Address				
Options (if any)				
Data				

Figure 3.2   IPv4 datagram format.

addresses in this field. Within the header, the first four bits gives the IP protocol version number, such as version 4. The next four bits specifies the header length in 32-bit words. The type of service (TOS) field occupies one octet, which is used to indicate how the datagram should be handled in terms of precedence, delay, throughput, and reliability. The original usage of TOS is:

- the first three bits for precedence
- one bit flag for delay sensitivity
- one bit flag for throughput sensitivity
- one bit flag for reliability sensitivity
- two unused bits.

Since IP provides a best-effort service, the TOS field only provides a hint to routers, not a demand. Recently, the TOS field has been used to support the IP QoS model, *diffserv*, where:

- the first 6 bits are used to represent a codepoint (i.e. a differentiated service), referred to as DSCP: total length gives the length of the datagram in octets;
- the next three fields: identification, flag, and fragment offset, are used for fragmentation and reassembly.

Fragmentation is needed for the following reasons:

- As indicated in the total length field, an IPv4 datagram has a maximum size of 65,535 octets.
- An IP datagram needs to be embedded into a layer 2 transmission frame. Different network types have different physical frame sizes. For example, an Ethernet frame is limited to 1500 octets, and a FDDI frame is limited to 4470 octets. These limits are referred to as Maximum Transfer Unit (MTU).
- Individual hosts and routers usually support even smaller MTU, for example, a router interface could have the MTU size of 576 octets (RFC 1122).

The two-octet identification field stores an identifier that uniquely represents the original datagram. The three-bit flag field represents the fragmentation flag. This flag can be used to indicate the existence of datagram fragmentation and also the fragment tail segment. The 13-bit fragment offset field specifies the position of this fragment in the original datagram measured in units of eight octets. That is a datagram must be fragmented on an 8-octet boundary.

Figure 3.3 shows an example of datagram fragmentation and reassembly. In the example, the original packet is divided into four fragments. Each fragment has the same original datagram header except the value in the flag field and in the fragment offset field. The first fragment has the offset 0; the second fragment has the offset equivalent to the MTU; the third fragment has the offset 2*MTU; the last fragment has the offset 3*MTU. The fragmented datagrams can be reassembled into the original packet by removing datagram headers and merging data sections according to their offset.

The IP datagram also contains a one-octet time-to-live (TTL) field, which specifies the time in seconds that a datagram can remain in the network. When a datagram arrives at a router, the router will decrement the TTL by the number of seconds that

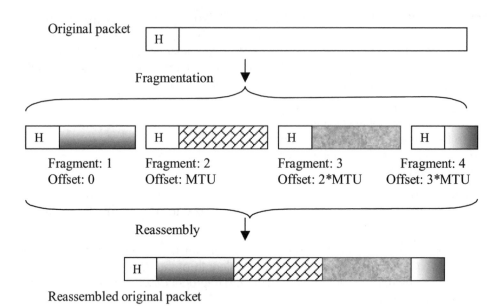

Figure 3.3 Fragmentation and reassembly.

the datagram remained inside the router waiting for service. The new TTL is written back to the header before the datagram is sent. Once the TTL reaches 0, the datagram is discarded and an error message is sent to the source. The next one-octet field represents the high-level protocol to receive the packet. Protocol field encoding is listed in Table 3.1.

Table 3.1   Protocol field encoding

Value	Protocol name
1	ICMP
2	IGMP
4	IP (used for IP-IP encapsulation)
6	TCP
8	EGP
17	UDP
46	RSVP
75	PVP
89	OSPF
201-254	Unassigned

The header checksum field uses two octets to ensure the header integrity. The source and destination IP address are the datagram's sender and receiver. IPv4 uses four octets for an IP address. These addresses are never changed during transmission although same source and destination packets may travel different paths and intermediate hops.

IP by itself does not guarantee latency for packets. Nor does it guarantee delivery, or notify hosts if a datagram is not delivered. IP only covers the integrity of the datagram header not the packet payload. IP does not maintain a session context so each packet is forwarded independently of any other packets.

3.1.2 QoS Queuing Models

Certain types of applications, for example, real-time video and interactive voice, require QoS with minimal delay. Too much jitter in voice applications and not enough bandwidth with real-time video applications will result in unacceptable communication across the network. IP provides best-effort routing so providing any level of QoS better than best-effort needs regulated and managed resource sharing across the network.

A service contract or service level agreement (SLA) can be honoured only when the underlying network(s) understands and implements QoS policies and mechanisms to administer access to the network resources. Hence, this requires sender, receiver, and all the interconnecting devices between them to be equipped with agreed precedence mechanism and QoS policy.

Queuing techniques are used for providing QoS. Queues implemented by buffers play an important role in a packet switched network, since a packet network has higher network utilisation compared to a circuit switched network. But high level of resource sharing raises the problem of contention. In fact, all current packet switches have buffers to match input rate to service rate to avoid packet loss. For arriving traffic, each packet is examined and dispatched to the corresponding queue (see Figure 3.4).

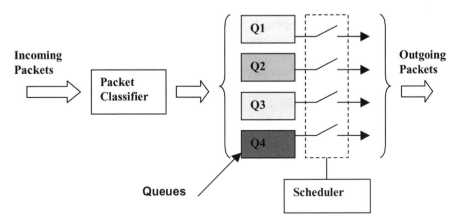

Figure 3.4 Router packet queuing.

A router has a scheduler that implements a queuing algorithm to schedule and control which queue goes next. There are several QoS queuing models developed:

- *First-in, first-out (FIFO) queuing:* the default and the simplest queuing method, where all packets are queued according to their arrival time. This is also called FCFS (First Come First Served).

- *Priority queuing:* each packet is assigned to one of several queues, for example, high, medium, normal, and low priority queues, according to, for example, network protocol, incoming interface, packet size, or source/destination address. During transmission, a higher priority queue has absolute preferential treatment over low priority queues. This queuing model can be unfair since low priority queues can be starved.
- *Custom queuing:* this is designed for applications with bandwidth or latency requirements to share the network effectively. A specific amount of queue space is assigned to each class of packets. Then, the queues are served in a round-robin fashion.
- *Weighted fair queuing (WFQ):* a flow-based queuing method, where interactive traffic is scheduled to the front of the queue and the remaining bandwidth is shared among high-bandwidth flows. If there is no 'higher priority' flows in the queues, bandwidth is used to forward 'lower priority' flows. This queuing method has several advantages. First, it ensures that queues do not starve for bandwidth. Second, traffic gets predictable service. For example, if multiple high-volume conversations are active, their transfer rates and inter-arrival periods are predictable. Third, low-volume flows that are the traffic majority can receive preferential service.

3.2   Internet Traffic Engineering

Traffic engineering is the aspect of network engineering concerned with the design, provisioning, and tuning of operational networks. It aims at specific service and performance objectives including the reliable and expeditious movement of traffic through the network, the efficient utilisation of network resources, and the planning of network capacity. Internet traffic engineering is targeted at performance optimisation of traffic handling in the Internet to minimise over-utilisation of certain network component capacity when other capacity is available in the network. Capacity is the maximum rate at which the network can transfer data. Throughput is the actual data a network can transfer for a given time frame. For a given time unit, throughput is always less than or equal to capacity.

3.2.1   Shortest Path Routing

An IP routing algorithm exploits the shortest path paradigm. Each link in an IP network is assigned a weight. The routing protocol ensures that an IP router forwards an IP packet to the next-hop towards the destination, such that the next-hop is on the shortest path to the destination. This algorithm is simple and works well when the network is lightly loaded. However, a drawback to this algorithm is that it is traffic independent, and there are no diverse routes. Therefore, depending on traffic conditions, the shortest path routing paradigm can lead to traffic congestions while non-shortest paths to the same destination are being underutilised (or even unused). If two or more shortest paths exist to a given destination, the algorithm will choose one of them.

3.2.2   Equal Cost Multi-Path (ECMP)

Some IP routing protocols, such as OSPF, do allow multiple equal cost paths to a destination, and can split traffic between the multiple paths. Each one of the multiple routes is of the same type (intra-area, inter-area, type 1 external or type 2 external) and cost, and has the same associated area. However, each route specifies a separate next hop and advertising router.

This operation is called equal cost multipath (ECMP) operation. ECMP splits traffic among multipaths in one of two ways. First, it distributes on a packet basis (i.e. packet-by-packet) to paths in round robin fashion. Second, it distributes on a per flow basis (i.e. flow-by-flow) to paths (for example, according to a hashing function on certain field(s) of the IP header). The first approach gives better load balancing as packet level distribution has finer granularity than the flow level. But the first approach will generate out-of-order packets. The second approach may be effective since it splits traffic in flow level but flow-based load sharing is desirable only when the number of flows is relatively large and the hash function can distinguish the flows. However, ECMP is still traffic independent in the sense that there is neither load balancing between the multiple paths, nor feedback between the traffic loading on the network links and the routing algorithm. Hence, they are still incapable of traffic engineering. An enhancement to ECMP to perform load balancing, also known as Optimised MultiPath, has been proposed.

3.2.3   Optimised Multi-Path (OMP)

Optimised MultiPath (OMP) relies on a link state routing protocol, such as IS-IS and OSPF, to periodically broadcast link loading information. The routing algorithm utilises the link loading information to adaptively split traffic load among multiple equal cost paths. Packet forwarding computes a hash on the $<$ *source address, destination address* $>$ tuple, and the hash space is split among the available paths. Load adjustment is dynamically performed in accordance to link load conditions, which is achieved by adjusting the hash boundaries among different paths.

However, to use OMP for traffic engineering effectively, the network administrator has to tune link-weights systematically to ensure a sufficient number of alternate paths between node-pairs. For large sized networks, this can be a difficult task. OMP uses relaxed shortest path criteria when calculating paths to a destination. In this case, a node considers all next hops that are closer to the destination than the current node of the shortest path. This implies that the only constraint on searching an alternate path is to avoid creating forwarding loops. Nevertheless, the number of alternate paths between a node-pair is limited and it is independent of traffic conditions. To support alternate paths on-demand, other techniques such as MPLS are needed.

3.2.4 MPLS OMP

Multiprotocol Label Switching with Optimised Multipath (MPLS-OMP) combines MPLS for its capability of creating on-demand label switched paths (LSPs) with OMP for load balancing. It assumes that all nodes are MPLS-capable routers so that the number of alternate routes between a router-pair can grow and shrink depending on the traffic demand, and the alternate routes are chosen in a load-dependent fashion. MPLS-OMP attempts to balance loads among different possible routes between a router-pair. Initially, at low loads, only the shortest path is being used. As traffic load between a router-pair increases (or if loads on links of the path increases), MPLS-OMP selects alternate routes, and splits the traffic load between the alternate routes. The alternate route selection process works by computing the shortest path on an auxiliary graph formed by deleting heavily loaded links of the topology.

3.3 TCP Traffic Policing

TCP is a layer 4 protocol providing end-to-end reliable data transport service. TCP has the following features:

- Connection-oriented, two applications using TCP must establish a connection with a three-way handshake before they exchange data. Connection termination also requires a three-way handshake.
- Reliability:
  - TCP ensures in-order delivery of application data;
  - TCP uses positive acknowledgement with retransmission to ensure lossless reception.

- Byte-stream oriented:
  - data flows between the sender and receiver as a stream of unstructured 8-bit data;
  - sending process writes bytes in arbitrary quantities;
  - TCP packs bytes into segments to send over IP;
  - Receiving process reads bytes in arbitrary quantities;
  - Byte ordering is preserved by end-systems.

- Full-duplex connection:
  - two independent data streams can pass in opposite directions on a single connection;
  - allows ACKs to be piggybacked on data segments in the opposite direction;
  - if one direction closes, the connection becomes half-duplex.

Figure 3.5 shows the TCP segment header format. Within the header, the source port and destination port specifies the port number of the source and destination hosts; the sequence number field identifies the first data octet in the segment; the acknowledgement number field gives the next sequence number that the sender of

01234567   01234567   01234567   01234567

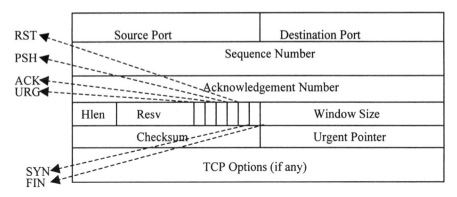

Figure 3.5 TCP segment header format.

the message expects to receive; the 4-bit header length specifies the TCP header length in 32-bit words; the 6-bit reserved field is reserved for future use. There is a 6-bit control field, among them:

- URG bit: if set, indicates the urgent pointer is valid;
- ACK bit: if set, indicates the acknowledgement number is valid;
- PSH bit: if set, the receiver passes the data to application immediately;
- RST bit: if set, the connection is reset;
- SYN bit: if set, during connection establishment, used to synchronise the sequence number;
- FIN bit: if set, indicates no more data from sender.

The 16-bit window size field specifies the number of bytes that the sender of the message is willing to accept beginning with the one indicated in the acknowledgement number field. The 16-bit checksum field gives the checksum of the entire TCP segment including TCP header and an IP pseudo header with source and destination addresses. The 16-bit urgent pointer field is used together with the URG field. When URG bit is set, this field indicates the offset from the first byte in the segment where the urgent data is located.

TCP option information can be carried in the TCP option field. An example of the TCP option is the Maximum Segment Size (MSS) that specifies the maximum segment size the sender can receive.

3.3.1 TCP Flow Control

TCP flow control is an end-system oriented feature trying to prevent a fast sender overwhelming a slow receiver. TCP employs a sliding window protocol for flow control. Between the sender and the receiver, the sliding window protocol adjusts dynamically a window size in the number of segments that can be transmitted at one time. TCP window size specifies the maximum number of bytes that the source can send in a set of segments for which it has not received acknowledgements. The sender must wait for acknowledgement for the transmitted segments explicitly. The receiver can acknowledge only the recent segment if it receives several segments in order. Once the sender receives a segment acknowledge, it will reposition its sliding window. If the sender does not receive acknowledge within a certain time interval (i.e. timeout), it will resend the segments in its current window.

As indicated in the current TCP segment header, the window size is in the range of $(0, 2^{16})$. This window size is a dynamic variable, which should be adjusted accordingly so that the sender gets its fair share of the network bandwidth. If the window size is too large, the router could backlog; if the window size is too small, the link capacity is not fully utilised.

The window size (w) has the following relationship with transmission rate (r):

$$r = w/RTT$$

where RTT is the round trip time for sending a packet and receiving its acknowledgement.

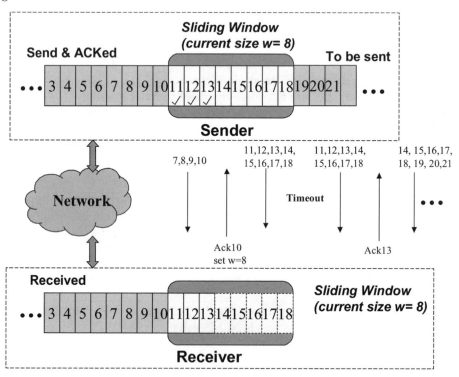

Figure 3.6   TCP flow control.

Figure 3.6 illustrates the TCP sliding window protocol. In the figure, the receiver acknowledges the receiving of segment 10 (or up to segment 10) and suggests the new window size to be 8. The sender sets its sliding window size (w) to 8, and then sends the segments (11, 12, 13, 14, 15, 16, 17, 18) in its window to receiver. For some reason, the sender may not receive an acknowledgement during the timeout period. Hence, it resends the segments in its window. The sender receives an acknowledgement for segment 13. It assumes the receiver has received the segments 11, 12, and 13, so it repositions its window to the next 8 segments and sends them (i.e. 14, 15, 16, 17, 18, 19, 20, 21).

3.3.2   TCP Congestion Control

Compared with flow control, TCP congestion control is a network-oriented feature trying to prevent from overloading the network or a portion of the network. In a dynamic network, congestion may occur which triggers the host timeout and packet retransmission. Fluctuations of traffic can result in overloads in certain network links. TCP congestion control lets individual routers decide how many packets it can send by observing the network so that network capacity is highly utilised but not over-loaded. Network congestion is perceived using timeouts in TCP. The assumption of conventional TCP is that packets are rarely lost due to an error during transmission.

TCP congestion control algorithms can be classified into two categories: reactive congestion recovery such as TCP Reno and proactive congestion prevention such as TCP Vegas. The first approach probes the bandwidth limit by increasing the sending window size continuously until a packet is lost. Once the timeout occurs, the sender slows down its window size. The second approach tries to detect congestion in its early stage, for example, through observing the expected throughput and the actual throughput. Once the incipient congestion is detected, the sender additively decreases its window size. If the incipient congestion is not detected, the sender either keeps the current sending rate or additively increases its window size.

In contrast, the first approach is easy to implement but packet loss is likely to occur and the window size changes dramatically during transmission. In the second approach, the packet loss may be avoided, but accurate congestion prediction is not an easy task and its implementation is complex.

TCP Reno

Several reactive congestion control algorithms are implemented in TCP Reno [Jaco88]. The algorithms start by increasing the window size exponentially to discover the available transmission rate. When the source fails to receive an acknowledgement, it suspects that a segment has been dropped so it reduces the window size. Then, the source gradually increases the window size, for example, by one unit every *RTT*.

Slow start

The TCP slow start algorithm can be described as follows:

- TCP maintains a new state variable for each connection, congestion window (*cwnd*), which is used by the source to limit how much data is allowed to transmit at a given time.
- Once a new connection is established, set *cwnd* = *1*(TCP segment).

- For each acknowledgement received, then *cwnd* + + (increment by 1 segment, i.e. exponential increase).
- When a timeout is received, set *cwnd* = 1 (TCP segment).
- Send *min*(advertised window, *cwnd*).

Fast retransmission and fast recovery

Fast retransmission means the receiver always sends an acknowledgement for every packet arrived even if according to sequence numbers this packet is already answered. So when an out-of-order segment is received, a duplicate acknowledgement is generated and sent to the sender, which informs the sender what segment number to expect.

If the sender receives 3 duplicate acknowledgements, it assumes that the segment has been lost and immediately retransmits without waiting for timeout and performs window size linear increase (instead of window size exponential increase).

Slow start with congestion avoidance

In addition to the exponential increase in the window size to probe the network bandwidth, one can also consider linear increase in the window size for TCP probing. The TCP slow start and congestion avoidance algorithm can be described as below:

- In addition to *cwnd*, TCP maintains another state variable for each connection, a threshold size (*ssthresh*), which is used to switch between the exponential and linear increase algorithms.
- Once a new connection is established, set *cwnd* = 1(TCP segment) and *ssthresh* = advertised window.
- For each acknowledgement received, if *cwnd* < *ssthresh* then *cwnd* + + (increment by 1 segment, i.e. exponential increase); else *cwnd* = *cwnd* + 1/ *cwnd* (linear increase).
- When a timeout is received, set *ssthresh* = *cwnd*/2 and then *cwnd* = 1 (TCP segment).
- Send *min*(advertised window, *cwnd*).

Retransmission timeout

Timeout function and retransmission play an important role in TCP congestion control. TCP is an end-to-end transmission control protocol so TCP traffic is likely to pass through an internetwork that may be composed of many different types of networks with different transmission delays. In addition, segments may take different routing paths through the network.

To overcome the varying network delays, TCP uses an adaptive retransmission scheme and defines the timeout as a function of measured *RTT*. *RTT* represents the best current estimate of the round trip time to the destination. When a segment is sent, a timer is set till the segment acknowledgement is back or the timeout is reached. When an acknowledgement is received, TCP computes/estimates a new

RTT, represented as *R*, using the following equation:

$$R = \alpha \times R + (1 - \alpha) \times M,$$

where *M* refers to the measured *RTT* for the last packet acknowledged and α represents a smoothing factor, a filter gain constant, which determines how much weight is given to the old value. In the recommendation, α takes the value of 0.875. In addition, *R* and the variation in *R* increase quickly with load. The mean deviation from mean variance *D* can be estimated using:

$$D = D + h(|Err| - D),$$

where *Err* represents the error in the estimation. As such, the round trip timeout, *RTO*, is given as:

$$RTO = R + 4D.$$

TCP Vegas

TCP Vegas prevents congestion through detection in early stages. TCP Vegas detection uses throughput changes as a measure for congestion. The assumption is the measured throughput should be very close to the actual throughput if there is no congestion. The difference between the measured throughput and the actual throughput, *DIFF*, defined as:*DIFF* = *cwnd*/*BaseRTT* − *cwnd*/*RTT*, can be used as the trigger for the congestion control algorithm.

TCP Vegas congestion prevention algorithm works as follow:

- If (*DIFF* × *BaseRTT* < φ)
- then *cwnd* + + (increment by 1 segment, i.e. exponential increase)
- Else if (*DIFF* × *BaseRTT* > τ)
- then *cwnd*– – (decrement by 1 segment)
- Else *cwnd* = *cwnd*(remain unchanged).

φ and τ are two constants to represent the lower threshold and the upper threshold in order to achieve high throughput and avoid overload.

3.4   Internet Traffic Characteristics and Models

In this section, we examine the characteristics of the information transfers of different applications over the Internet. These characteristics describe the traffic generated by the applications as well as the acceptable delay and loss by the delivering network. We will also examine the traffic behaviour on representative Internet backbone links and illustrate their properties from operational networks.

The IETF Integrated Services Architecture (ISA) divides traffic into two categories: elastic and inelastic. The former includes the conventional data traffic that is flexible to delays. It tries to use the available bandwidth but will lower demands if congestion is apparent. The latter includes interactive voice and real time video traffic that are delay sensitive. The inelastic traffic requires QoS assurance.

ISA defines three types of QoS services:

- guaranteed service
- controlled load service
- best-effort service.

To describe traffic generated at a source, we use the term peak rate (α) referring to short-term measurement for a number of packets and average rate (β) referring to the long-term average. The rate is expressed in bits/second for physical transmission links and packets/second for frame-based protocols. Bursty traffic is generated at a source when the source sends traffic at markedly different rates at different times. Non-bursty traffic means the source always sends traffic at the same rate.

A simple measure of the burstiness (b) over a period of time can be defined as $b = \alpha / \beta$. A similar measurement of the burstiness is the source activity probability (p), which specifies the frequency that the source sends packets. The p can be expressed as $1/b$, so $p = \beta / \alpha$. However, the traffic from two sources can have the same ratio, b, but may possess quite different characteristics in terms of the timescale on the traffic burst. So, in addition to b, burstiness must also be expressed in a time window, i.e. burst duration, expressed as d. For an instance, a network may limit the maximum duration, d, of a burst, b, to ensure that it can deliver the desired service quality.

3.4.1   Internet Traffic Statistics

The study of Internet backbone traffic shows strong time-of-day correlations, and the traffic volume fluctuations that arise due to time-of-day effects are significant [Thom97]. In this section, we examine the daily, weekly, monthly, and yearly traffic statistics on several IP links that belong to research and education networks, NCREN and Abilene:

- The North Carolina Research and Engineering Network (NCREN), http://www.ncren.net, is a metropolitan network interconnecting a number of university campuses.
- Abilene, http://www.abilene.iu.edu, is a high-speed national backbone network linking major cities in US.

Metropolitan Area Network

NCREN (see Figure 3.7) is a private telecommunication network interconnecting universities, research institutions, and medical and graduate centres in North Carolina. Network performance data can be obtained from their web site, http://skippy.ncren.net (data can be download in a graphical format). In this section, we will examine three links in the NCREN network.

The colour code in each of the following performance figures is as follows: (a) green represents incoming traffic in bits per second, (b) blue represents outgoing traffic in bits per second, (c) dark green represents the maximal 5 minute incoming traffic, and (d) magenta represents maximal 5 minute outgoing traffic.

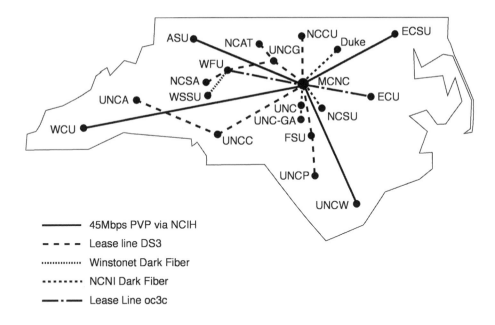

Figure 3.7 NCREN data network topology.

Link UNCG

Interface: Fast Ethernet 4/0/0 (7)
IP: ncren-gw.uncg.edu (152.13.254.253)
Link capacity: 12.5 Mbytes/s
Statistics last updated, Tuesday, October 9, 2001, 4:13 pm.
http://jazz.ncren.net/stats/uncg-gw.ncren.net.7.html

Figure 3.8 shows the link UNCG traffic statistics. This IP link is an access link with 12.5 Mbytes/s bandwidth and fast Ethernet interface. In the daily graph, one can observe traffic asymmetry between the incoming and outgoing link. Also, there is an increase of over 250% in outgoing traffic volume between 4 am and 5 am. The link is highly utilised in the outgoing direction during the afternoon rush hours. It shows the peak volume traffic at 4 pm on the outgoing link. In the weekly graph, the weekend shows a reduced traffic volume in comparison to the weekdays on the outgoing link. There are large volumes of traffic on Monday on the incoming link. The monthly pattern again illustrates the 24-hour periodicity. In the past month, there is more incoming traffic than outgoing traffic. In the yearly graph, the traffic volume picks up significantly from August especially for the incoming traffic. For the first three months of the year (Jan to March), the traffic volume on the outgoing link doubles that on the incoming link.

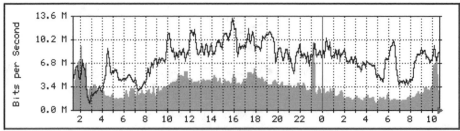

(a) Daily graph (5 minute averages).

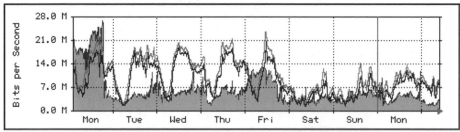

(b) Weekly graph (30 minute averages).

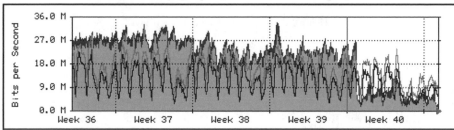

(c) Monthly graph (2 hour averages).

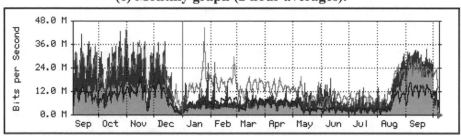

(d) Yearly Graph (a day average).

Figure 3.8   Link NCREN-UNCG traffic statistics.

Link ECU

Interface: Gigabit Ethernet 1/0/0 (1)
IP address: 150.216.243.1
Link capacity: 125.0 MBytes/s

Statistics last updated, Tuesday, October 9, 2001, at 4:34 pm.
http://jazz.ncren.net/stats/ecu-gw.ncren.net.1.html

Figure 3.9 shows the link NCREN-ECU traffic statistics. This link represents a core link in the metro network and has a bandwidth of 125 Mbytes/s with Gigabit

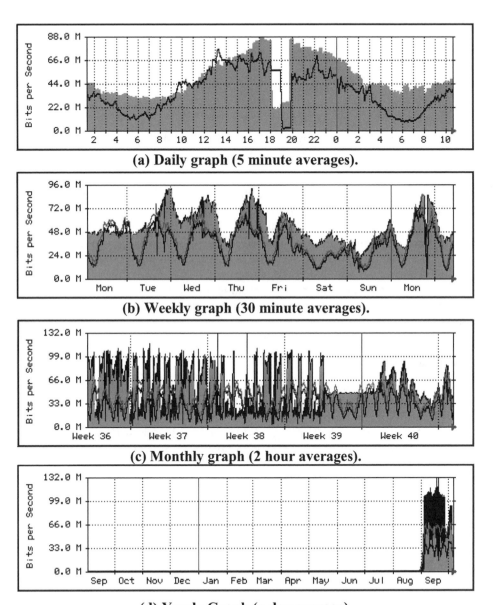

(a) Daily graph (5 minute averages).

(b) Weekly graph (30 minute averages).

(c) Monthly graph (2 hour averages).

(d) Yearly Graph (a day average).

Figure 3.9 Link NCREN-ECU traffic statistics.

Ethernet interface. The daily graph shows symmetric traffic in respect of traffic volumes in each direction. The incoming traffic is peaked at 5 pm, while the outgoing traffic is heavy from 12 pm to 6 pm. Both directions show traffic significantly reduced between 7 and 8 pm. The weekly traffic shows the link is currently underutilised. The traffic volume on each direction is still symmetric. In the monthly graph, there is more incoming traffic than outgoing traffic in the first three weeks. In fact, in the 1st and 2nd weeks, the incoming traffic is triple the outgoing traffic.

Link Abilene

> Interface: POS0/0 (2), propPointToPointSerial
> IP address: 198.86.16.61
> Link capacity: 77.8 MBytes/s
> Statistics last updated, Tuesday, October 9, 2001, at 4:36 pm.
> http://jazz.ncren.net/stats/abilence-gw.ncren.net.2.html

Figure 3.10 shows the link Abilene traffic statistics. This link provides a point-to-point connection to the Abilene long haul backbone network. The daily graph shows similar time-of-day fluctuations as link ECU. However, the utilisation of this link is much higher. This link is also symmetric with respect to traffic volumes in each direction. The weekly graph shows evenly distributed, symmetric traffic. The monthly figure again shows evenly distributed, symmetric traffic. From the yearly figure, one can conclude that traffic volume is comparatively low in the summer and in the month of December, and high in the month of September. Overall, this link is heavily loaded.

Long Haul Backbone Network

Abilene is an advanced backbone network that connects regional network aggregation points, known as gigaPoPs, to support the work of Internet2 universities (http://www.internet2.edu) as they develop advanced Internet applications. Figure 3.11 shows the Abilene network topology and traffic conditions in October 2001. The network nodes are located at these US cities: Seattle, Sunnyvale, Los Angles, Denver, Houston, Kansas City, Indianapolis, Atlanta, Cleveland, Washington DC, New York City.

In this section, we examine traffic performance on four links of the Abilene network. Traffic graphs presented in this section use the same notation and colour scheme as those of the previous section.

Point-to-point link between New York City and Cleveland

> Source: http://monon.uits.iupui.edu/abilene/clev-nyc-bits.html
> Statistics last updated Tuesday, 9 October 2001 at 5:35 pm

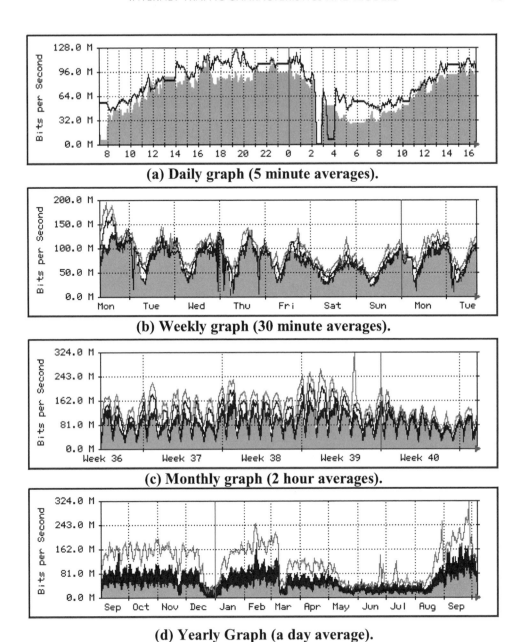

(a) Daily graph (5 minute averages).

(b) Weekly graph (30 minute averages).

(c) Monthly graph (2 hour averages).

(d) Yearly Graph (a day average).

Figure 3.10   Link NCREN-Abilene traffic statistics.

Figure 3.12 shows the point-to-point link between NYC and Cleveland. In the daily graph, the outgoing traffic from Cleveland to NYC is evenly distributed over 24 hours. The incoming traffic from NYC is peaked at 4 pm and the traffic volume is dipped to

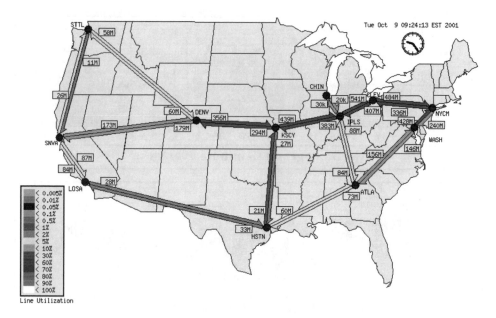

Figure 3.11   Abilene network traffic map (October 2001).

the lowest point at 7 am. The weekly, monthly, and yearly graph all indicates the symmetric traffic volume on the incoming and outgoing links. Traffic surged in September after the slowdown in the summer months. The weekly traffic is evenly distributed except that weekends have slightly lower traffic volume than weekdays.

Point-to-point link between Cleveland and Indianapolis

Source: http://monon.uits.iupui.edu/abilene/clev-ipls-bits.html
Statistics last updated Tuesday, 9 October 2001 at 5:35 pm

Figure 3.13 shows the point-to-point link between Cleveland and Indianapolis. Overall, this link shows similar performance as the link between NYC and Cleveland. The figure also shows the recent traffic surge in September.

Point-to-point link between Denver and Sunnyvale

Source: http://monon.uits.iupui.edu/abilene/snya-denv-bits.html
Statistics last updated Tuesday, 9 October 2001 at 5:39 pm

Figure 3.14 shows the point-to-point link between Sunnyvale and Denver. In the daily, weekly, and monthly figures, the outgoing traffic from Sunnyvale outweighs the incoming traffic from Denver.

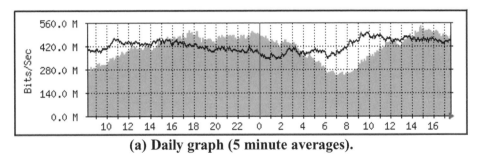

(a) Daily graph (5 minute averages).

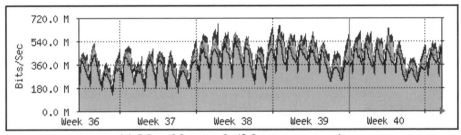

(b) Weekly graph (30 minute averages).

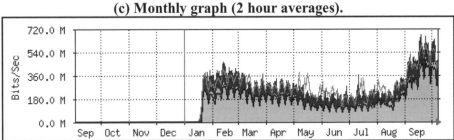

(c) Monthly graph (2 hour averages).

(d) Yearly Graph (a day average).

Figure 3.12 Link Abilene-NYC-Cleveland traffic statistics.

Point-to-point link between Sunnyvale and Seattle

Source: http://monon.uits.iupui.edu/abilene/snya-sttl-bits.html
Statistics last updated Tuesday, 9 October 2001 at 5:40 pm

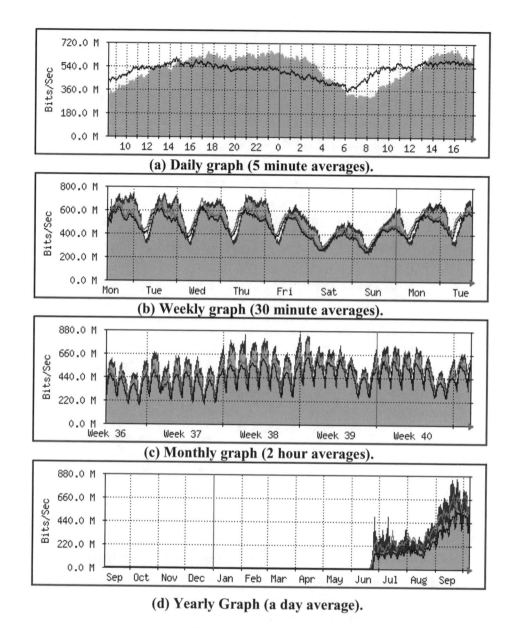

(a) Daily graph (5 minute averages).

(b) Weekly graph (30 minute averages).

(c) Monthly graph (2 hour averages).

(d) Yearly Graph (a day average).

Figure 3.13 Link Abilene-Cleveland-Indianapolis traffic statistics.

Figure 3.15 shows the point-to-point link between Sunnyvale and Seattle. This link shows strong traffic volume asymmetry. The outgoing traffic from Sunnyvale is nearly doubled the incoming traffic from Seattle. The yearly figure also shows a traffic surge to four times of the average traffic volume from Seattle to Sunnyvale.

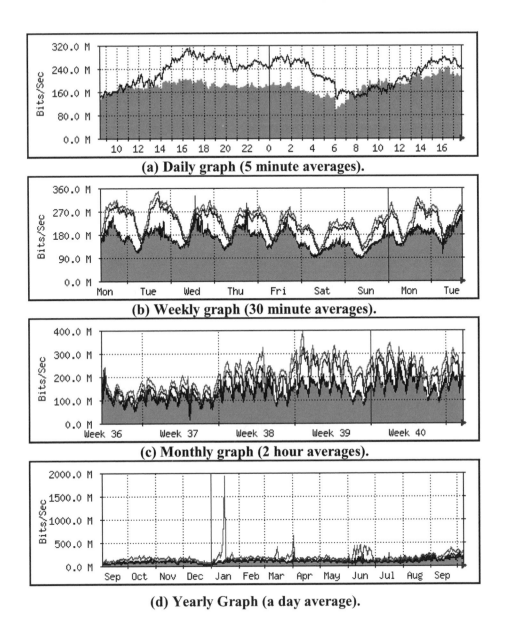

Figure 3.14 Link Abilene-Denver-Sunnyvale traffic statistics.

3.4.2 Traffic Models and Long Range Dependence

Traditionally, traffic arrival processes, for both connections and packet arrivals have been modelled as Poisson processes. The Poisson arrival model assumes that each of the N users generates traffic at an average rate of λ arrivals every T/N seconds. Thus, the probability that an arrival occurs from a specific user within an interval T is

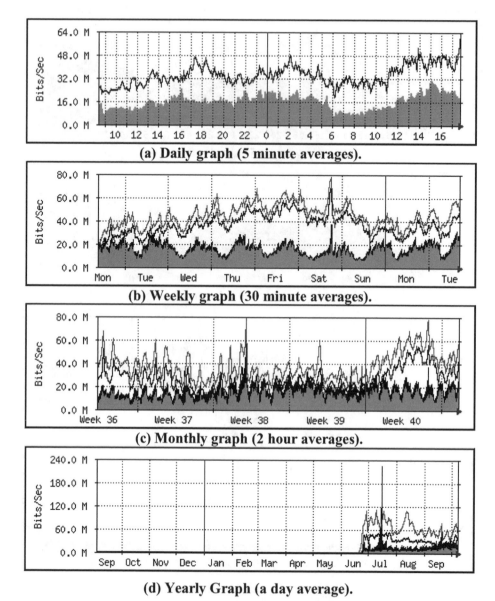

(a) Daily graph (5 minute averages).

(b) Weekly graph (30 minute averages).

(c) Monthly graph (2 hour averages).

(d) Yearly Graph (a day average).

Figure 3.15   Link Abilene-Sunnyvale-Seattle traffic statistics.

$\lambda \times T/N$. So the probability density for the random variable K refers to the probability, where K arrivals occur within a T interval from the population of N sources. Specifically, the probability of K arrivals in a T second interval follows:

$$P(K) = \frac{(\lambda T)^{K}}{K!} e^{-\lambda T}$$

where $K = 0, 1, 2, \ldots$ The average number of arrivals in a T second interval can be expressed as:

$$E(K) = \sum_{K=0}^{\infty} KP(K) = \lambda T$$

Poisson models have been considered suitable for modelling the aggregate traffic of a large number of similar and independent sources. However, there is significant experimental evidence [Lela94, Paxs95] that packet traffic arrival processes are not Poisson, and that they exhibit correlations over a wide range of timescales.

This type of traffic in fact poses a sharp contrast with the Poisson and Markovian models, where the traffic tends to become smoother and more predictable when considering long-term averages. The self-similarity of network traffic was first demonstrated for Ethernet traffic traces in [Lela94]. Subsequently the non-Poisson nature of packet arrival processes in wide-area Internet traffic was pointed out in [Paxs95], and the long-range dependence for ATM VBR traffic traces in [Bera95]. Recently work has demonstrated evidence that Internet web transactions are self-similar [Crov97]. The work in [Will95] provided a physical explanation for long-range dependence by demonstrating that the multiplexing of ON-OFF sources with the ON and OFF period lengths having heavy tails leads to self-similar traffic.

Self-similar can be defined as follows [McDy00]. A discrete process x is self-similar if for all values of a, the process $ax_k^{(a)}$ is statistically identical to the process of $a^H x_k$. The discrete process x is second-order self-similar if the following conditions are true:

$$E(x_k) = E\left(x_k^{(a)}\right)$$

$$a^{-2(1-H)} Var(x_k) = Var\left(x_k^{(a)}\right)$$

$$r(k) = r^{(a)}(k), \quad \text{as } a \to \infty.$$

To recall that the expected value $E(Y)$, of a discrete random variable Y with a probability mass function $p(y)$, is a weighted average of the possible values that the variable Y can take on, each value being weighted by the probability that the variable Y assumes. That is:

$$E(Y) = \sum_{y:p(y)>0} y \times p(y).$$

Variance is a measure of the dispersion of a distribution around its expected value: the larger the variance, the greater the dispersion. The variance of Y, denoted by $Var(Y)$ is defined by:

$$Var(Y) = E\left[(Y - \mu)^2\right] = E\left[Y^2\right] - \mu^2 = E\left[Y^2\right] - (E[Y])^2.$$

A continuous process $x(T)$ is self-similar if for any $0.5 \le H < 1$, the process $b^{-H}x(bT)$ has the same statistical properties of $x(T)$. The $x(T)$ is second-order self-similar if the expectation and covariance satisfy the following conditions:

$$b^H E(x(T)) = E(x(bT))$$

$$b^{2H} Var(x(T)) = Var(x(bT))$$

$$b^{2H} C_x(T, S) = C_x(bT, bS)$$

where $0.5 \le H < 1$. Note the expected value $E(Y)$, of a continuous random variable Y with a probability density function $f(y)$, is defined as:

$$E(Y) = \int_{-\infty}^{\infty} yf(y)dy$$

The variance of a continuous random variable is defined exactly as it is for a single variable. Since it is difficult to attach meaning to a squared percentage, the standard deviation σ is often used as an alternative measure of dispersion about the expected value. To interpret properly the implications of standard deviations as measures, for example, of the relative risks of investments whose expected values are different, coefficient of variance C, is defined as follows:

$$C_Y = \frac{\sqrt{Var(Y)}}{E(Y)} = \frac{\sigma_Y}{E(Y)}$$

A real-world application provides the value of the Hurst parameter H.

Long range dependency is one of the most important properties of self-similar processes. Traffic engineering needs to take the self-similarity of network traffic into account because one of the impacts of self-similarity of traffic is that the buffers (or predicted link capacities) at routers must be larger than those predicted using traditional Poisson models. Predicting link capacity based on conventional Markov models assuming Poisson arrival processes can result in under provisioning. To simulate self-similar processes, in random variable modelling, a number of ON-OFF sources need to be generated and related to the number of active sources. A heavy-tailed distribution, such as a Pareto distribution, can be used to describe the pattern of the times between ON and OFF transitions.

Another way to generate self-similar traffic uses a $M/G/\infty$ queue, where the arrivals are Markovian (not the general service time distribution) and the state variable is the number of users in the $M/G/\infty$ itself.

Let us consider a queue with a constant service rate and use first Poisson arrivals. Let ρ be the ratio between the average arrival rate and the service rate. The average queue length for this queue (M/D/1) diverges as $(1 - \rho)^{-1}$ as ρ approaches unity. On the other hand, consider the queue with the same service rate but with a self-similar arrival process such as the Fractional Brownian Motion (FBM) process with Hurst parameter H. The average queue length in this case diverges as:

$$(1 - \rho)^{-\frac{H}{1-H}}$$

when ρ tends to unity. When $H = 0.5$, this is the same as $(1 - \rho)^{-1}$ as predicted by the traditional model. However, for higher values of H, e.g., $H = 0.8$, the average queue length diverges much faster as $(1 - \rho)^{-4}$. This example illustrates that the use of standard queuing approximations may not be applicable for traffic engineering.

3.5 Internet Routing

The Internet is organised into regions, known as Autonomous Systems (AS). Each AS is an administrative domain and usually supports a single routing policy. An AS consists of one or more routing domains, each of which comprises a set of inter-connected routers under a single instance of an Interior Gateway Protocol (IGP). Within an AS, routing domains may use different IGPs such as OSPF or RIP. An Exterior Gateway Protocol (EGP) is used for communication among AS.

The Internet, essentially, is a peer model, in which routers have connections to their neighbours. Packet routing over such a large peer-to-peer network is a complex and weighty task. For scalability, routing hierarchies are introduced. In addition, routers are managed and configured by network administrators in each AS. The use of an AS also draws boundaries based on network ownership, policy enforcement, and administrative authority. A Stub AS is an AS that has a single exit point to reach networks outside its domain. A stub AS does not have to learn Internet routes from its provider network. All outgoing traffic can be directed to the provider network.

A multi-homed non-transit AS is an AS that has at least two exit points to the outside network but does not allow transit traffic. This type of AS only advertises its own routes and not the routes learned from other AS. A multi-homed transit AS allows transit traffic. It has at least two exit points. This type of AS advertises not only its own routes but also those routes learned from other AS.

Figure 3.16 shows three ASs, each of which has multiple routing domains. AS *1* and *3* have only one IGP instance, but AS *2* has two IGP instances. An example of AS *2* configuration uses OSPF for one routing domain and RIP for another routing domain, and manual (or static) configuration to interconnect the two domains.

Figure 3.17 illustrates the difference between the concept of a routing table and a forwarding table. A routing table is maintained by the IP routing protocol, whereas a forwarding table is obtained from local mechanisms (such as ARP) in the presence of a data link layer. In fact, whether the routing table and forwarding table are separate data structures is purely an implementation decision. A forwarding table can even be implemented in specialised hardware.

An Internet routing protocol computes routing paths according to the routing information it maintains and updates entries in the local routing table. There are two types of routing protocols: distance vector (including path vector) and link state routing protocols. The distance vector protocol works by letting each node tell its neighbours its best idea of distance to every other node in the network. Once a node receives the distance vectors from its neighbours, it compares these with its own distance vector, each destination and, if necessary, computes/re-

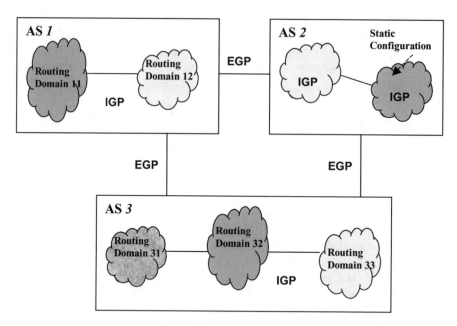

Figure 3.16   Internet routing.

computes its best path to each destination and the next hop for that destination. A distance vector protocol has these features:

- Each node maintains a set of triples < destination, cost, next hop > .
- Nodes exchange updates with directly connected neighbours periodically or triggered by updates (in fact, a triggered update is an amendment to the early distance protocol for better convergence).
- Each update is a list of pairs < *destination, cost* > .
- Each node updates its local table when a cheaper route is learned from neighbours.
- Each node uses timers to refresh and delete the existing routes.

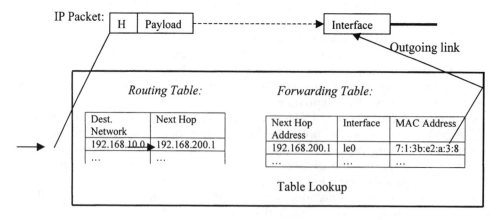

Figure 3.17   Routing vs. forwarding.

The distance vector protocol has the advantage of simplicity, and with amendments, it also supports route aggregation. However, a distance vector protocol suffers from:

- slow distance vector convergence
- formation of routing loops during convergence
- the problem of counting to infinity.

An example of a distance vector protocol is RIP, where to address the counting-to-infinity problem, a destination is declared as unreachable when the path cost to the destination reaches 16.

A link state protocol requires that each router stores the entire network topology and computes the shortest path by itself. A link state database is maintained at each router and link state information is exchanged by the means of link state update packets. In contrast to a distance vector protocol, a link state protocol provides:

- faster network convergence
- the ability to support multiple routing metrics.

A link state protocol is more complex than the distance vector protocol and has a higher computational overhead. An example of a link state protocol is OSPF. We will describe OSPF in the next section.

3.6 Open Shortest Path First Protocol (OSPF)

A scalable and widely deployed IGP in the IP protocol suite is Open Shortest Path First protocol (OSPF). OSPF is a distributed, link state protocol, by which each router keeps a link state database locally and performs route computation based on the link state database. The link state database is synchronised using OSPF messages with a reliable flooding mechanism. For better scalability, hierarchical routing is used with the introduction of the OSPF area. OSPF supports ECMP for load balancing and CIDR and subnetting. All OSPF message exchanges are authenticated.

Due to its popularity, the OSPF update can also be used to disseminate traffic engineering information and externally derived routing data, for example, those routes learned from BGP.

The main functionality of OSPF can be summarized as below:

- *Topology discovery:* automatically discover neighbours in the case of initial start or network changes.
- *Link state construction and maintenance:* each router has an internal data structure to represent the link state database. The structure is populated with operational network data during the initial network topology discovery stage, and then updated with network link state changes.
- *Reliable flooding:* OSPF runs directly over IP. OSPF Link State Advertisement (LSA) messages are encoded and transmitted in such a way that reliable transmission is guaranteed.
- *Route computation:* each router computes a shortest path tree starting from this router. The computation is based on a selected routing metric and the link state database, and is subject to dynamic routing constraints.

- *Routing table update:* Once route computation is completed, OSPF interfaces with the kernel to update the routing table.

All OSPF packets have the following header format (see Figure 3.18). Within the 24-byte header:

- the version field specifies the OSPF version number;
- the type field indicates the type of this message;
- the packet length field gives the total packet length including header in bytes;
- the router ID field specifies the packet's source;
- the area ID field identifies the OSPF area that this packet belongs to;
- the checksum field gives the checksum of the entire packet;
- the AuType field specifies the authentication procedure to be used for the packet;
- the authentication field uses 64 bits to encode the authentication scheme.

```
0 1 2 3 4 5 6 7     0 1 2 3 4 5 6 7     0 1 2 3 4 5 6 7     0 1 2 3 4 5 6 7
```

Version	Type	Packet Length	
Router ID			
Area ID			
Checksum		AuType	
Authentication			
(cont.)			

Figure 3.18 OSPF header format.

3.6.1 OSPF Messages

OSPFv2 defines five types of message:

- hello (type 1)
- database description (type 2)
- link state request (type 3)
- link state update (type 4)
- link state acknowledgement (type 5).

Hello messages are sent periodically on all interfaces to discover and maintain neighbourhood relationships. Database Description (DD) messages are used during the adjacency initialisation to describe the link state database. Once a router finds a part of its link state database is out of data, it sends Link State Request (LSR) messages to request pieces of the neighbour's link state database. Link State Update (LSU) and Link State Acknowledgement (LSAck) messages implement the OSPF reliable flooding. LSU carries one or more Link State Advertisement (LSA) one hop away from their origin. When a router receives non-self-originated (bypass) LSA, it will check the LSA for aging and integrity. If the LSA is alive (i.e. within the specified time period) and valid, the router will continue the flooding process by sending a LSA to its neighbours except the one that the LSA originated from. LSAck messages are used to explicitly acknowledge the receipt of the flooded LSA. Both LSU and LSAck can deal with multiple LSAs in one message.

OSPF flooding is discussed in Chapter 6. Figure 3.19 shows message flows between two routers during initialisation.

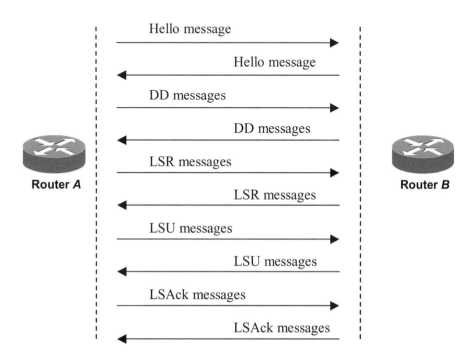

Figure 3.19   OSPF messages exchange during initialisation.

3.6.2   Link State Advertisement (LSA)

OSPF link state database is a collection of LSAs. Figure 3.20 shows the 20-byte LSA header. Within the LSA header:

- the LS age field is the time in seconds since the LSA is originated;
- the options field indicates the router capability on supporting optional features;
- the LS type field specifies the type of the LSA;
- the link state ID field identifies the portion of the routing domain described in the LSA;
- the advertising router field represents the router ID that generates the LSA;
- the LS sequence number is used to distinguish and select the most recent LSA in the case of receiving duplicate LSAs;

01234567   01234567   01234567   01234567

LS Age	Options	LS Type
Link State ID		
Advertising Router		
LS Sequence Number		
Checksum	Length	

Figure 3.20   LSA header format.

- the checksum field specifies the checksum for the entire LSA including the header but excluding the LS age field;
- the length field specifies the LSA length in bytes.

There are five types of standard LSAs:

- **Type 1 – router LSA:** used to describe the state of the router interfaces and originated by the router. All the interfaces to the area must be encoded in a single router LSA. There are four types of router links: point-to-point interface, broadcast and NBMA interface, virtual link, and point-to-multipoint interface.
- **Type 2 – network LSA:** used to describe the set of routers attached to the network and originated by the network Designated Router (DR). DR is a network representative router, which is elected by the hello protocol. Introducing DR reduces the amount of adjacency in the network and therefore reduces control traffic and the size of the link state database. DR is used in the broadcast network or the NBMA (Non Broadcast Multi Access) network.
- **Type 3 – network summary LSA:** used to describe the routes to networks and originated by Area Border Router (ABR).
- **Type 4 – ASBR summary LSA:** used to describe the routes to AS boundary routers (ASBR) and originated by ABR.
- **Type 5 – AS external LSA:** used to describe the external routes and destinations for the AS and originated by ASBR, for example, those routes learned from an EGP.

3.6.3   Routing in OSPF

OSPF routing performs in three layers(see Figure 3.21):

- intra-area routing
- inter-area routing
- inter-AS routing.

Intra-area routing refers to a single area routing, where each router maintains a link state database of the area. An area consists of a set of routers that are administratively configured to exchange link state information. Router and network LSAs are flooded

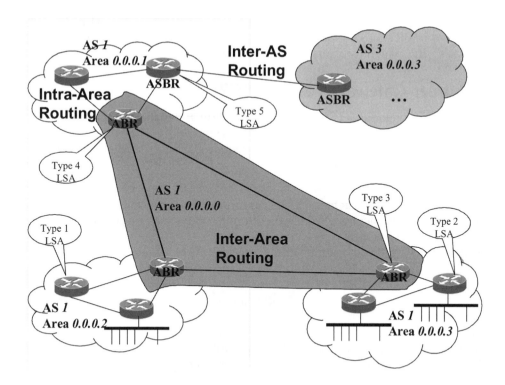

Figure 3.21 Routing in OSPF.

within the area. Multiple areas can be configured within an AS, in which they are interconnected through ABRs. All the ABRs form the backbone area, i.e. area 0.0.0.0. OSPF requires that all areas must attach to the backbone area directly but not necessarily through physical links. For inter-area routing, ABR originates and floods network summary LSA and ASBR summary LSA.

Comparing with other types of LSAs, summary LSAs are encoded in a condensed manner so that LSA traffic is reduced. This also supports the growth of single routing domains (i.e. areas) without overburdening the intra domain routing protocols. Inter-AS routing describes interfaces to external AS. The external routing information is incorporated into the AS by ASBR, which originates AS external LSA.

The intra-area routes are computed by building the shortest-path tree for each attached area. The inter-area routes are computed by examining the summary LSAs. The routes to external destinations are computed by examining the AS external LSAs. Using hierarchical routing, OSPF is able to confine the router's routing table size so that it can scale up to a large number of routers. However, the hierarchical routing may affect routing efficiency. Since certain levels of OSPF routing only have partial or condensed routing information, this will hinder the ability to make perfectly optimal routing decisions. An example of the shortest path routing algorithm is Dijkstra's algorithm, which computes a shortest path tree from one node to all other nodes in the graph. OSPF hierarchical routing can be viewed as multi-area routing, according to which, a router forwards a cross-area packet to ABR. Then, the source area's ABR sends the packet through the OSPF backbone area to the destination area's ABR, which forwards the packet to its final destination.

3.7   Border Gateway Protocol (BGP)

As discussed in previous sections, for administration and management purposes, the Internet is divided into regions, known as AS. Within an AS, IGP is used as an intra-AS routing protocol. An EGP is the inter-AS routing protocol. A widely implemented and deployed EGP is the Border Gateway Protocol (BGP). BGP is designed with the following features:

- Exchanging network reachability among AS. This information is used to construct a graph of AS connectivity from which routing loops may be pruned and policy decisions at AS level may be enforced.
- Support CIDR and route aggregation.
- Support policy-based filtering and routing and load balancing among equal cost multipaths.
- Support backup paths so in the event of failure over the primary path, it need not wait for network convergence.
- Employ path vectors to support multiple routing metrics instead of the conventional distance-vector in a distance vector protocol.
- BGP speakers are peers. The exchanged messages between peers are delivered using a reliable transport protocol, i.e. TCP.
- A BGP peering relationship is manually configured so that a BGP speaker can choose to either support or deny a peering request.

3.7.1   Internal and External BGP

BGP requires peering between BGP speakers. There are two types of peering, Internal BGP (IBGP) and External BGP (EBGP) peering. Within an AS, IBGP is responsible for collecting network reachability information and redistributing BGP information into the IGP of the AS. IBGP speakers may not be directly connected; for example, there are intermediate routers (or routing hops) between IBGP speakers. Thus, they need not be next hop neighbours.

External routes, learned using BGP, can be injected into OSPF, which distributes them to routers within the AS using AS-External LSA. EBGP is used between different ASs. EBGP speakers may not be physically directly connected, but virtually they must be neighbours. BGP messages are carried in a reliable transport protocol, TCP. Figure 3.22 illustrates the IBGP and EBGP peering.

BGP confederation and route reflectors are used to address the explosion of IBGP peering within an AS. BGP confederation divides an AS into multiple AS and assigns the entire group to a single confederation. Each AS by itself will be fully IBGP meshed and has connections to other AS inside the confederation. Although these ASs will have EBGP peers to ASs within the confederation, they exchange routing updates as if they were using IBGP. To the outside world, the confederation (i.e. the group of ASs)

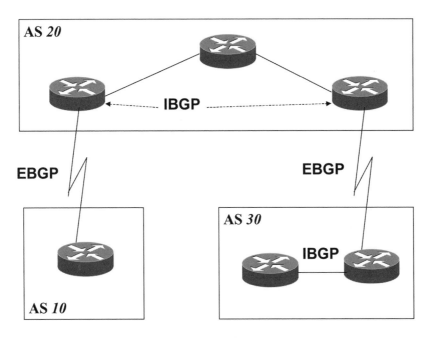

Figure 3.22 IBGP and EBGP peering.

will still look like a single AS. Without route reflectors, BGP does not allow IBGP to pass routing information learned from an IBGP peer to a different IBGP peer. With route reflectors, this restriction is relaxed. Hence, there is no longer a need to have a full meshed IBGP within an AS.

3.7.2 BGP Messages

A BGP peering relationship is configured manually. Two BGP speakers initially exchange their full BGP routing tables. After this exchange, incremental updates are used to synchronise the copies. BGP messages have a common header shown in Figure 3.23. Within the header, the 16-octet marker field is used to detect loss of synchronisation between a pair of BGP peers and authenticate the incoming messages. This field is computed based on the authentication mechanism used. For OPEN messages, there is no authentication, so the marker field has all ones.

The two-octet length field gives the total message length including the header. The one-octet type field specifies the message type. Four messages are defined in the BGPv4:

- **Type 1 – OPEN:** after a TCP connection is established between the BGP speakers, OPEN message is sent to initiate the communication between peers. The receiver confirms receiving an OPEN message by sending back a KEEPALIVE message. So OPEN and KEEPALIVE messages work as a handshaking process during the initial contact. After that, peers can communicate using other messages. This message has the following fields:

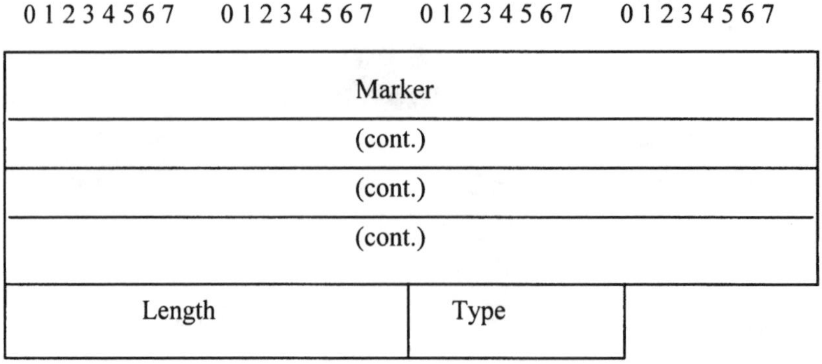

01 2 3 4 5 6 7   01 2 3 4 5 6 7   01 2 3 4 5 6 7   01 2 3 4 5 6 7

Marker
(cont.)
(cont.)
(cont.)

Length	Type

Figure 3.23   BGP header format.

- *Version:* 1-octet field to indicate the version of the BGP.
- *My Autonomous System:* 2-octet field to specify the BGP speaker's AS number.
- *Hold Time:* 2-octet field to propose a value for the Hold Timer in number of seconds. This defines the maximum number of seconds that may elapse between the receipt of successive KEEPALIVE or UPDATE messages.
- *BGP Identifier:* 4-octet field to indicate the sender's ID.
- *Optional Parameters Length:* 1-octet to specify the length of the optional parameters.
- *Optional Parameters:* a list of the optional parameters, each of which is encoded using a TLV (Type Length Value) triplet.

- **Type 2 – UPDATE:** this is the main BGP message to advertise a single feasible route to a peer or withdraw multiple routes. Three blocks of information are carried in one UPDATE message: unreachable routes, path attributes, and network layer reachability information (NLRI). The first block has two fields:

  - *Unfeasible Routes Length:* 2-octet field to specify the total length in bytes of the withdrawn routes.
  - *Withdrawn Routes:* variable length fields to list the withdrawn routes, each of which is encoded using a < *length, prefix* > format.

  The second block also has two fields:

- *Total Path Attribute Length:* 2-octet field to specify the total length in bytes of the path attributes.
- *Path Attributes:* variable length fields to list the path attributes, each of which is encoded using a < *attribute type, attribute length, attribute value* > format.

  The third block, NLSI, contains a list of the networks being advertised in the form of IP prefix route. Each network reachability information is encoded using a < *length, prefix* > format.

- **Type 3 – NOTIFICATION:** this message is used to report an error condition with these fields: 1-octet error code, 1-octet error subcode, and 6-octet data.
- **Type 4 – KEEPALIVE:** this message contains only the header, sent periodically to

determine whether the peers are reachable. If the Hold Time has the value of zero, the periodic KEEPALIVE message will not be sent. The specification recommended the KEEPALIVE message should be sent every one third of the Hold Time interval.

3.7.3 Path Attributes

BGP is a path vector protocol. Path attributes representing route-specific information are carried in the UPDATE message. The Attribute Type is a 2-octet field: 1-octet Attribute Flags and 1-octet Attribute Type Code. The first bit of the Attribute Flags field indicates whether the path attribute is optional or well known. The second bit of the flags field indicates whether the optional path attribute is transit or non-transit. The third bit represents whether the optional transit attribute is partial or complete. The fourth bit indicates whether the following Attribute Length field is one octet or two octets.

All BGP speakers must implement the well-known attributes. Mandatory attributes must be included in every UPDATE message that contains NLRI. In addition to mandatory attributes, each path may carry one or more optional attributes. BGP speakers pass along unrecognised transit optional attributes, but unrecognised non-transit optional attributes are ignored and not passed further. Table 3.2 lists some common path attributes.

Table 3.2 Common path attributes[a]

Name	Type	EBGP	IBGP	Explanation
ORIGIN	1	M	M	The origin of the path information
AS_PATH	2	M	M	The list of the ASs that message has passed
NEXT_HOP	3	M	M	The next hop (border-router) IP address
MULTI_EXIT_DISC	4	D	D	Used on external links to discriminate among multiple exit or entry points to the same neighbour AS.
LOCAL_REF	5	N/A	M	Allowing AS border routers to indicate the preference they have assigned to a chosen route when advertising it to IBGP peers.
AUTOMIC_ AGGREGATE	6	D	D	Used to ensure that certain NLRI is not deaggregated. If a BGP speaker selects the less specific route, the local system must attach this attribute when sending to other BGP speakers.
AGGREGATOR	7	D	D	A BGP speaker performing route aggregation may have this field to specify its own AS number and IP address

[a] Note: 'M' stands for Mandatory and 'D' refers to Discretionary.

3.7.4   Policy Filtering

BGP is a policy-based routing protocol. It provides the capability, through its mechanisms, to enforce policy routing related constraints and preferences, which can be input or updated by configuration. Policy-based filtering can be applied to information dissemination as well as traffic forwarding.

Route filtering includes two phases: identifying the route and taking actions against the route. The actions include permitting, denying, or manipulating attributes. Route filters can be applied to the UPDATE message. Routes can be identified through its NLRI and/or the AS_PATH list. Once a route is identified and permitted, its attributes may need to be modified to affect routing decisions. For example, all routes from a particular neighbour are ignored; the cost on a route from a specific neighbour is doubled before propagating to the AS. Path filtering defines the access list on both inbound and outbound UPDATE messages based on AS path information. Path filtering supports the same actions (permitting, denying, or manipulating attributes) as route filtering. In Cisco Routers, for example, route maps, distribute-lists and filter-lists are used to configure update policies.

A BGP peer group is a group of BGP neighbours sharing the same update policies. Instead of defining the same policies for each separate neighbour, we can define a peer group name and these policies to the peer group. As the number of networks and ASs increases, there is a growing concern of the scalability of BGP. Today, a backbone router routinely carries routes for over 110,000 different network addresses. To address this problem, BGP also supports different levels of route aggregation:

- aggregate only but suppress the specific routes;
- aggregate together with the specific routes;
- aggregate with only a subset of the specific routes;
- loss of information inside aggregates by summarising all paths with certain attributes;
- changing the aggregate attributes;
- forming the attribute from a subset of the specific routes.

3.7.5   BGP Routing

BGP routing is shown in Figure 3.24, where a BGP speaker maintains not only an IP routing table but also a BGP routing table, input and output policies. When routes are received from the peers, the BGP speaker will apply the input filtering policies to identify routes or paths and take specified actions such as path attribute manipulation.

A BGP routing table is kept separately from the IP routing table. According to the BGP routing table, a decision process is performed to select the best route. BGP uses the following rules, also known as tiebreakers, to choose between two equal BGP routes:

- Choose the route with the lower route weight.
- Choose the route with the higher Local Preference attribute.

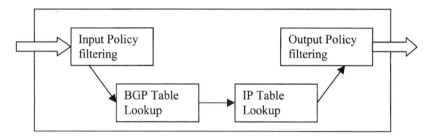

Figure 3.24 BGP routing flows.

- Choose the route with the lower Inter-AS Metric attribute (if both routes include this optional attribute).
- Choose the route with the lower interior cost to the Next Hop.
- Choose external BGP over IBGP.
- Choose the route with the lower BGP identifier.
- Choose the route with the lower BGP connection remote address.
- Choose the route with the lower BGP connection local address.

The decision process based on the BGP table will decide how this router passes routing information to peers. As external AS information is injected into the AS, the IP routing table entries reflect both internal and external route information. The outgoing messages or packets look through the IP routing table for the next hop destination address. The messages are filtered through output policies before sending out.

3.8 IPv6

With the continuous growth of the Internet, there is a global push towards next-generation IP, IPng. The global demand for next generation IP comes from a number of sources, such as:

- the inefficient use of IP addresses in IPv4;
- the insufficiency of the IPv4 address space;
- the emerging market of Internet-enabled appliances;
- the popularity of wireless Internet;
- the security and potential performance concerns over the current Internet;
- the inefficiency introduced by the use of Network Address Translation (NAT);
- the advent of technology and infrastructure deployment in developing countries.

IPng or IPv6 offers several key advantages over the current version of IP, i.e. IPv4, which includes bigger address space, support for mobile devices and Internet appliances, and built-in security. IPv6 uses 128 bits for addresses that means 2^{128} addresses in total. As a result, there is no longer the need for workaround technologies such as NAT. It promises a global, unconstrained peer-to-peer IP network including mobile devices and Internet appliances such as PDAs, mobile phones, and home networking devices.

Supporting mobility is one of the design requirements in IPv6. With IPv6, roaming between different networks can be performed efficiently and effectively with possible global notification on entering or leaving a network. IPv4 has no support for security (IPsec is optional for IPv4), so applications have to employ their own mechanisms to ensure the data security. Although possibly encrypted, the data in transport across the network poses considerable security concerns. The IPv6 protocol stack includes IPsec, which allows authentication and encryption of IP traffic.

Figure 3.25 shows the IPv6 header format. Within the header, the four-bit version field specifies the IP protocol version, i.e. version 6; the four-bit priority field indicates the packet *diffserv* class; the 24-bit flow label field represents the flow identifier that this packet belongs to; the 16-bit payload length gives the total length of the packets in octets excluding this header but including the IPv6 extension header. The 8-bit next header field indicates which higher layer protocol is in use after the IP header. The value of this field can be TCP, UDP, or the IPv6 extension header. The hop limit field specifies the packet lifetime in terms of router hops. Every router decrements the value of this field before forwarding the packet to the next hop. When it reaches zero, the packet is discarded. The 128-bit source and destination address gives the packet's source and destination IPv6 address. An IPv6 address is usually written in groups of 16 bits using four hex digits, separated by colons, e.g. fa57:0:0:0:3a1:d3ee:cbb6:d18c.

IPv6 allows three types of address:

- unicast
- multicast
- anycast.

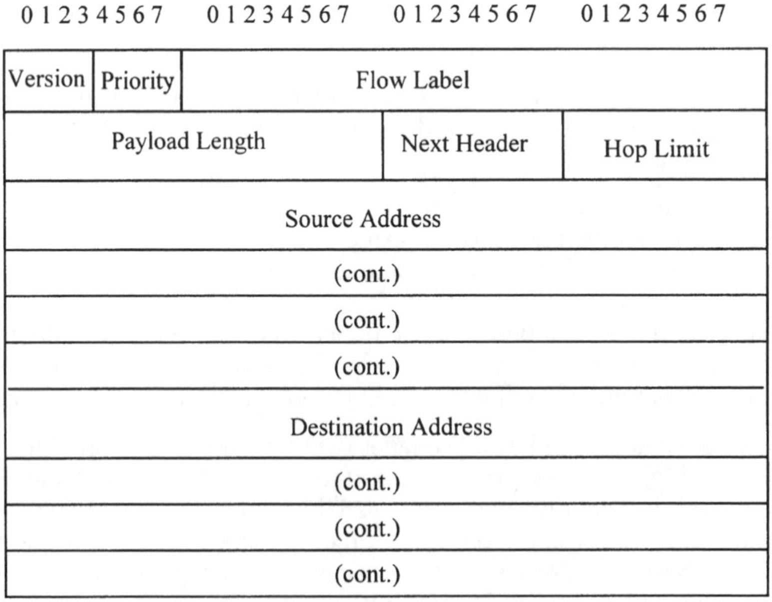

```
0 1 2 3 4 5 6 7    0 1 2 3 4 5 6 7    0 1 2 3 4 5 6 7    0 1 2 3 4 5 6 7
```

Version	Priority	Flow Label		
Payload Length			Next Header	Hop Limit
Source Address				
(cont.)				
(cont.)				
(cont.)				
Destination Address				
(cont.)				
(cont.)				
(cont.)				

Figure 3.25   IPv6 datagram header format.

A unicast address specifies an identifier for a single interface. Both multicast and anycast give an identifier to a set of interface typically belonging to different nodes. A packet sent to a multicast address is delivered to all the interfaces in the multicast group. However, a packet sent to an anycast address is only delivered to the nearest interface of the anycast group.

IPv6 has a fixed size (i.e. 40-octet) standard header. It also has the flexibility to carry any number of extension headers between the initial header and the higher layer protocol headers (see Figure 3.26). Example extension headers include:

- hop-by-hop options
- destination options
- source routing
- fragmentation
- authentication
- IPv6 encryption.

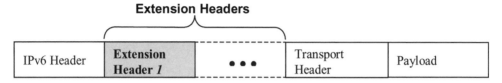

Figure 3.26   IPv6 extension header.

Each extension header typically occurs only once with a given packet, except for the destination options header that can occur twice.

In addition, IPv6 supports host autoconfiguration. First, it supports serverless host address configuration, which lets hosts construct their own address, for example, by using their MAC address. Second, it allows auto-reconfiguration, which supports cost-effective numbering and renumbering of large-number IP hosts. In terms of routing, IPv6 uses IGP and EGP developed under IPv4, but the protocols need to be modified with the address bits extension.

IPv6 products have already been supported by network operators such as Worldcom, AT&T, BT, NTT, and developed by network equipment vendors such as Cisco, and software vendors such as Microsoft, Sun Microsystems, and Linux vendors. IPv6 information and updates can be found in the following Web sites:

- www.ipv6forum.org
- playground.sun.com/ipng
- www.6bone.com (field trail IPv6 network deployment).

4

WDM Optical Networks

- Optical modulation
- Optical switching component and technology
- WDM NC&M framework
- WDM network information model
- WDM NC&M functionality
- WDM NE management
- WDM signalling
- WDM DCN
- WDM network views
- Discussion

In this chapter, we describe optical modulation, optical components, switching technology, and software development in WDM optical networks. Optical modulation formats and methods used in optical networks are presented. Optical networking components and switching technology are described. Next, an overview of WDM network control and management (NC&M) is presented, where the Telecommunication Management Network (TMN) framework and the WDM optical network management and visualisation framework are introduced. An information model for WDM optical networks is presented to abstract the WDM optical network resources. We present a detailed discussion on WDM NC&M functionality including connection management, connection discovery, WDM client topology reconfiguration, optical signal quality monitoring, NE management, and fault management. We will also describe the procedure of WDM network signalling. Finally, we present a section on network management and visualisation.

Throughout the chapter, the MONET network and its NC&M system is used as an example WDM optical network to illustrate the NC&M functionality. The MONET DC testbed network consists of two rings, where the west ring network comprises three O-E-O WADMs and the east ring network comprises three all-optical WSXCs.

4.1 Optical Modulation

An optical network transfers data in digital form, i.e. in 1s and 0s. It represents 1s by a pulse of light and 0s by little or no light. There are two formats developed for optical modulation. They differ in the transition between bits. The Non-Return-to-Zero (NRZ) modulation transmits a sequence of 1s by continuing to give out light, and only stops giving out light once it needs to transmit a 0 again. The Return-to-Zero (RZ) modulation represents successive 1s, each given out as a pulse of light. Hence, RZ switches the light to 0 even in between the transition of two 1s (see Figure 4.1).

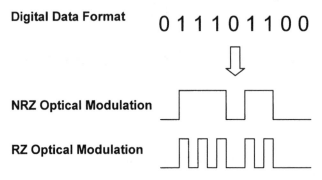

Figure 4.1   Optical modulation formats.

NRZ is in common use because of its simplicity and ease of its implementation; RZ may have advantages for long distance applications since it is less susceptible to chromatic dispersion and polarisation mode dispersion.

There are two approaches to modulate the laser light. They differ in whether the modulation is conducted using the laser itself or performed in an external modulator. Direct modulation takes the electrical current representing the digital data and directly outputs the optical signal by switching on and off the lights rapidly. During external modulation, the laser's output is constant. The modulation is performed in an external modulator, which works as a shutter to effectively switch the light output on and off (see Figure 4.2). Direct modulation is simple and cost-efficient, but it suffers from poor chirp performance leading to dispersion and has poor performance

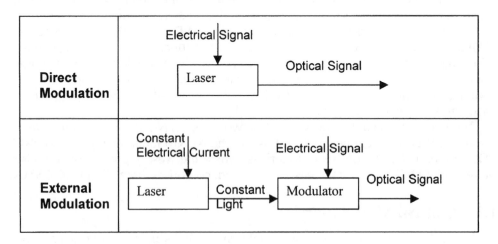

Figure 4.2   Optical modulation methods.

at high bit-rates. External modulation has good chirp performance and is suitable for very high bit-rate applications, but this approach is complex and expensive.

There are two main types of external modulators used in optical networks: electro-optic and electro-absorption modulators. Electro-optic modulators works by splitting laser light into two paths and then using phase shift to either cancel out the two waves (to give a 0) or combine them (to give a 1). Electro-absorption modulators are reverse-biased lasers that can absorb or transmit light.

4.2   Optical Switching Components and Technology

We describe various optical networking equipment in this section. We introduce a taxonomy of the switching fabric. We discuss various network equipment including amplifiers, OADMs, OXCs, transponders, and optical switches/routers.

4.2.1   Optical Amplifier (OAMP) and Repeater

An optical signal attenuates as it propagates. If the signal is too weak, for example, over a long distance without regeneration, the receiver cannot detect the signal. An optical amplifier is used to strengthen the optical signal through amplification.

A related device to the amplifier is the repeater, which has been used extensively in the past to completely regenerate the signal after a certain distance. A repeater is an O-E-O (optical-electrical-optical) device, so the signal is not only regenerated but also reshaped and retimed, i.e. 3R regeneration. A repeater-regenerated signal is clean but its processing is expensive. A repeater has an O-E-O converter, receiver circuits, the signal regenerator, transmitter-driven circuits, and a light source. An optical signal is converted into an electrical signal on entry to the repeater; the regenerated electrical signal is converted back to an optical signal through transmitter circuits.

There are several types of optical amplifiers, such as Erbium-Doped Fibre Amplifier (EDFA), Praseodymium-Doped Fibre Amplifier (PDFA), and Semiconductor Optical Amplifier (SOA). Among them, EDFA is the most widely used optical amplifier. An EDFA is designed by equipping a common single-mode fibre with erbium in fabrication.

Figure 4.3 shows the EDFA. The amplifier fibre is about 10 m long. To compensate for the weak input signal, a strong wavelength, e.g. 980 nm, is pumped into the amplifier section and mixed with the arriving input signal. The erbium ions in the EDFA fibre are promoted to a higher energy state by the pump laser. When the input signal, e.g. 1550 nm, reaches the promoted erbium ions, the energy carried by the erbium ions is transferred to the input signal. The erbium ions finish with a lower energy level but the arriving signal has been strengthened. To prevent the EDFA fibre itself from becoming a laser, an optical isolator is used to avoid reflections. Also, amplifier 'noise' will occur at both ends of the fibre amplifier section. This noise must be removed to reduce the cumulative noise in the signal.

In contrast, an OAMP is cheaper and simpler than a repeater and is also more robust in terms of its environment. A repeater requires a specific bit rate and modulation format, but an OAMP is insensitive to the bit rate or signal format. An OAMP can

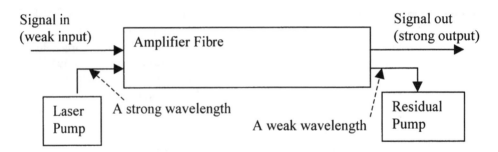

Figure 4.3 Erbium-Doped Fibre Amplifier (EDFA).

simultaneously amplify several WDM signals since it has fairly large gain bandwidth. A major drawback of OAMP is that it will amplify everything within the signal including the noise and the distortion. Therefore, OAMPs must be placed at a greater frequency along the fibre than would repeaters.

4.2.2   Optical Add/Drop Multiplexer (OADM)

OADM and OXC (Optical Crossconnect) are the two most common optical networking devices. In fact, the functionality of OADM and OXC can be implemented in a single piece of network equipment. We will introduce OADM and OXC separately for illustration purposes.

As the name implies, the main functionality of OADM is to access, drop, or pass-through wavelength channels in a WDM enabled optical network. Figure 4.4 shows a possible structure of an OADM. In the figure, there are four input and output fibres, each of which supports *n* wavelengths. An incoming optical signal over the input fibre is demultiplexed through a wavelength demux. Each of the wavelength channels matches one fabric port. The demuxed signal can propagate directly through the fabric without changing wavelength. Or it can be dropped onto one of the fabric drop ports through physical configuration of filters. Likewise, a wavelength can be added through an add fabric port and directed to a wavelength port by configuring corresponding filters. The outgoing wavelengths are multiplexed unto outgoing fibres through wavelength mux. Fabric add/drop ports represent the WDM network entry and exit points. To handle certain client signal formats, the corresponding interface cards are employed at the add/drop ports, which are also known as the OADM client interfaces. These client interfaces represent the implementation of the corresponding transmission framing technology.

4.2.3   Optical Crossconnect (OXC)

An OADM isolates wavelengths to selectively access a wavelength channel. However, another useful function is to rearrange wavelengths from fibre to fibre within a WDM network. This is provided in an OXC. An OXC provides wavelength-level switching.

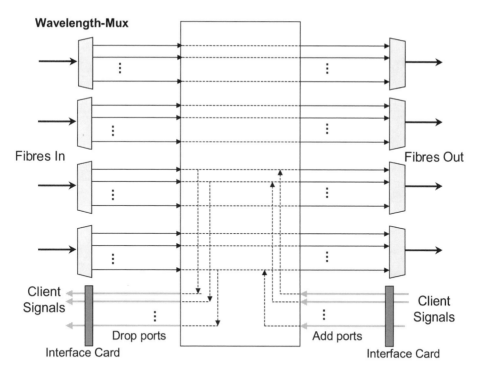

Figure 4.4 OADM.

Figure 4.5 shows a possible structure of an OXC. In the figure, there are four input and output fibres, each of which has a number of wavelengths. Through demux, signals can reach fabric ports. Depending on the switch setting, a signal over a certain wavelength from one fibre can be connected to the same wavelength but on a different outgoing fibre. In fact, it is likely that more than one signal will compete for a wavelength channel on one outgoing fibre, which causes outgoing fabric port contention. To ease this problem, wavelength interchange is introduced, according to which, a wavelength can be directed to a fibre with a different optical frequency. Wavelength interchange is expensive and may degrade the signal quality in all optical lambda conversion, so it should be used only when necessary.

In addition to wavelength switching, OXC can also support waveband switching and fibre switching. Waveband switching connects a subset of wavelengths simultaneously from an incoming fibre to an outgoing fibre. Fibre switching switches an entire fibre including all the wavelength channels to an outgoing fibre. Wavelength switching provides finer granularity switching than that of waveband and fibre switching.

Another issue of OXC is its architecture design. To support a large-size switching fabric with a number of ports, it requires OXC architecture scalability. Scalability should also be addressed in terms of connection data rate for each port.

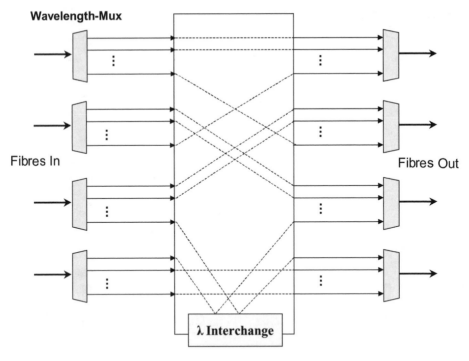

Figure 4.5   OXC.

4.2.4   Transponder

So far, our discussion on OADM and OXC does not consider signal quality or degradation. In fact, there is usually a transponder associated with each input or output fabric wavelength port to automatically correct optical imperfections accumulated during transmission. An electrical domain transponder regenerates signals in 3R or 2R. Since regeneration involves re-modulation, the transponder can generate a signal with a different wavelength, i.e. supporting wavelength interchange.

Optical networking hopes to process optical signals purely in the optical domain except for the control functions, which have to be implemented in the electrical domain (based on the current commercial technology). Recently, an optical transponder has also been developed. An optical transponder takes an intensity-modulated optical signal on one wavelength and another unmodulated carrier wavelength as inputs. Through optical waveguides (e.g. in silicon), the two wavelengths are brought together in the optical transponder so that the intensity modulation on one wavelength can be picked up by the carrier wavelength. The optical transponder outputs the modulated carrier wavelength that carries the same bit sequence as the input wavelength (see Figure 4.6).

Optical transponder technology is mostly under research and the use of electrical transponders is cheaper and more efficient.

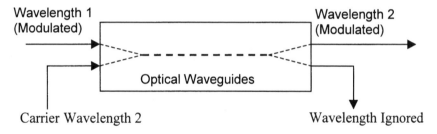

Figure 4.6 Optical transponder.

4.2.5 Switching Fabric

One of the most important optical components in optical networking is the switching fabric. Incumbent carriers prefer to use switches implemented by electrical switching fabrics because they are more mature and reliable than their all-optical counterparts. In addition, management information can be easily obtained through the O-E-O network equipment. However, there are competitive carriers who are deploying all-optical networks.

Figure 4.7 shows a taxonomy of switching fabrics. An electrical fabric is also known as an opaque fabric whereas all-optical fabric is referred to as a transparent fabric. In opaque fabrics, light pulses are converted into electrical signals so that their routes across the switch can be handled by conventional ASICs (application-specific integrated circuits). Since electrical fabrics, which can be found in almost all levels of switches and routers, have been used for decades, and electrical processing has been driving the information revolution, standards and, more importantly, commercial products are widely available. This translates into lower cost and

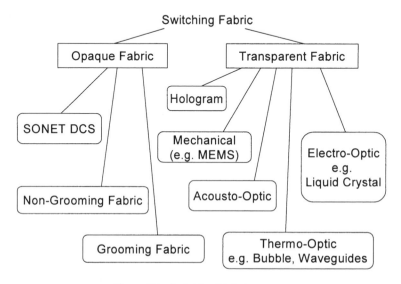

Figure 4.7 Switching fabric taxonomy.

ease of development for optical network equipment vendors. Also, networks formed by opaque fabrics are easier to manage and control as most of the management information can be obtained from the electrical domain. However, opaque switches are difficult to scale up in terms of the number of ports (i.e. fabric size) and the data rate.

High-speed applications using opaque fabrics require very high power consumption and introduce high processing costs. Opaque fabrics require a certain signalling format and so demand extra effort in the case of protocol update and multiprotocol support. By avoiding the O-E-O conversions, transparent fabrics, theoretically, can scale up to a large number of ports without adding lots of complexity.

Opaque Fabrics

First-generation optical networks use SONET DCX (Digital Crossconnect), which still represents the majority of switches deployed today. SONET DCX is designed primarily for voice traffic and can be used to support multiservice environments. Once wavelengths are separated by the demux, SONET ADMs are employed to convert the optical signal into an electrical signal and then feed it to the electrical fabric. The output signal from the fabric is also in electrical format. Through a SONET ADM, the electrical signal is converted into an optical signal. This signal is multiplexed onto a fibre for transportation. A SONET DCX can integrate ADM functionality so it can add or drop traffic, e.g. DS3.

Recently, opaque fabrics brought about the feature of arbitrary concatenation, which represents a key issue in bandwidth efficiency for data traffic. Arbitrary concatenation in grooming can be used to switch subwavelength or TDM traffic (see Chapter 2). Both non-grooming and grooming fabrics use transponders for signal regeneration (see Figure 4.8). In the figure, all the ports on the fabric have SONET interfaces. The switching fabric uses the optic with the 1310 nm centred waveband; transportation uses the long-reach waveband centred on 1550 nm. A non-grooming fabric is protocol and bit-rate independent and is less flexible than a grooming fabric. A non-grooming fabric is scalable to a large number of ports so it is suitable for network nodes where a large number of WDM channels terminate. Generally, non-grooming traffic does not eliminate the need for SONET ADMs. Grooming fabrics integrate DCX and ADM functionality. They are protocol and bit-rate dependent. Grooming fabrics can support mesh topologies.

Transparent switch fabric technologies

A transparent switching fabric tries to eliminate the repeated O-E-O conversions in the network, so eventually it hopes to create a high performance and low cost network with simplicity and scalability. We describe several optical switching fabric technologies below. These technologies can be applied to a number of applications based on the fabric characteristics. We list below several criteria by which different fabric technology can be compared.

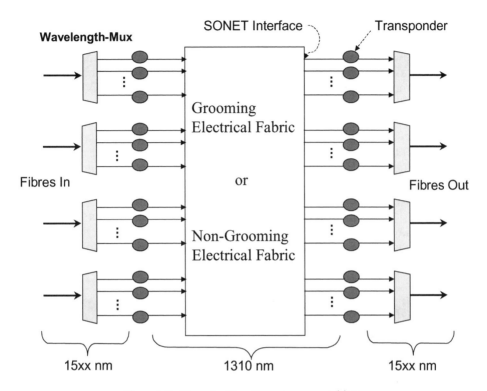

Figure 4.8 Grooming/Non-Grooming opaque fabric.

- **Scalability:** how scalable is the technology? How many ports, i.e. wavelength and/ or fibre ports, can be supported on the fabric? Are there any potential obstacles of the technology to prevent it from supporting a large number of ports?
- **Switching speed:** how fast can the fabric switch, for example, in microseconds or in nanoseconds? This represents the time to switch light from an input port to an output port. Fast switching is required in optical packet switching.
- **Power consumption:** what is the power consumption level of the switching fabric when it is operating?
- **Reliability:** how reliable is the technology? When the traffic granularity is small, for example, packet switching, the fabric needs to perform switching operations a million times per second. Is the technology robust enough to support this kind of 'heavy duty' switching?
- **Signal power loss:** as an optical signal attenuates during transmission over the fibre, its power level is reduced after crossing the fabric. The power loss is related to the fabric material and the coupling of the fibre to the switch fabric. Higher power loss requires signal regeneration, so it introduces extra cost in its solution.
- **Signal quality degradation:** except signal power loss, are there other factors that will cause signal quality degradation during fabric switching? Example factors include cross talk, polarisation dependence, and path-dependent loss.

MEMS switch

The current leading optical switching fabric technology uses the Micro Electro-Mechanical Systems (MEMS), which relies on silicon manufacturing processes to construct ultra-small arrays of switches. These switches use tiny silicon mechanical structures to move small mirrors, so that the switching function can be performed by reflecting the light from the input source to the output destination. If the mirror has only two positions, on and off, the MEMS forms a 2-D architecture, also known as a digital switch. If the mirror is able to adjust its positions according to configuration, i.e. more than two positions, the MEMS forms a 3-D architecture, also known as an analogue switch. A 2-D switch is easier to implement, more reliable and has lower insertion loss, but it is not as scalable as the 3D switch. An example of the MEMS mirror is shown in Figure 4.9 (left).

To support large switch configurations, a number of these mirrors are interconnected to relay the reflected light. There are several interconnection structures: Crossbar, n-stage Planar, two stage, and three-stage Clos. Figure 4.9 (right) shows an example of 8 × 8 MEMS with a chip size of 1 cm × 1 cm.

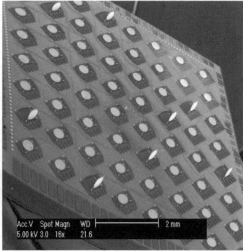

Figure 4.9 MEMS technology (source: Lucent).

Bubble switch

Inkjet printers use bubbles to blow ink out of the holes of the inkjet printhead. This is a simple and inexpensive technology due to the bubbles' high reflectivity. The same technology can be applied to light switch circuits. Figure 4.10 shows a bubble switch circuit with multiple optical paths, for example, waveguides or light pipes. These paths intersect at several cross points where the light travels through a fluid. An optical pulse will travel straight through the junction, unless it needs to be switched along another route. Using inkjet technology, a bubble can be inserted into the junction, which causes the light to be reflected along a new path, thereby switching

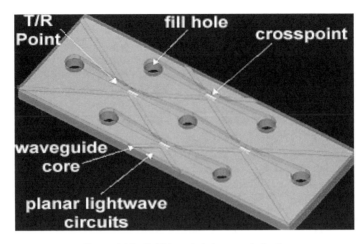

Figure 4.10 Bubble switch (source: Agilent).

the information it contains to the right destination. The optical bubble switch has the same reflective index properties as the waveguides. The bubble is inserted or removed by the printhead mechanism. A bubble switch is small, fast, and efficient but it is not as scalable as MEMS.

Thermo-optical waveguide switch

As we recall, waveguides refer to paths on an integrated circuit chip that have all the properties of optical fibres, so they can be regarded as 'fibre on a chip'. A waveguide also has a core and cladding, which are made of glass with different reflective indices. Thermo-optical effects occur when temperature changes are applied to the waveguide, triggering the change of the phase of the travelling light. Changing the phase in turn can cause the light to switch to a different wave-guide pathway.

Figure 4.11 (bottom) shows a switching unit. It uses a Mach-Zehnder Interferometer with thermo-optic phase shifters that function as switching components. The unit has one input waveguide and two possible output waveguides. In between, there are two other internal waveguides, which will split the input signal first and then couple the two interval waveguides. Through heating the splitter, the refractive index is changed to alter the way in which it divides wavelengths between one output and another. Figure 4.11 (top) shows an 8 × 8 optical matrix switch comprising 64 switching units.

Thermo-optical waveguides are small. Since such waveguides can be made in large batches as in mass production of integrated circuits, the cost can be very low. However, this type of switching fabric is not scalable. Currently the largest switch of this type only has a 16 × 16 fabric including over 500 switching elements on a silica wafer.

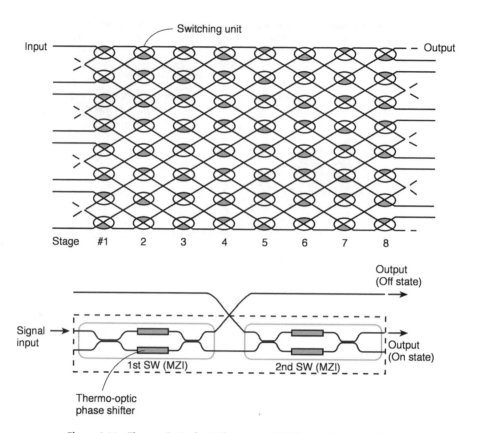

Figure 4.11 Thermo-Optical switch (source: NTT Electronics Corporation).

Liquid crystal switch fabric

A liquid crystal fabric switches light according to the change in the polarisation of optical signals when crystals are applied with an electrical voltage. The same technology is used in a desktop screen display, where by applying an electrical voltage, the change of the orientation of long molecules results in the update of the optical properties of the material in each pixel.

A liquid crystal switching element has two active components: a liquid crystal cell and a displacer. The cell is formed by placing the liquid crystals between two plates of glass. The glass is transparent and coated with oxide materials, so that it can be applied with a voltage. The function of the cell is to reorient the polarised light entering the cell when a voltage is applied. The displacer is formed by a composite crystal, which is used to guide the polarised light leaving the switch element. In the absence of a voltage, input light passes through the cell and then hits the displacer. The light polarisation is unchanged and output through the default path in the displacer. In the presence of a voltage, the input light polarisation is changed, for example, by 90° at the cell. The polarised light enters the displacer and uses a second output path.

Unlike MEMS (mechanical switch) and thermo-optical switches, a liquid crystal switch is a kind of electro-optic switch. The electro-optic switch is capable of changing its states very rapidly, and therefore, it has fast switching time.

Acousto-optic switch

Acousto-optic switches use sound waves to deflect a light beam, for example, from one fibre to another, to provide switching. This kind of switch does not have any mechanical parts, so, to some extent, it is reliable. The switch uses the electronically and computer-controlled acousto-optic deflector set, which can randomly connect the channels from the array to the single fibre-optic line, resulting in fast switching times of a few microseconds.

Holograms

A hologram switch has an electrically energised Bragg grating inside the crystal, where the Bragg grating comprises a series of stripes of different refractive index materials, each of which reflects a specific wavelength of light. In the presence of a voltage, the Bragg grating deflects the light to the output port. With no voltage, the light can pass straight through the crystal. Each input fibre requires a row of crystals, one for each wavelength on the fibre. A hologram switch provides nanosecond switching speeds so it can be used for packet-by-packet switching in optical routers.

Table 4.1 shows a switching fabric comparison in respect to scalability, switching speed, power consumption, reliability, signal power loss, signal quality degradation, and cost (or potential cost).

4.2.6   Optical Switch/Router

Combining the switch fabric, the client and network interfaces, and the transponders allow us to form an optical network switch. The optical switch can be designed to connect layer 2 frames, e.g. an OXC. An OXC utilises a circuit-switched paradigm, in which there is a shadowed data communication network (DCN) functioning as the control channel, and there are connection setup and tear-down phases during a connection. Once a connection is set up, it works as a link or pipe, and the traffic can flow in and out of the pipe without any examination. However, this scheme does not fully utilise the network capacity since the level of resource sharing and usage is very low.

Optical burst switching is introduced in particular to address the network efficiency. A burst itself can be considered as a packet, but it has a larger size than that of a regular IP packet. In IP terms, a burst can be treated as a flow or 'fat packet'. However, a burst can be a layer 2 concept, and, therefore, it is not necessarily related to IP. Optical burst switching requires a separate control channel to transport control messages.

The optical switch can also be designed to switch layer 3 packets, i.e. an optical packet router (OPR). An OPR no longer needs pre-established circuits (for example, setup using signalling protocols). It forwards packets to a next hop according to a

Table 4.1 Switching fabric comparison

Switch fabric name	Scalability	Switching speed	Power consumption	Reliability	Signal power loss	Signal quality G! degradation
MEMS-2D	Medium	Slow (10ms)	Low	Low to Medium	Low	Low
MEMS-3D	Very Good	Slow (10ms)	Medium	Low	High	Low
Bubble switch	Medium	Slow (10ms)	High	Low	Medium	Medium
Electro-optic (e.g. liquid crystal)	Good	Medium (6 ms)	Very low	High	Medium	Medium
Acousto-optic	Good	Fast (1ms)	Medium	Medium	Low	Low
Hologram	Very Good	Very fast (5 ms)	High	NA	Medium	Low
Thermo-optical waveguides (e.g. Silica)	Poor	Medium (6ms)	Low	Low	Low	Medium

routing table maintained locally. An OPR requires on-the-fly packet header processing to determine a packet path based on a packet destination. We will introduce optical burst and packet switching in the next chapter.

4.3 WDM NC&M Framework

NC&M-related functions can be described and classified using the OSI management framework, which defines five functional areas:

- **Fault management** is responsible for fault detection, root cause analysis, fault isolation, fault correction, and alarm/notification generation.
- **Configuration management** is responsible for maintaining an accurate inventory of resources including hardware, software, and circuits within the network, and with the ability to change/control the inventory in a reliable and effective manner.
- **Accounting management** is responsible for identifying costs associated with network communication resources and establishing charges for the use of these resources and services.
- **Performance management** is responsible for evaluating and analysing the effectiveness and efficiency of network resources and related communication activities.
- **Security management** is responsible for network and system security management.

This book will cover areas related to network control and management in the OSI and TMN management framework. We will not include discussions on accounting management and security management of OSI, and the business management layer of TMN.

4.3.1 TMN Framework

The ITU-T proposed and developed an infrastructure to support management and deployment of dynamic telecommunications services, known as Telecommunications Management Network (TMN). TMN is a framework for achieving interconnectivity and communication across heterogeneous operating systems and telecommunications networks. TMN principles are incorporated into a telecommunications network to send and receive information and to manage its resources. TMN adopts the object-orientated concepts to define communication interface between management entities in a network.

A telecommunications network is composed of switching systems, circuits, and terminals. In TMN terminology, these resources are referred to as network elements (NEs). TMN enables communication between operations support systems (OSS) and NEs. TMN is designed as a multivendor framework so it aims at network and system interoperability, extensibility, and scalability. Within the framework, telecommunication network management is described from several viewpoints: a logical model, a functional model (functional and physical architecture), and a set of standard interfaces.

TMN logical model

The TMN logical model employs a layered concept to group functionalities according to business practice. The following layers are defined in the model:

- *Business management layer (BML):* responsible for business planning and management for the enterprise. It includes budgeting, goal setting, business-level agreement, and communication management.
- *Service management layer (SML):* responsible for managing service-related tasks according to items defined in the service-level agreement, for example, QoS service assurance. A contracted service may involve several networks over different technologies administered by different operators.
- *Network management layer (NML):* responsible for managing the network, which is composed of various NEs. This layer has a network level view in terms of topological configuration and connectivity usage.
- *Network element management layer (EML):* responsible for managing the network element. Each element manager usually manages a set of NEs in terms of element data, logs, error, and activity.
- *Network element layer (NEL):* responsible for maintaining TMN-manageable information in an individual NE.

Figure 4.12 shows the layers in reference to functional blocks and reference points. Reference points can be regarded as the implementation of interfaces. Reference points are used in describing services, whereas interfaces are associated with protocol stacks. In most cases, interfaces and reference points have one-to-one relationship. However, no interface exists for the reference points that lay outside of the TMN and interconnect function blocks within a single building block.

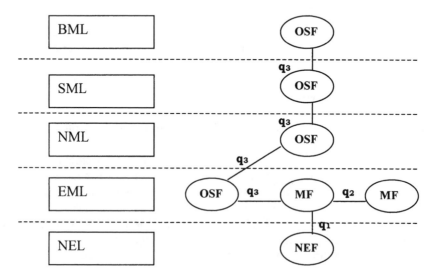

Figure 4.12   TMN logical model hierarchy.

TMN functional model

Figure 4.13 shows the TMN functional blocks:

- OSF (Operations System Functions)
- MF (Mediation Functions)
- QAF (Q Adaptor Functions)
- NEF (Network Element Functions)
- WSF (Work Station Functions).

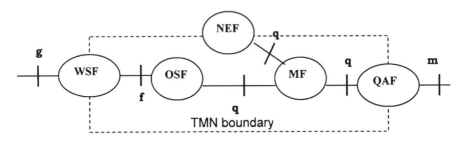

Figure 4.13 TMN function blocks (Functional Architecture).

It is unnecessary for all of these types to be present in each TMN configuration. Most TMN configurations support multiple function blocks of the same type. Within the figure, OSF, MF, and the 'q' and 'f' (also 'x') reference points are completed defined in the TMN; WSF, QAF, NEF, and the 'g' and 'm' reference points are partially defined in TMN.

A WSF provides the means to interpret TMN information for the management information user:

- a QAF connects those entities that do not support standard TMN reference points to TMN;
- a MF works as an intermediate entity to pass information between NEF or QAF and OSF;
- an OSF initiates management operations and receives notifications;
- a NEF implements the NE management functions;
- the 'q' reference point connects the function blocks NEF to MF, MF to MF, and MF to OSF;
- the 'f' reference point connects function blocks OSF, MF, NEF to the WSF;
- the 'g' reference point connects the user to the WSF;
- the 'x' reference point connects a TMN to other TMNs.

To implement the functional model according to the functional architecture, TMN also defines a physical architecture, which details how function blocks should be mapped to building blocks and also reference points to interfaces. TMN physical architecture has the following building blocks:

- Network Element (NE)
- Mediation Device (MD)
- Q Adaptor (QA)

- Operations System (OS)
- Work Station (WS)
- Data communication network (DCN).

It is possible to implement multiple function blocks of the same type or of different types into a single building block. For example, an OS can be used to implement OSF, MF, and WSF or even multiple OSFs.

A special kind of building block is a DCN, which does not relate to any specific functional blocks. A DCN presents a transport network for the control plane, upon which other building blocks exchange management information.

TMN application functions

In addition to the general functions defined by the above functional blocks, TMN supports the following application functions:

- *processing functions* for information analysis and manipulation;
- *retrieval functions* for information access;
- *security functions* for reading and updating information authorisation;
- *storage functions* for storing information over controlled amounts of time; *transport functions* for the movement of information among TMN Nes;
- *user terminal support functions* for input and output information.

4.3.2   WDM Network Management and Visualisation Framework

For scalability reasons, WDM networks can be organised topologically into a hierarchy of subnetworks. The subnetworks are categorised, for example, based on geographic proximity, equipment vendor identities, management boundaries, or network ownership. Figure 4.14 shows an example of the WDM network with subnetworks.

Figure 4.15 illustrates the network control and management system framework architecture that corresponds to the example WDM network shown in Figure 4.14. The functionality within the framework is grouped into four layers:

- Network Element Layer (NEL)
- Element Management Layer (EML)
- Network Management Layer (NML)
- Network Visualisation Service (NVS).

Within the framework, the NEL, EML, and NML are TMN concepts.

The NVS is introduced to provide management functionality to different levels of system and network users, possibly located at different remote locations. As a result, a NC&M user can control and monitor the network through the NVS including administration services, views, and system APIs. The administration kit includes a daemon, a query server, and a **Network Process Status** (NPS) server, and verifies user information and/or user profile with an **Authorisation and Authentication** (A&A) database.

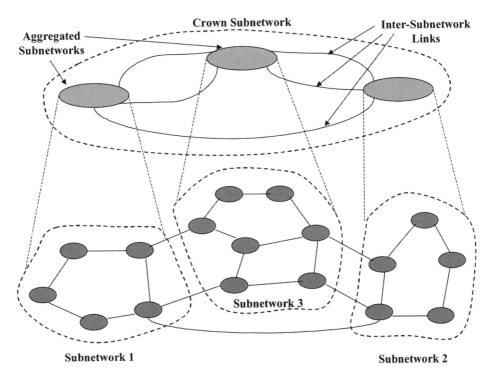

Figure 4.14 An example of WDM optical network.

The query server provides the searching facility over the entire WDM network (i.e. in its administrative domain). It is designed as the SQL-like interface to persistent network information and serves as the front end of the Network configuration and performance databases. The NPS is analogous to the 'ps' command in Unix. It keeps track of network process status and provides a GUI to spawn, terminate, or migrate network processes among hosts. For a legitimate user, various views can be queried and displayed and the views can be further customised through user profile updates. Conceptually, views are application orientated, so they are designed to serve the user by providing an information abstraction on the NML and/or EML functionality. Example views may include WDM map, WDM network spare capacity, performance monitor, configuration view, making and deleting connection window, and NE fabric view.

Application Programming Interfaces (APIs) are employed to interface with external applications or services. They perform as a reflection interface for the exported WDM management features. An *Interface Repository* (IR) can be used to maintain full information about the interface definitions.

The NML consists of Configuration, Connection, Performance, and Fault functional modules. It controls and monitors the entire network to provide end-to-end connection path creation, network level fault reporting/isolation, and end-to-end statistics collection. The CORBA event service can be used to implement event

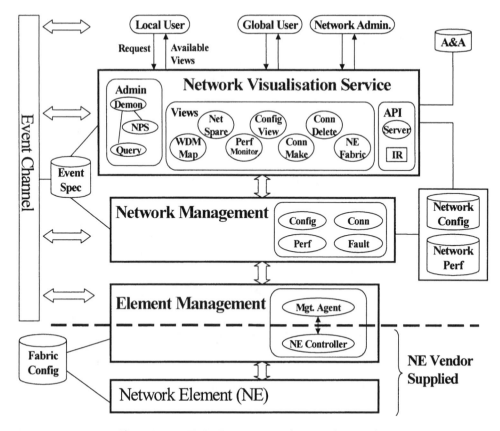

Figure 4.15   WDM NC&M system architecture framework.

channels to provide a unified messaging mechanism throughout the network. The definitions and meanings of events are given in the Event Specification database.

The EML is made up of two software components, a management agent and a NE controller. Each NE has a low-level hardware orientated NE controller, but a management agent (EML) may control and monitor one or more NE. NE inventory and its connectivity are stored and maintained by EML in a Fabric Configuration database. Since there is frequently *direct* and *heavy* traffic between an agent and the NE controller, a low-level interface (such as a socket) is the preferred technology for implementing the communication channel between them. Network level managers and visualisation servers are likely to be distributed. CORBA technology (ORB) may be used to implement the communication between network processes, for example, agents, network managers, and network visualisation servers. CORBA is gaining acceptance in the telecommunication community, and helps to simplify the implementation effort with location and platform transparency.

4.4 WDM Network Information Model

An important task of NC&M is to efficiently store, access, and update the WDM network information. From a software development point of view, network resources such as links and NE are objects with attributes, where the relationship between objects can be described in a data structure. Hence, the purpose of a NC&M system is to automate and computerise the process of management to simplify and to optimise the network resources usage. A network information model specifies the network abstraction and resource representation.

Figure 4.16 shows the connectivity representation of a point-to-multipoint trail in a WDM layer network in one administrative domain. A trail is a logical representation of network resources for an end-to-end lightpath, which may include signal splitting such as multicasting. To a client, each trail is perceived as two or more Network Trail Termination Points (TTPs) that represent the optical WDM add/drop ports. Within the WDM network, a trail consists of Subnetwork Connections (SNCs) and Link Connections (LCs). A LC represents a single wavelength channel interconnecting two network Connection Termination Points (CTPs) in different NEs. A link typically supports more than one LC. A SNC has two end points, each of which is bound to a CTP. (This is one representation of a SNC. Other models allow SNCs to be multipoint to multipoint.) The connection manager keeps track of the connectivity graph.

In a testbed WDM network, partial connections are common (a partial connection is a network fragment of a complete trail and may have one or none TTPs).

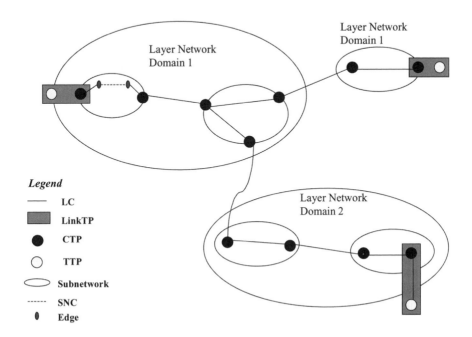

Figure 4.16 A Point-Multipoint trail in a WDM layer network.

4.4.1   WDM Object Model

Data in a NC&M system is composed of persistent and transient information. The persistent information is used for resource configuration. An example of the transient information is the connectivity information in connection management. The status of network resources indicating faults/alarms and their severity and probable cause, is another group of transient information.

Figure 4.17 shows an object model to implement a WDM optical network. A layer network (hereafter in this section simply referred to as 'network') is made up of components of a specific transmission and switching technology that transports information of a specific format and coding. For example, WDM and ATM networks are two different layer networks. A network physically consists of NEs and topological links. A WDM line, a logical concept also known as an optical multiplex section, comprises one or more links interconnected by WAMPs that regenerate the optical signal to support long distance transmission. The line representation of the network can simplify the user's view on network topology and allow the user-articulated manually routed trails.

A Subnetwork (SN) recursively contains smaller subnetworks and links interconnecting subnetworks. The smallest subnetwork is the NE itself. In Figure 4.17, objects have two main relationships, *has-a* and *is-a*. The former represents a Whole-Part object connection and the later indicates a Generalisation-Specialisation object connection.

In WDM optical networks, there are three types of NE: WAMP, WADM, and WSXC.

A port can be either a physical **Transport Interface** (TI) or a **Client Interface** (CI), but only the CI can add or drop an optical signal.

A network may have 0 or more trails. In Figure 4.17, a trail orientated object subtree represents transient objects. 'Transient' here does not mean these objects are never written to a database or stored in some way. It simply indicates these objects are more dynamic than objects associated with resource configuration because they represent the network usage. Resource configuration objects such as NE also need to be updated. For example, a network manager may wish to remove a circuit pack so that it may be used elsewhere in the network. However, this happens less frequently and the objects are static in nature in a 'perfect' WDM network. Separating transient from persistent objects not only enhances system performance but also addresses issues of scalability and extensibility.

A trail consists of a sequence of SNCs and LCs, where a high level SNC contains LCs and other SNCs recursively. The trail can also cross different domains and have one or more inter-domain LCs. A trail may be protected by employing two SNCs, one primary and one backup. Fault or alarm information indicates the status of the object, and may be summarised as the object attributes. For instance, a NE object can support these fault-related attributes: Alarm Status, Current Problem List, Probable Cause, and Perceived Severity. An alarm could be one of these types: Critical, Major, Minor, Under Repair, and Alarm Outstanding.

The following is a list of key objects in a WDM network.

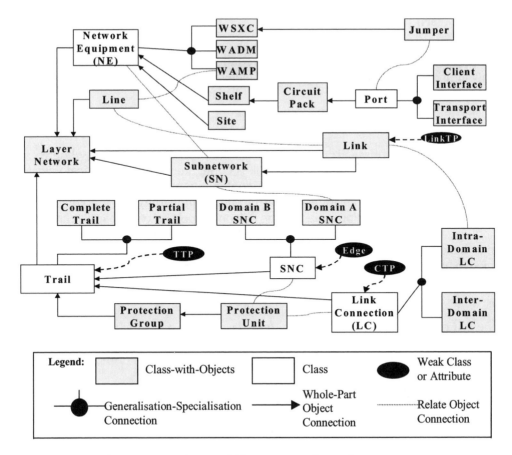

Figure 4.17 An object model for WDM optical network management.

Layer network

The concept of the layer network was originally defined by ITU-T to describe the functional model of the SDH transport model. Layer network can be defined as a transport network that carries a particular technology. Generally, a layer network is closely tied to a specific type of network transmission and switching technology. For example, in the MONET network, there is a WDM layer network. The introduction of the network layer concept allows each layer network to be defined independently of the other layers, so adding or changing the technology and/or structure of a single layer network may not affect other layers.

Trail

In the layer network, the network resource that transfers client information between two or more endpoints is a trail. A trail has one of these configurations: point-to-point unidirectional, point-to-point bi-directional, or point-to-multipoint unidirectional. A trail is an end-to-end concept, so its endpoints are the interface to client networks.

Subnetwork (SN)

A layer network is decomposed into subnetworks, which in turn may be further decomposed into smaller subnetworks interconnected by links. A subnetwork represents an interconnected group of network elements or subnetworks that is entirely within a layer network.

Link

A link represents the physical connectivity between two subnetworks or a subnetwork and a customer premises equipment. A link is configured using topological links in one of these ways: 1:1 configuration, N:1 configuration, or 1:N configuration. 1:1 configuration is implemented by assigning the entire bandwidth of the topological link to the link. N:1 configuration assigns to the link only a portion of the bandwidth of the topological link. 1:N configuration assigns the entire bandwidth of a set of topological links to the link. In a WDM network, a link can represent an optical fibre. The WDM link can be further categorised into access link and transport link.

Line

Line is a concept used in some optical transmission networks such as SONET. Line has one or more links and terminates at line termination points. For example, a line can consist of several links interconnected by optical amplifiers or repeaters.

Customer premises equipment (CPE)

This represents the client terminal equipment such as a computer.

Link connection (LC)

The network resource that transports information across a link between two subnetworks or a subnetwork and a CPE is called a link connection. In a WDM network, a link connection is a wavelength channel on the link. A trail is composed of one or more subnetwork connections and link connections.

Subnetwork connection (SNC)

The network resource that transports information across a subnetwork between two or more end points in the subnetwork is called a subnetwork connection. As defined in subnetwork, a subnetwork connection can in turn consist of one or more subnetwork connections and link connections.

Trail termination point (TTP)

A termination of a trail is called a trail termination point. Since a trail supports point-to-multipoint connections, a trail has two or more trail termination points. This is the point that information is accepted or delivered in a layer network.

Link termination point (LinkTP)

A termination of a link is called a link termination point. After a link termination point is configured, one or more connection termination points can be created.

Network connection termination point (CTP)

A termination of a link connection is called a network connection termination point. Since a link connection is always a point-to-point connection, a link connection has only two network connection termination points.

Line termination point

An end point of a line is called a line termination point.

Edge

A termination point of a subnetwork connection is called an edge. Each edge is bound to a network connection termination point.

NE

This represents the general object of network equipment, which will be inherited by all types of NEs. There are 3 types of NE in WDM networks: WADM, WSXC, and WAMP. Each NE has one or more shelves, each of which consists of multiple circuit packs.

4.4.2 An Example of WDM Network and Connection MIB

As an example to illustrate how these information objects are used to describe a network and its connections, we consider the network and connection example depicted in Figure 4.18.

The example network in this figure consists of three subnetworks. Two of the subnetworks are single ring networks each consisting of multiple WADMs and WSXCs. The other subnetwork is a double ring network with WADMs and WSXCs. The WSXCs are used to connect to their counterparts in the other subnetworks.

Figure 4.19 shows the resulting MIB (Management Information Base) formed from the *bottom* portion of this example network (i.e. the resources associated with the lightpath). The MIB shown consists of:

- A WDM Layer Network consisting of a CROWN subnetwork that in turn consists of the two ring subnetworks (Subnetworks II and III). They are interconnected by link B.
- Ring subnetwork II consisting of six network elements of which only two (WADM1 and WSXC1) are shown. They are interconnected by link A.

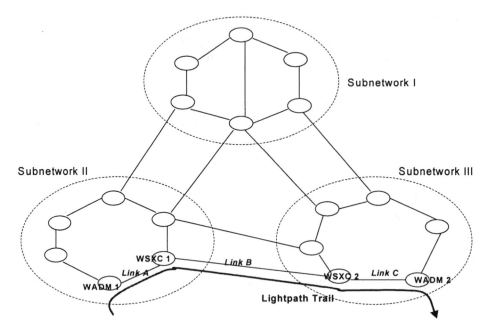

Figure 4.18   Sample network and connection example.

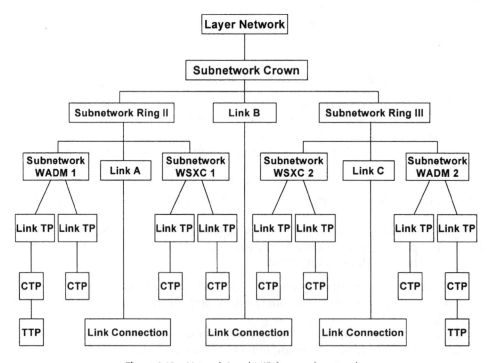

Figure 4.19   Network-Level MIB for sample network.

- Ring subnetwork III also consisting of six network elements of which only two (WADM2 and WSXC2) are shown. They are interconnected by link C.
- Each fibre link consisting of eight link connections (wavelength channels). At each of the network element, a subset of the LinkTPs, CTPs, and TTPs are shown. For example, at WADM1:
  - WADM 1 has three LinkTPs denoting three connecting fibres. In Figure 4.19, we show only two of them: one denoting the incoming access link, and the other denoting the outgoing link (A).
  - The LinkTP of WADM 1 associated with link (A) contains eight CTPs representing termination points for the eight wavelengths supported on the fibre. In Figure 4.19, we show only a single CTP that denotes an end point of a particular link connection (wavelength connection). These network resource objects are organised into NML and EML layer entities, and are maintained by their corresponding NML and EML resource configuration managers.

When the end-to-end lightpath connection depicted in Figure 4.18 is provisioned across the network, it is represented in the MIB by a trail object, and a corresponding set of supporting subnetwork connection objects (SNCs). Figure 4.20 shows the corresponding MIB view with the new information objects created as a result of provisioning this end-to-end trail:

- A WDM layer end-to-end trail is supported by an NML-level subnetwork connection (SNC0) over the crown subnetwork.

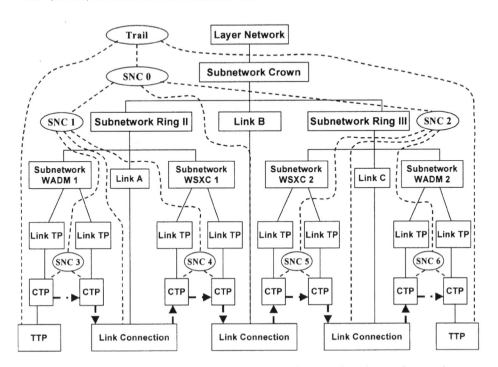

Figure 4.20 Network-Level resource and connection information base for sample network.

- The NML-level SNC consists of two EML-level SNCs, one over the ring II subnetwork (SNC1), and the other over the ring III subnetwork (SNC2), and a link connection over link B.
- The EML-level SNCs are further decomposed into link connections and NEL-level SNCs (cross-connects).

The management information base is shared by all management applications that can navigate and retrieve the information needed for their processing. For example, starting from a trail object, a manager application can trace out the route of an end-to-end trail, by enumerating all the SNCs and link connections used to support the trail. As another example, a network administrator running a GUI can start at a link object and navigate to discover which end-to-end trails are routed over the selected link.

Figure 4.21 (left) shows an example of NC&M information model browser, where object classes are drawn as rectangles and object inheritances are drawn as lines. Clicking in the box on the figure opens an object information window for that box. The browser is designed for administrators and application developers, and can also be used for trouble shooting and debugging. Figure 4.21 (right) shows the WDM link information (i.e. one link) that lists each fibre detail such as the link-fibre ID, source and sink NE, directionality, signal type supported over the fibre, and current fibre status.

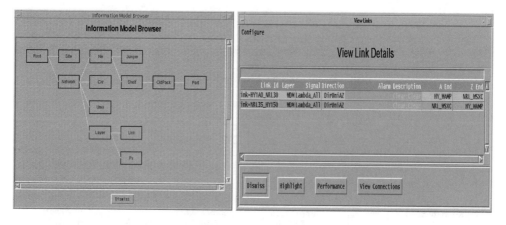

Figure 4.21   WDM information model browser and link configuration (source: MONET NC&M).

4.5   WDM NC&M Functionality

This section describes some examples of management functionality for WDM networks. This discussion is not intended to be comprehensive but, rather, it illustrates some of the many management functions useful for WDM.

4.5.1   Connection Management

Connection management computes the routing path based on a routing algorithm, and then reserves the wavelength at the corresponding NE. With regard to connection establishment, there are two approaches. One approach is **User Signalling,** in which the end user through the signalling interface sets up a connection. Another approach sets up the connection using the layered **Network Management Interface** (NMI).

MPλS is an industry approach to leveraging the existing control plane techniques developed for MPLS. MPλS is an example of the user signalling approach, and uses the IP-based **Label Distribution Protocol** (LDP) or RSVP (Resource Reservation Protocol) for wavelength signalling. NMI employs a hierarchical layered approach, where the route is computed by a connection manager and established by a connection setup interface. A WDM connection setup request specifies the source and destination client termination points that the connection management system has to connect. The connection provisioning process has two phases:

- **Route selection phase:** connection route is computed from the crown subnetwork and proceeds downward towards the NEs. As indicated in the NC&M system architecture, both NML and EML have connection managers that are responsible for connection computation and reservation in their managed subnetwork.
- **Connection commit phase:** connection setup involves a connection commit phase. There are two styles of connection commit, fast connect commit and verified connect commit. In the case of fast connect commit, reservation messages can be sent to NEs in parallel. Once the message arrives at an NE, the cross-connection is set up even in the absence of a client signal. In the case of verified connect commit, an input signal must be presented and detected for verification before the cross-connection is installed; otherwise, the request will be rejected. Hence, the presence of a valid client signal at the time of connection setup can be used as a criteria to select the connect commit style.

There are several important issues associated with lightpath routing. Our discussion focuses on circuit switched WDM networks.

Routing Metrics

Routing metrics define the routing objective function such as maximising the network throughput, minimising the link congestion, and minimising the connection blocking rate. Wavelength routing and assignment has received attention in recent years. Our discussion is based on a shortest path algorithm. There are several routing metrics that can be applied to the shortest path algorithm:

- **Path length:** this represents the default best effort routing service. The path length is given in terms of the number of intermediate hops.
- **Cost:** different links and different NEs have different connection and usage cost. Using this metric, route computation gives the cheapest connection path between the source and destination NE.
- **Path quality:** different wavelength channels have different signal quality. For example, signal degradation is likely to occur over an all-optical wavelength channel, i.e. no O-E-O conversions on the intermediate hops. Path quality can also be interpreted as protection and restoration capability. High level of QoS usually requires 1:1 or 1 + 1 path protection.

- **Connection rate:** based on fabric scalability, each port can support up to the maximum connection bit rate, e.g. OC-48 or OC-192, of the fabric. The connection rate may need to be explicitly configured during cross-connect setup, a process known as **clocking**. When a connection manager computes a route for a trail, it needs to assure that each hop on the trail path can support the data rate requested by the client.

Static routing and wavelength selection

Lightpath route computation can be either performed by the centralised connection managers or conducted through a distributed routing protocol. A distributed routing protocol enables all WDM network elements to establish a consistent view of the network, and outputs a routing table. The main functionality of a wavelength routing protocol includes physical network topology discovery, WDM link state information maintenance, and a constraint-based wavelength routing algorithm. A wavelength signalling protocol fulfils wavelength routing decisions and conducts these activities: wavelength assignment, setup/tear-down optical light paths, and priority arbitration with pre-emption. Using the distributed WDM routing protocol, the control system can cope with dynamic lightpath connection requests between end nodes. In addition, the lifetime of these dynamic lightpaths is flexible, for example, it can be relatively short lived (in days or even hours). However, the route of the lightpaths could be less optimal in comparison with that of centralised connection managers. The dynamic routing and wavelength assignment issues are discussed in Chapter 6.

This section focuses on static routing and wavelength selection that is widely used in network provisioning. In a TMN framework, a connection manager is responsible for route and wavelength selection based on the routing metrics between the user-specified end points. The connection manager is logically centralised although its copies can be distributed to improve availability. In such an environment, the lightpath connection requests are known in advance, for example, as a result of network planning. Moreover, the lightpaths remain in service for a relatively long time (in months or years). With respect to connection setup, this approach uses the NMI that can guarantee wavelength channel QoS during setup, for example, using verified connect.

Figure 4.22 shows an example of WDM wavelength routing. For simplicity, we assume that all switches belong to a single subnetwork and do not have wavelength interchange capability. Certain switches or all switches supporting wavelength interchange can be considered as a special case of wavelength continuity. In the example, there are two unidirectional fibres between any two switches, and each of the fibres supports four wavelengths. Within the figure, each line represents a bi-directional wavelength link: solid line indicates the wavelength channel is available; dotted line indicates that the wavelength channel is in use or has been reserved. For example, the connection manager received a connection request between the Source and Sink switch as indicated in the figure.

As we can see, two imperative issues in WDM routing are wavelength availability and wavelength continuity.

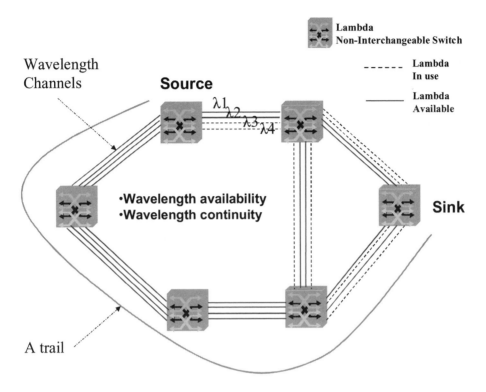

Figure 4.22   WDM wavelength routing.

Problem formulation

The static routing and wavelength selection problem can be formulated using Integer Linear Programming (ILP) with the objective function as maximising the carried traffic in the network (i.e. maximising the number of lightpath requests accommodated in the given physical topology). According to [Waut96], two types of formulation can be considered: route formulation and flow formulation. In route formulation, all possible routes for all node pairs are listed and the number of times a route is used is also determined. In flow formulation, the flows for each node pair are used as the decision variable.

Assume the set of all possible cycle-free routes between node pairs are determined and represented by $A = \{a_1, a_2, \ldots a_a\}$, the ILP variables are represented by a route-wavelength matrix $B = [b_{ij}]$, where b_{ij} is the number of paths using route a_i and wavelength j, and L is the number of wavelength channels on each fibre. The objective function for the route formulation can be expressed as:

$$\max \sum_{i=1}^{a} \sum_{j=1}^{L} b_{ij}$$

subject to these constraints:

$$C_L \times B \times Q \le D$$

and

$$B \times E \le C_L \times F/L$$

where C_L is a $C \times L$ vector in which all elements are 1; Q represents a route node pair incidence (with A) matrix (each element is set to 1 only if a route is incident to the node pair, otherwise it is set to 0); D represents the lightpath request demand vector; E represents a route fibre link incidence (with A) matrix (each element is set to 1 only if the edge is used on the route, otherwise it is set to 0); F represents the fibre link capacity vector.

In the flow formulation, assume the route-wavelength matrix is $Z = [z_{ijk}]$, where z_{ijk} indicates the amount of traffic flow from node pair k on link j on wavelength i, and the routed traffic is represented by $Y = [y_{mk}]$, where y_{mk} is the number of lightpaths routed between node pair k on wavelength m. The objective function can be expressed as;

$$\max \sum_{m=1}^{L} \sum_{k=1}^{s} y_{mk}$$

subject to these constraints:

$$\sum_{m=1}^{L} y_{mk} \le D_j,$$

$$\sum_{k=1}^{s} z_{mjk} \le F_j/L$$

and

$$\sum_{\in \alpha(v_i)} z_{mjk} - \sum_{\in \beta(v_i)} z_{mjk} = \begin{cases} y_{mk}, \text{if } v_i = \text{ source of node pair } k \\ -y_{mk}, \text{if } v_i = \text{ sink of node pair } k \\ 0, \text{otherwise} \end{cases}$$

where s represents the number of node pairs in the topology; $\alpha(v_i)$ represents the set of links for which v_i is the source; $\beta(v_i)$ represents the set of links that v_i is the destination.

Heuristic algorithms

Since the above ILP formulation is NP-complete, heuristics-based algorithms can be developed for static routing and wavelength assignment. There are two correlated tasks in routing and wavelength assignment: path selection and wavelength assignment. One group of heuristics divides the two tasks into separate phases. These heuristics perform path selection first in respect to physical fibre topology and then wavelength assignment on the selected path.

An example of the path selection metric is to minimise the number of hops across the network, i.e. the shortest path for a given node pair. A common metric for wavelength assignment is to minimise the number of wavelength channels (or colours) used in the network.

In [Baro97], a heuristic algorithm is proposed for lightpath allocation in an arbitrarily connected WDM networks. Every node pair is assigned to a single lightpath through the network and the algorithm tries to minimise the number of wavelength channels required in fibres to route the traffic over the given topology. In the algorithm, the shortest path routes are computed and assigned to the lightpath requests first. Alternate paths may be used to substitute the primary candidate path if the number of channels of the most loaded link in the alternate path is lower than that of the primary path.

Wavelength assignment is conducted through certain indexes, where lightpaths with the longer paths are assigned to smallest index wavelengths available through its route before the other lightpaths. This approach implies it is harder to find a free wavelength on a longer lightpath with more links. [Waut96] proposed a heuristic algorithm to further minimise the number of wavelength channels used for requested lightpaths. The algorithm starts with computing the shortest paths for each node pair and then searching the shortest paths to minimise the wavelength requirement. Then, the algorithm enters a loop trying to lower the wavelength channels required through rerouting some of the lightpaths. The algorithm stops with the shortest path of minimal wavelength requirements.

Another group of heuristics makes use of the notation of wavelength topology that details the wavelength availability on each fibre link. An edge on the wavelength graph is identified using the wavelength ID and the fibre ID in the physical WDM network. In addition, each wavelength can be assigned a weight to indicate the wavelength or the usage cost. In the absence of wavelength conversion capability, for a given node pair, the heuristics exhaustively search the minimal weight wavelength path. In the presence of wavelength interchange, for a given node pair, the heuristics can either still search the minimal weight wavelength path or look for the minimal weight wavelength path requiring the least number of wavelength conversions.

Instead of colouring the wavelength for routing, an approach is presented below to colour the switch according to its wavelength interchange capability. This approach is useful when the wavelength interchangeable ports/nodes are fixed in the network and they co-exist with non-wavelength interchangeable ports/nodes. A WDM switch is either wavelength interchangeable or non-wavelength interchangeable. We model the non-wavelength interchangeable WDM switch into a group of logical nodes at the same location (for example, these logical nodes can be regarded as wavelength switches located on isolated shelves). Each logical node interconnects a bi-directional wavelength channel, and can add, drop, or pass through traffic.

Figure 4.23 shows an example of WDM node colouring to model WDM switches. The example network is composed of six switches, half of which are wavelength non-interchangeable. As shown in the figure, the wavelength non-interchangeable switches, B, D, and F, are modelled into a group of nodes, where the number of nodes is equal to the bi-directional wavelength channels. Hence, this scheme translates the WDM network with mixed wavelength interchange capability into a network where every node has the same functionality, i.e. switching any of the input wavelengths to one of the output wavelengths across a wavelength graph.

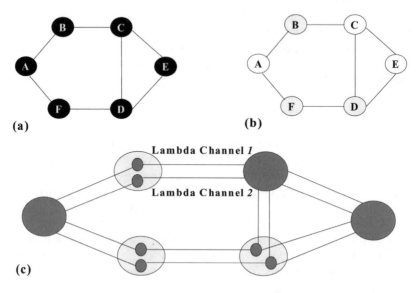

Figure 4.23   WDM node colouring.

Figure 4.23(c) shows the modelled WDM network with two bi-directional wavelength channels. For a given node pair, the minimal weight wavelength path can be obtained through exhaustive searching. This scheme can be applied to reconfigurable WDM networks or OXCs, where network topology is relatively simple and wavelength transparency is desirable.

In principle, wavelength topology is more complex to represent than the fibre topology. Employing node colouring schemes (as presented above) to wavelength topology introduces extra complexity into the modelled network graph.

Several wavelength-specific scheduling heuristics will be presented later in this chapter. Dynamic routing and wavelength assignment issues are discussed in Chapter 6.

WDM Routing Constraints

There are two groups of WDM routing constraints: client-imposed and network-orientated routing constraints. Clients may prefer specific paths or have special requirement for certain reasons. **Client-imposed constraints** include:

- *Source and sink end points:* these are usually given in terms of switch names. The NC&M will also accept port names assuming each port name is unique in the WDM optical network.

- *Connection directionality:* bi-directional or unidirectional.
- *Disjoint path:* there are many scenarios that a client may demand a disjoint path. For example, a client asks for two bi-directional connections between the same source and sink switch with disjoint paths. Another example is requesting a bi-directional wavelength channel with a disjoint fibre path.
- *Preferred lambda:* different lambdas on the same fibre may have different signal quality.
- *Number of hops:* in all optical networking, in particular with all optical wavelength conversions, end-to-end signal quality may not satisfy user requirement if too many hops are used. Hence, a client would like to enter the number of hops as an explicit routing constraint.
- *Wavelength protection:* this indicates whether the client requires wavelength protection. This can be a Boolean value. Alternatively the NC&M system may support a number of wavelength protection schemes.
- *Wavelength restoration:* if a client prefers a wavelength restoration scheme, the request can be entered in this field. This can be a Boolean value.
- *Client signal type:* an OADM may support several client signal formats, such as SONET or Gigabit Ethernet.
- *Line rate:* the preferred connection data rate.
- Other physical link QoS requirements:

  - Optical Signal-to-Noise Ratio (OSNR),
  - Wavelength dispersion.

In addition to client-imposed constraints, there are routing constraints that are determined by the WDM network properties. These **network-orientated constraints** include:

- *Fabric type:* a rather simple classification based on fabric transparency, according to which a fabric can be either opaque (O-E-O) or transparent (all optical). Certain fabrics imply certain routing constraints. For example, minimum number of hops is desirable in a transparent subnet.
- Number of wavelengths per TI/CI fibre.
- Wavelength availability.
- Wavelength interchange capability.
- Wavelength signal quality:

  - better OSNR, less dispersion,
  - path stability, e.g. avoiding a fast switching fibre,
  - preferred wavelength with less cross talk.

- Protection and Restoration capabilities of network/EMS.

Figure 4.24 (left) shows an example of connection request GUI provided by the NC&M, which allows explicit input of the client-imposed routing constraints. Sometimes, users prefer explicit routed connections. This can be used for testing and analysing the performance of optical networks. Or there are external routing entities that compute the complete connection path. An example testing connection originates at a NE, routes completely around the network and returns to the origin. To

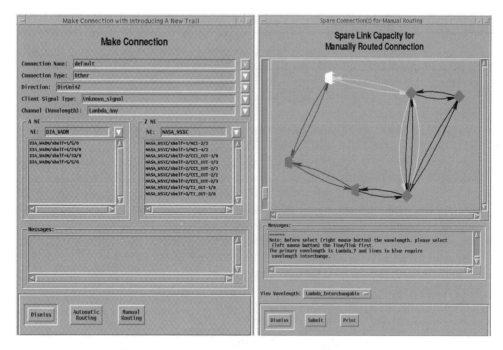

Figure 4.24   Connection request (source: MONET NC&M).

support this, the NC&M system has an explicit routing interface. The manual routing GUI to assist creating such connections is shown in Figure 4.24 (right). This capability allows a user to completely specify how a connection is routed in the network. The approach taken for this feature is as follows:

- The user first selects the trail origin and destination TTPs as in automatic routing.
- The GUI then invokes a network spare wavelength feature in the Trail Manager.
- The spare wavelength information is returned to the GUI. The GUI displays the spare information in a map and allows the user to select the desired path of link connections across the network.
- The GUI then invokes a manual routing feature in the Trail Manager.
- The Trail Manager then directs the Connection Manager to create the specified connection.

This capability places no restrictions on the connection that may be made beyond what the equipment supports. For example, some equipment does not support wavelength interchange and all equipment places restrictions on what wavelengths are reachable from certain client interface ports.

Wavelength scheduling algorithms

An important part of the connection management is the wavelength scheduling that is based on a routing scheme or metric to provide traffic control by assigning wavelengths and add-drop nodes (WADM) to a given set of lightpaths [Rama98]. In the

MONET WDM network, the east ring, consisting of Lucent WSXCs, is highly transparent so the connection rate can be set dynamically upon request up to the rate of
OC192 over all channels on a fibre and the quality of the signal is crucial to optical
data transmission. The connection manager computes the lightpath not only based
on current network usage, history data summary, and network topology but also
relies heavily on QoS monitoring. For example, it is highly likely that channels in
the middle of the spectrum produce better signal quality than that of channels using
the spectrum extremes. If a fibre supports eight wavelengths numbered 1 to 8, wavelength 4 tends to support a higher connection rate than those of wavelength 1 and 8.

Figure 4.25 shows the problem of wavelength scheduling that can be described in
three dimensions. At a given time t, there are v add-drop nodes and each link
supports n channels in the network. For a given topology, the network receives a
series of connection request (a_i, z_i). In the figure, on the time d, there are four intra-NE
connections on Lambda 4; Lambda 5 supports a loop connection both starting and
ending at Node 4 (usually used for testing purposes); there is a wavelength interchange on Lambdas 1 and 2 between Nodes 1 and 3 for the connection originating at
Node 6 and heading to Node 5.

Intra-connection scheduling

For a connection request, we present three methods below using path length, rate,
and load as the scheduling criteria. Path length is calculated in terms of both the

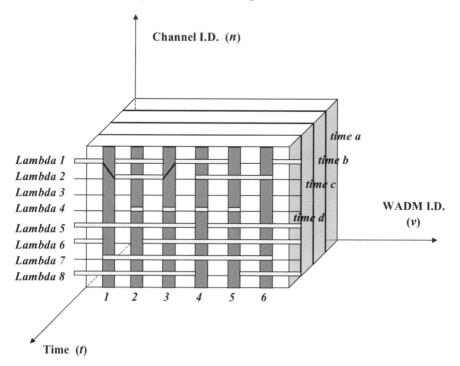

Figure 4.25   Wavelength scheduling.

number of WADMs and the wavelength. As the cost of bypassing a WADM is far more expensive than that of the wavelength, this method can be simplified to treat the number of WADMs as the path length. The maximum supported rate for each signal varies, although theoretically the channels on a fibre support the same rate. This implies that for a given time, a channel has a maximum connection rate depending on the QoS monitoring report, in particular the measurements of wavelength registration and signal-to-noise ratio degradation. For example, the Lucent WSXC supports connection rate setting, up to the highest level of signal quality, but the connection rate cannot be tuned to a higher rate during the connection if a lower rate is set at the beginning. Both WADM and wavelength have a current traffic representation, i.e. the NE fabric view and the spare wavelength view, so it may be wise to allocate the least congested road to the new connection.

K-shortest path method (KSP)

KSP has been treated as a static method because the calculation for the shortest path is always conducted off-line. However, the topology information used for the shortest path calculation is updated from time to time and this increases the level of feasibility of primary and alternate paths. A KSP method works as below:

- given a source and its destination, generate the feasible paths according to the current network usage;
- sort the paths in increasing order of path length;
- select the first one on the path list as the primary path and store others on an alternate path list.

Best-fit rate routing method (BFR)

For a given network $G(V, E)$ where V is the number of nodes and E is the number of links, a connection is represented by:

r, rate requirement,
Q, the set of admissible paths,
Q_i, the number of links on the ith path in the set Q,
P, the set of feasible paths, $\{p_1, p_2, ..., p_i, ...\}$,
P_i, the number of links on the ith path in the set P, $\{p_{i1}, p_{i2}, ..., p_{ij}, ...\}$,

and a link, l, is characterised by R_{ij}, the maximal rate that the link can offer to a connection where i and j are the indexes for the path and the link on that path.
   The BFR method can be implemented by:

- generating the feasible paths set P from Q if the path in Q satisfies this condition:

$$\left(\left(\min_{j \in Q_i} R_{ij} \right) - r \right) > 0$$

- sorting the paths in P into a list according to:

$$\min_{i \in P} \left(\sum_{j \in P_i} \left(R_{ij} - r \right) \right)$$

- selecting the first path on the list as the primary path and storing the other paths on the list as alternate paths.

Least loaded routing method (LLR)

Assuming:

F_{jn}, the number of fibres on the jth link,
π, the set of wavelengths in the network, $\{\lambda_1, \lambda_2, ...\lambda_k, ..., \lambda_N\}$,
λ_n, the number of wavelengths on each fibre,
A_{jk}, the total number of channels on the jth link for which wavelength k is active i.e. either the channel using wavelength k is available or the channel supports wavelength interchange, the LLR method has the following steps:

- generating the admissible paths set Q;
- sorting the paths in Q into a list according to:

$$\max_{i \in Q} \left(\max_{\lambda_k \in \pi} \left(\min_{j \in Q_i} \left(\lambda_n \times F_{jn} - A_{jk} \right) \right) \right)$$

- selecting the first path on the list as the primary path and storing the other paths on the list as alternate paths.

Inter-connection scheduling

When the scheduling application receives a number of connection requests simultaneously, the complexity of the scheduling increases dramatically and a globally optimal solution may be very computationally expensive if not impossible to calculate. When there are sufficient wavelengths and add-drop nodes in the network, the connection requests can be processed as in intra-connection scheduling. When there are more requests that the network can accept due to current capacity and topology, heuristic algorithms such as a greedy approach can be developed to organise requests into groups and process them group-by-group.

Multiple-connection scheduling is based on one or more objective functions, for example, network utilisation, throughput, or latency. In this chapter, maximising network throughput has been used as the scheduling objective. When there are more requests than the network can satisfy due to current capacity and topology, a multiple-connection scheduling algorithm that has the following phases is introduced.

Phase1 – Grouping

During this phase, connection requests are grouped and thereafter the group resource request is considered instead of individual connection requests. An example of a group request is a pair of connection requests between two nodes, where the connections in the pair have opposite directions and one's starting node is the other's ending node. This represents a bi-directional link between the two end nodes. Grouping criteria can also be based on a per-customer basis and QoS. At the end of the grouping phase, a queue is constructed where each element on the queue has a link list recording the individual connection requests. This queue is named the 'blocked' queue.

Phase 2 – Sorting on the blocked queue

In this phase, the elements on the blocked queue are sorted in ascending order, according to:

- the number of lightpaths in the group
- the average number of nodes per lightpath in each group.

Phase 3 – Constructing the ready queue

Starting from the top of the blocked queue, compute the lightpath(s) based on the selected single connection algorithm and the current network resource availability. If the group request can be granted and does not turn the resultant topology into a disconnected graph, the group request is moved to the ready queue from the blocked queue and the network resource capacity is updated. A disconnected graph is one from which any one node cannot reach the other nodes, which can be verified using a depth-first search (DFS) graph algorithm. During the construction of the ready queue, the blocked queue is accessed twice:

- First, the group request that can be granted and does not result in a disconnected topology is inserted into the ready queue and deleted from the blocked queue. The topology used for further computation is updated.
- Second, the remaining requests on the blocked queue are processed. If any of the group requests can be granted, the group request is inserted into the ready queue and removed from the blocked queue.

The group requests on the ready queue are forwarded to the connection manager and consequently the lightpaths are set up. The group requests on the blocked queue are rejected or moved to a waiting queue. Only when an existing connection terminates, can we reconsider the connection requests on the waiting queue. This heuristic algorithm favours short-trip connections (i.e. a small number of hops) so in an extreme case, a long-trip connection may never get accepted or wait on the blocked/waiting queue forever causing starvation. This effect can be reduced by giving higher priority to requests on the waiting queue than those on the blocked queue.

A connection request can also explicitly ask for protection. The scheduling application will interpret the request as an alternate path. This will be implemented as end-to-end service level protection. At the link level, the SONET ring is self-healing and offers 1 + 1 and 1:1 protection.

4.5.2 Connection Discovery

Inconsistency may occur over the connectivity information, especially in a testbed network. For example, users can directly make connections through the NE controller for testing and optical signal evaluation. When connection notification fails or relies on a polling scheme, the NC&M system may enter a state, where the network management system and the NE Controller have different views on local fabric connectivity. Connection discovery is a network recovery operation by which connection managers and NE are forced to reconcile so that all network and element information is consistent according to the information model.

The connection inconsistency is notified to the human operator, who can make a decision on taking corrective actions. Connection discovery can also be considered as an auditing approach. Hence, it can be invoked at any time by a network administrator to perform periodic consistency checks. Incorporating and utilising such an operation will enhance the consistency and reliability of the NC&M system and the connection service it provides.

The MONET NC&M developed the following connection discovery algorithm [Wils00]. Within the context of a subnetwork, every CTP can be considered to be an *internal* CTP or a *boundary* CTP. An *internal* CTP is a CTP that is contained in a LinkTP whose supported link is contained in the subnetwork. All other CTPs are considered *boundary* CTPs.

The connection discovery algorithm has the following phases:

- **Phase 1:** The EML Connection Manager retrieves all child subnetwork connections (i.e., subnetwork connections in its children subnetworks).
- **Phase 2.1:** For each subnetwork connection whose destination CTP is a boundary CTP, the connection is extended as follows:

    - If the origin CTP is a boundary CTP, then the Connection Manager has discovered a valid, complete, EML connection across the subnetwork and the connection is recorded.
    - If the origin CTP of the connection is an internal CTP, then the Connection Manager consults the resource model to determine the link connection bound to the CTP and obtains the link connection's far end CTP. The Connection Manager then finds the child subnetwork connection whose destination CTP is the just-calculated CTP (This child subnetwork connection is in a different child subnetwork than the current child subnetwork connection. Also, there should only be one such child subnetwork connection since the equipment supports merging connections only at drop CTPs.) This child subnetwork connection is pre-pended to the route of the EML connection being discovered and the process continues.

- **Phase 2.2:** Any child subnetwork connection whose destination CTP is not a boundary CTP is ignored. These subnetwork connections will either be included as part of the route of some other EML subnetwork connection or are a dangling connection that do not support any higher level connection.

After the EML Connection Managers have discovered their connections, the NML

Connection Manager uses the equivalent process to discover NML subnetwork connections. After the NML Connection Manager completes its discovery, the Trail Manager then determines trails for each of the discovered connections. The Trail Manager is able to resolve point-to-point trails, multicast trails, and protected trails. Only network resource model errors can prevent all NML connections from being resolved into trails. Possible errors occur when the origin and destination CTPs cannot be associated with TTPs. In such cases, the configuration manager needs to re-synchronise the management system's internal resource information database with the actual network state.

All trails discovered by this process are manageable by the NC&M system. In particular, discovered trails may be deleted, protected, or rerouted. The NC&M system provides an interface that allows the user to find and delete dangling NE subnetwork connections. The user may view the fabric of a specific NE that illustrates the connections in the NE. Any connection that does not have a supported trail is a dangling connection. Connection discovery can be viewed from Figure 4.29, where the solid lines on the fabric represent the connections made by the NC&M system, and the dotted lines on the fabric represent the discovered connections possibly made through craft interfaces.

4.5.3   WDM Client Topology Reconfiguration

In an overlay WDM network, client layer reconfiguration can be realised in the WDM server layer. For a WDM optical network, the client layer topology reconfiguration can be interpreted as setting up and tearing down optical lightpaths by the server layer. This section focuses on the enabling mechanism in a WDM NC&M system to support client layer lightpath topology reconfiguration. WDM reconfiguration comes in two flavours, lightpath reconfiguration and network reconfiguration.

Lightpath reconfiguration is typically a reactive approach triggered by network faults or maintenance. For example, when a fibre is cut or a NE is down due to power loss or burned circuit, an alarm is generated by the management agent. The fault may last a while before it is repaired. Through reconfiguration, the faulty components are isolated and contained, while the rest of the WDM network is available for service. The affected lightpaths may be rerouted, but the virtual lightpath topology is intact. Alarms are processed by the fault manager that determines the cause (root) of the alarms and informs other management modules when a reconfiguration is needed.

In the case of reconfiguration, the configuration manager takes the lead by updating the WDM network topology and configuration database, and the connection manager purges out-of-date connectivity information and reroutes connections affected by the faulty components. Other management modules, for example, fault and performance managers, also update their view of the information model. As soon as the WDM network enters a stable state, the NVS informs users and applications, and updates their views of the network.

Network reconfiguration is a proactive approach by which a new WDM network topology is generated for the purpose of adapting to a new lightpath traffic pattern projection. Network reconfiguration results from WDM network planning since reconfiguration and convergence have a deep impact on the entire network. Following a WDM network topology change, the management system may choose to reroute existing lightpaths to take advantage of more efficient routes in the new network. Relatively short-term traffic surges and/or skews may be handled by the management system using alternate routes and, possibly, lightpath topology reconfiguration.

Figure 4.26 shows the information flow in a reconfiguration. Lightpath traffic patterns for network planning may be based on probabilistic modelling together with on-line sampling since the problem of traffic prediction is non-deterministic in nature.

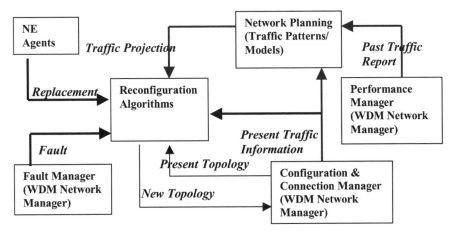

Figure 4.26 Reconfiguration in WDM network.

The design of reconfiguration algorithms and the construction of traffic models or a demand matrix are presented in Chapter 7.

4.5.4 Signal Quality Monitoring

An important and unique aspect of WDM optical networking is signal performance monitoring, which is responsible for monitoring and isolating optical signal quality related soft faults. As indicated in the previous chapters, signal degradation may occur for a number of reasons. In particular, all optical networking requires end-to-end optical signal transmission without O-E-O conversion of intermediate hops.

Without signal quality monitoring, once a trail is set up, NC&M system has little information on wavelength channel performance over time. It is possible that a wavelength channel fragment is interfered with by internal or external factors so that the signal is unrecognisable at the receiver. A signal monitor may need special hardware to split the signal and therefore, it will also have adverse impact on signal quality.

By introducing the optical signal monitor, the NC&M system can provide an interface to allow administrators to query a connection's signal quality, such as OSNR and signal power readings. This in fact provides a trouble-shooting tool for the administrators. If the NE supports threshold setting on certain performance parameters, a notification scheme can be developed. Consequently, clients or administrators can provide the thresholds for ongoing connections. Once the threshold is exceeded, the corresponding user will receive a notification. Furthermore, the optical signal performance information collected over time can be sent to the connection manager.

By periodically updating the optical performance attributes in the related objects, variations in these parameters can be taken into account in subsequent provisioning computations.

Figure 4.27 shows the wavelength SNR and power monitoring and threshold-setting GUI for an existing connection. The left picture indicates the SNR obtained in a 60-second polling interval from the two connection termination points. The right-hand picture shows the connection wavelength power in 'dBm' obtained from the NE ports in a one-second polling interval. As expected, the wavelength power is quite volatile in a very short interval, i.e. one second, but it always falls into the acceptable range. If the wavelength power exceeds the thresholds, the NC&M generates a notification.

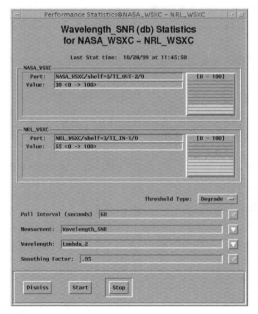 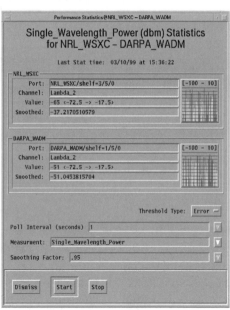

Figure 4.27   Wavelength SNR and power monitoring and threshold setting (source: MONET NC&M).

4.5.5 Fault Management

Faults may occur in a number of places for a number of reasons. For example, a circuit pack is burned or an EML connection manager is down. Faults will trigger alarms, which will be routed to and processed by the corresponding managers or the system administrator.

The MONET NC&M system implemented three tasks associated with fault management:

- *Alarm collection:* fault manager maintains a list of active alarms. During startup, the fault manager polls the NE to gather a list of outstanding alarms.
- *Alarm correlation and root cause analysis:* certain component failures may generate several alarms. The fault manager is responsible for analysing alarm correlation and determining root cause of the alarm based on the current topology and connection information. The logic behind correlation and root cause analysis can be implemented as rules, which can be added or modified based on the flexible and modular fault management architecture.
- *Alarm log:* this provides a trace facility on the alarms, which can be output to external files or databases. The alarm log can be switched on/off and can be applied to filters.

The Fault Manager receives alarms from NEs in the network. It collects alarms and correlates them with the network topology and connection information to locate the root cause of the alarms. Once it detects and identifies the failed resources, it alerts the human network administrators through an alarm notification window. The corresponding icons on the network map also change colours to highlight the failures. The network administrator can then initiate corrective actions.

4.6 WDM NE Management

The management agents are designed to control the managed objects (through a virtual NE resource abstraction) and monitor their behaviour by reporting extraordinary events to network level managers. Figure 4.28 shows the agent software architecture. The switching manager is responsible for element level resource management such as resource reservation and QoS assurance.

Users, applications, and connections might compete for resources. When resource concurrency occurs, a priority co-ordinator or scheduler arbitrates resource allocation. The priority scheme takes into account network, user, and task priorities. This scheme may be used whenever network resource is constrained or in the case of a network emergency. In an operational WDM network, events include many different kinds of alarms and QoS statistics, but a management application is only interested in a subset of these events and related resource state changes. The management application would be overwhelmed by the huge amount of event-related information. Hence, an event filter is designed to suppress the information at the element level. Agents are also equipped with signalling protocols if they are used for optical path setup.

Both event filters and priority schedulers employ predicates that can be defined by the functional managers, and installed and configured by the agent. By doing so, not only is application and user-specific information conveyed into the network but also the NC&M is able to oversee the amount of management information flow and the level of the information detail at any time. For example, if a Circuit Pack (CP) on a WADM will be replaced in three hours due to failure, the management application can configure the WADM agent so that the CP replacement-related alarms are suppressed temporarily. These alarms include CI and TI ports on the CP and communication links using the CP.

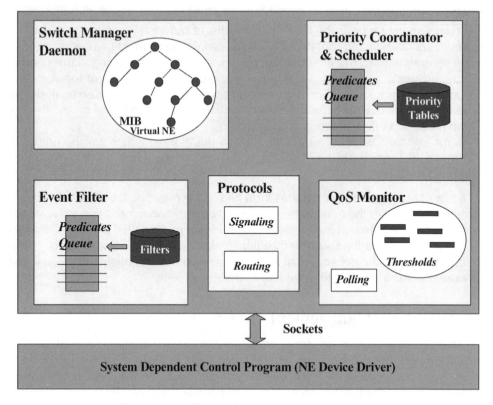

Figure 4.28   Software architecture of management agent.

In a high-speed WDM network, using traps is a preferred monitoring mechanism to polling as setting up the polling interval is a non-trivial issue. A performance manager can create a threshold object, which is enforced on the CTP (or other resource) by the agent. The agent generates an alarm whenever the threshold is exceeded. Examples of QoS alarms and their severity are:

- response time exceeded – major
- queue size exceeded – minor
- performance degraded – major
- congestion – major
- resource at or nearing capacity – major
- loss of signal – critical
- degraded signal – critical.

Figure 4.29 shows an example of NE fabric connectivity. In the figure, all the wavelength ports are listed on the fabric. A port is in use if there is a line starting or terminating at it. Ports in black such as those on shelf 1 and 5 represent wavelength ports on the transport interfaces. Ports in blue such as those on shelf 4 represent wavelength ports on the client interfaces. Arrows indicates the directionality of the wavelength channel. A dotted line on the fabric represents an incomplete trail, for

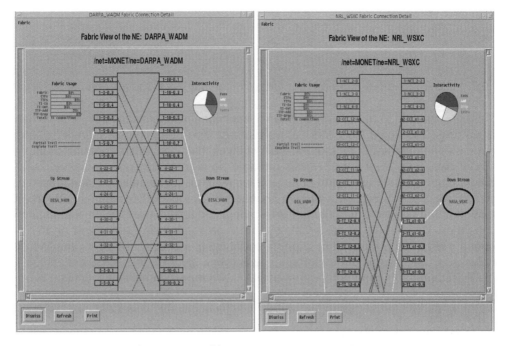

Figure 4.29   NE fabric view (source: MONET NC&M).

example, a dangling connection. Clicking on a connection on the fabric displays the connection's upstream and downstream NE. An add/drop connection may not have an upstream/downstream NE. The NE level traffic and fabric usage is also displayed in the figure.

4.6.1   NE MIB

Each NE implements a MIB to model the resources. The NE MIB can be implemented in software or in firmware and software. A standard NE MIB indicates the maturity of optical networking industry and allows communication activities across multi-vendor networks.

Three groups of managed objects are discussed below. The discussion is not intended to be comprehensive but, rather, it illustrates some of the useful information for WDM optical networks.

System group

This group is used to describe the agent software or the operating system that controls this NE. Example objects include:

- System description
- system object ID
- system-up time

- system name
- system location
- system services.

Configuration group

This group is used to model the hardware networking resources. Each NE may have a multi-chassis arrangement, and each chassis can be modelled as a single routing node. A chassis may consist of multiple shelves, each of which contains multiple circuit packs. Each circuit pack has multiple ports. A WDM port can be either a physical transport interface port or a client interface port. The transport port is used to switch the optical signal (wavelength channel) within the WDM optical network. It may support complete wavelength interchange or subset wavelength interchange capability. Depending on the switching fabric, it is either an O-E-O port or an all-optical port. The client port is used to add/drop a wavelength to/from the WDM network. Client ports may implement specific link layer interfaces such as SONET or GbE/Ethernet. Example objects associated with a port include:

- Port description
- port type
- reservation status (free, in-use)
- line rate
- interchangeable wavelength set
- signal requirement
- physical address
- admin status (up, down, testing)
- operational status (up, down, testing)
- inbound errors
- outbound errors.

Fabric connection group

This group is used to record the connections on this NE fabric. A fabric connection has this information:

- connection ID
- lightpath/trail ID
- connection rate
- upstream NE ID
- upstream wavelength channel (or ID)
- incoming port ID
- outgoing port ID
- downstream wavelength channel (or ID)
- downstream NE ID
- up time
- estimate connection duration.

4.6.2 NE Interfaces

NEs export control and management interfaces (i.e. APIs) to network managers. IP routers use SNMP (Simple Network Management Protocol) for this interface. WDM optical networks do not have a standard NE management protocol, but vendors have proprietary solutions.

We present a discussion on WDM NE NC&M protocol in Chapter 6.

4.7 WDM Signalling

WDM signalling can be used during connection setup and is invoked by the connection manager. WDM signalling can conduct one or more of the following activities:

- wavelength channel selection and reservation;
- wavelength channel status maintenance;
- setup/tear down light paths;
- priority arbitration with pre-emption;
- adaptive QoS with QoS negotiation.

In a circuit switched WDM network, the connection must be reserved and there are preserved connection setup and tear-down phases. Even in the case of MPλS, the lambda path is not a 'virtual path' as is the case with MPLS. The main functionality of signalling is path setup and tear-down.

Signalling paths can be predetermined, for example, by the connection manager, or dynamically obtained from a local routing/switching table. Signalling can be responsible for wavelength selection, in which case the connection manager only computes a fibre path. Introducing wavelength continuity issues such as wavelength interchange complicates the exchanged messages between signalling peers. In addition, signalling can support adaptive QoS during the course of path setup. For example, if a segment of the routing path does not support the required bit rate, so QoS issues must be negotiated/renegotiated using the signalling mechanism.

4.7.1 Wavelength Signalling and Routing

Wavelength routing and signalling could be considered together instead of separated control components. Figure 4.30 shows two approaches of wavelength routing and signalling in WDM networks. The left-hand side of the figure provides a *Source Routing with Destination Initiated Signalling*. This approach is based on explicit path routing so only the source (or the edge) has the need to equip with the wavelength routing algorithm but every node in the network is able to access the signalling protocol and the resource discovery protocol.

The routing path is computed once and is specified in fibres and wavelength channels. Then, the signalling protocol runs through node by node. First, the source sends a request message towards the destination. Only when the sink receives the request, does it start the reservation phase by reserving a wavelength and then sending a reservation message to its upstream. In this way, the reservation message is propagated back to the source. Once the source reserves its channel, it can declare

the end-to-end connection is up and ready for traffic. If a request message is rejected
or a reservation message cannot be fulfilled, an error message will be sent to the
source. In this scheme, the message itself can have timers, so it will be expired
without further update.

The right-hand side of Figure 4.30 shows a *Hop-by-hop Routing with Intermediate
Signalling*. This approach requires a closer co-operation between routing and signalling. All nodes in this approach are assumed to have access to the routing algorithm.
We focus on the control not the forwarding. After the source computes the path, it
shifts the decision making to the next hop on the routing path. Signalling takes place
between the source and the first hop to reserve a wavelength. Then in turn the first
hop computes the routing path towards the sink and forwards the decision-making to
the next hop. Signalling runs again to assign a wavelength. When the sink has
reserved a wavelength, it sends an end-to-end acknowledgement to the source to
indicate the completion of path setup.

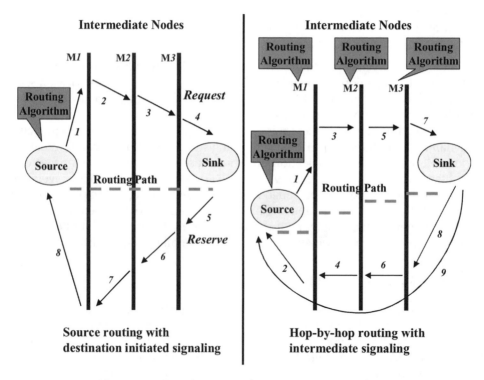

Figure 4.30   Optical circuit switching vs. Just-In-Time switching.

4.7.2   Circuit Switching vs. Just-In-Time Burst Switching

Optical circuit switching requires explicit circuit setup and release phases. During
circuit setup, the source/user has to wait (i.e. is blocked) until the end-to-end circuit
has been established. Just-in-time (JIT) burst switching has a single phase, also known
as the **tell-wait-go** approach. In this approach, after receiving a circuit setup request

from a user, the edge switch makes a delay estimate on the rest of the circuit setup time and sends the circuit setup delay estimate to the user who waits for the estimate delay interval and then starts to transmit data (see Figure 4.31).

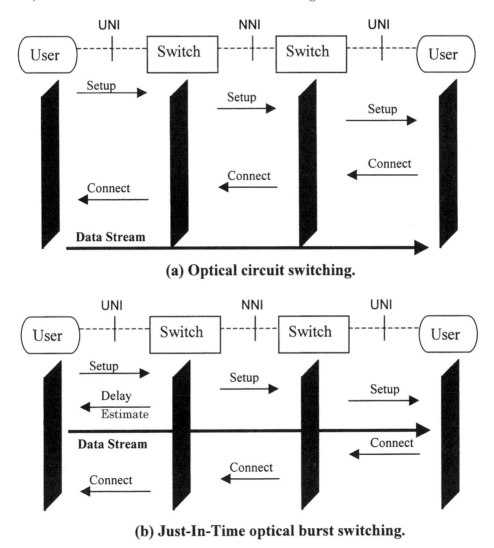

(a) Optical circuit switching.

(b) Just-In-Time optical burst switching.

Figure 4.31 Circuit vs. Just-In-Time signalling process.

Optical circuit switching is reliable but it has large round trip latency in circuit setup. If circuits are reconfigured from time to time, this approach results in inefficient use of bandwidth. By comparison, JIT burst switching has efficient use of bandwidth but it requires delay estimates and its performance relies on accurate delay estimation. It is less reliable since a distributed environment has many dynamic factors (and uncertainties).

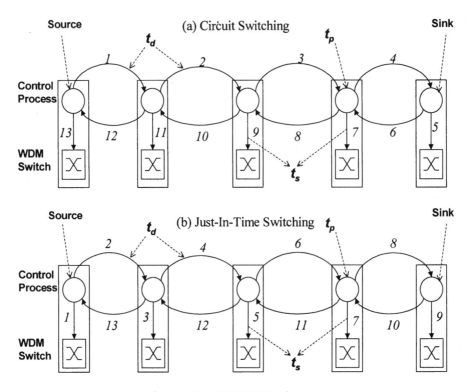

Figure 4.32 MONET DCN Subsystem.

Figure 4.32 shows the signalling process of a circuit with JIT switching, where the signalling protocol daemon runs as one of the control processes in the control plane. Signalling messages are transported over a control channel, for example, a dedicated wavelength channel. There are three types of time intervals involved in the signalling process:

- t_d: the propagation delay time between neighbour control processes.
- t_s: the switch setup time to send a command to the switch and make the connection over the crossconnect fabric. In the MONET testbed, the OXC accepts fast connect as well as verified connect. Verified connect takes more time than fast connect since it checks for signal quality before making connection.
- t_p: the control processing time once the signal message is received.

Figure 4.32(a) shows a destination-orientated circuit switching approach, where a setup message is sent hop-by-hop to the sink node. The sink starts reservation and sends a setup message to the switch. Then, it sends a confirmation message to its previous hop. Once the source control process receives the confirmation message and successfully sets up its switch, the signalling process completes.

Figure 4.32(b) shows a JIT burst switching approach, where switches are set up from the source and data streams are sent after the delay interval. The event sequences in the signalling process are indicated in the figure.

For illustration purposes, we use a just-in-time (JIT) signalling protocol developed in the MONET NC&M testbed. Details of the JIT signalling implementation in the MONET network can be found in [Wei00b]. MONET JIT is designed to support low latency transport of data-bursts across an all-optical network. It contains support for establishing fixed duration burst connections as well as variable duration 'circuit switched' connections. The MONET JIT capability support connection setup and connection tear-down capabilities using 5 messages involve:

- Setup (request for service);
- Call-Proceeding (acknowledgement from the network to the user);
- Connect (acknowledgement from the destination that the connection is complete);
- Release (request for end-of-service);
- Release-Complete (acknowledgement that the connection has been deleted).

The JIT protocol is implemented in a set of signalling messages that may be encoded in a single 48-byte ATM cell. The cell is transmitted from the client host to the agent managing the NE supporting the client host. The network uses pre-defined routing tables to determine the route of the connection across the network. The JIT wavelength channel can be configured through the NC&M system.

4.8   WDM DCN

WDM optical networks have dedicated control channels to provide a shadow DCN for transporting control messages. There are several ways to construct this DCN. First, is to reserve a wavelength channel from the data plane in a WDM network. This approach assumes every hop in the WDM network has add/drop ports, so control information can be converted into optical signals and inserted into the WDM network. Second, is to construct an independent shadow DCN, which may use other networking technologies, e.g. Ethernet. Third, is to use a sideband wavelength to carry the control information.

The third approach using the MONET network is illustrated, where a sideband wavelength centred on 1510 nm is used as the control channel [Ande00]. The use of an out-of-band DCN channel at 1510 nm is also recommended by ITU. The control channel does not occupy data plane wavelength channels and this sideband wave-length signal goes through O-E-O regeneration at each switch.

Figure 4.33 shows the MONET DCN subsystem. In the control plane, the control messages are transported using ATM/SONET/WDM fibre. The MONET DCN system operates at 155 Mbps data rate based on the ATM over SONET OC-3 protocol. As shown in the figure, an ATM switch is located at each NE. The ATM switch presents a layer 2 in the testbed. The ATM switching fabric offers multiple ports so that client signalling equipments and network management stations can be connected to the WDM network. Since available ATM switches only supports 1310 nm output ports, a transponder is needed to convert the 1310 nm wavelength of ATM traffic to that of the embedded channel at 1510 nm.

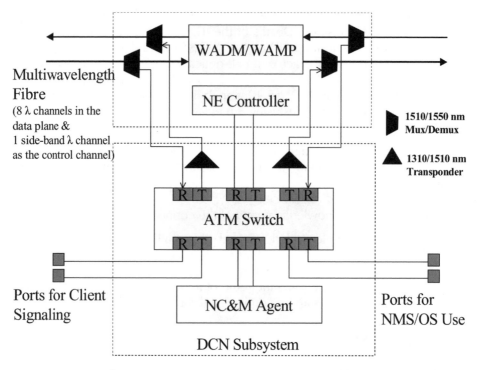

Figure 4.33   MONET NC&M GUI (source: MONET NC&M).

4.9   WDM Network Views

As shown in Figure 4.15, positioned between the network management applications and their users, NVS is designed for the following purposes:

- serving as a GUI to human users;
- presenting tools to assist the network administrator to launch management applications;
- providing view customisation;
- presenting the WDM NC&M APIs and an interface repository to other applications, external services, and client layer NM systems.

An end user may not be interested in each individual network element and might be overwhelmed and even confused by the vast amount of raw network and NE level information. More likely, a user prefers his/her individual view of the network and the user hopes to conduct network activities in a simplified way by the means of the NC&M system.

NVS views are organised to highlight NM functionality: configuration, connection, fault, and performance. Configuration management typically offers views on current network topology, individual NE information, and details of a network link. Likewise, connection management presents views for lightpath creation, deletion, and discovery, as well as subnetwork, lightpath, and NE fabric connection segment query. Fault management would provide notification views on general alarms, root cause alarms, and alarm query facility. Performance management provides views on lightpath monitoring and statistics collection, and SNC connection and NE port (interface) monitoring.

For a network administrator, NVS also provides a network process status view, which not only monitors the network managers and agents, but also controls these processes during their lifecycle. For instance, process migration may be utilised by a network-level manager to overcome platform failure or to address NC&M traffic congestion. The NC&M APIs may be provided in both CORBA IDL (Interface Definition Language) and scripts. CORBA IDL APIs are powerful but complex as applications have to initialise the connections to the ORB (Object Request Broker) and implement both client and server components for asynchronous communication. By comparison, script interfaces are easy to use. NVS also enhances the NC&M system stability by minimising direct exposure to casual users.

Figure 4.34 shows the WDM topology of the MONET DC network [Liu00b]. In the figure, a WADM is represented by a trapezoid, a WAMP is represented by a triangle, and a WSXC is represented by a rhombus. The co-ordinates of a NE are computed by scaling down the site location information in the configuration database. Lines indicate links and each line represents one or more unidirectional fibres between NEs. The colour of the NE gives the current status (e.g. faulty) of the resource. Each object

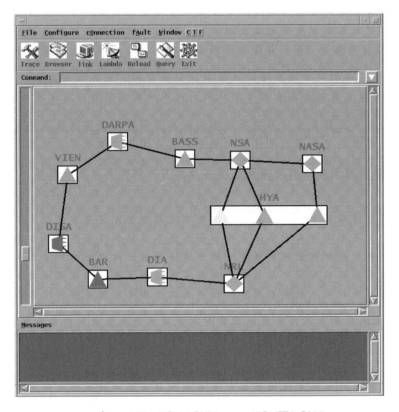

Figure 4.34 NC&M GUI (source: MONET NC&M).

on the map may be queried for additional information, such as a faceplate view of each NE or the lightpaths routed across a link.

Once a lightpath trail is established, its connection detail can be obtained and displayed as shown in Figure 4.34 [Liu00b]. The view shows the source, the sink, and all the intermediate hops' NE, shelf, slot, port, and wavelength ID associated with the trail. This view also allows a user to delete, add-leg, protect, print, and reroute the lightpath, or highlight the lightpath in the WDM map. Allowing add-leg to an existing lightpath increases both the functionality of connection management, for example, to emulate a physical layer multicasting in WDM networks, and the administration flexibility such as to insert connections to existing lightpath contracts. If the signal quality is below the threshold values or the client would like to order the transport path with certain QoS, the trail may be rerouted, which will bring up the connection request window. As requested by the client or the network administrator, an existing trail can be protected.

There are many other views supported by the network visualisation service such as lists of equipment and where it is deployed in the network, and faceplates of NEs. Together, these views provide a network administrator with clear and meaningful depictions of the network and enable efficient network management decisions and actions.

4.10 Discussion

Legacy optical network management systems do not provide interoperateability. TMN style management systems have been introduced and used to manage telecom equipments. However, a TMN system is complex and difficult to implement. Despite the ITU-T standard, the developed TMN management systems may not interoperate. In addition, these legacy systems require special trained workforces to monitor and maintain. Circuit provisioning across these legacy systems is conducted manually so it may take months to set up an end-to-end circuit. These manual operations catch errors easily and inventory control becomes a major problem. TMN systems/networks are simply too complex to be scalable.

Interoperability and routing across networks is the task of layer 3 in the OSI or TCP/IP model. IP has become the de facto standard of network layer protocol. IP community is well established such as IANA for global network addressing and IP control protocols are widely deployed and understood. A common control plane for computer and communication networks will be IP based. However, IP control protocols are developed for packet switched networks not for circuit switched networks in which circuits are always traffic engineered. In packet switched networks, network reachability implies resource availability, whereas in circuit switched networks, one has to explicitly describe resource availability in addition to reachability. IP networks can only scale to a certain size, e.g. an OSPF instance can control around 200 routers. A scaleable common control plane must scale up to at least tens of thousands nodes. Nevertheless, IP based control plane is the best candidate for a common control plane. Although GMPLS is immature, it is gaining momentum.

In the following chapters, we examine utilizing IP addressing and protocols to control WDM optical networks.

5

IP over WDM

- IP over WDM networking architectures
- IP/WDM internetworking models
- IP/WDM service models
- Summary

5.1   IP over WDM Networking Architectures

WDM networking technology can be classified into two categories: reconfigurable WDM and switched WDM. The former is employed in a circuit-switched WDM network, in which a lightpath topology formed by established circuits is reconfigurable in response to traffic changes and network planning. The latter is employed in a packet-switched WDM network, in which optical headers or labels are attached to the data, transmitted with the payload, and processed at each network switch. Based on the ratio of packet header processing time to packet transmission cost, switched WDM can be implemented using burst switching, label switching, or packet switching. A WDM technology comparison is presented in Table 5.1.

Table 5.1   WDM networking technology comparisons

	Reconfigurable WDM	Switched WDM		
Switching technology	Circuit switching	Burst switching	Label switching	Packet switching
Control traffic	Out of band	Out of band	Out of band/ In band	In band
Traffic granularity	Large	Medium	Medium	Small
Switching unit	Lambda channel/fibre	Data burst	Flow	Packet
Data forwarding	Established circuits	Virtual channels	Virtual channels	Next hop

Reconfigurable WDM is among the most promising technologies in the backbone transport networks. It mainly copes with large granularity aggregated traffic, which is less bursty than access traffic.

Switched WDM is still evolving especially for the metro and access networks. Switched WDM aims at switching medium or small granularity traffic, which requires flexible network architecture and needs comprehensive and scalable network control functions. Switched WDM can support burst switching, label switching, and packet switching. Label switching is similar to burst switching except that it follows the MPLS approach, i.e. using local labels and forwarding traffic over label switch paths. Optical packet switching copes with small sized optical packets. In this section, we introduce optical burst switching and optical packet switching first and then present IP over WDM networking architecture.

5.1.1 What is Optical Burst Switching?

In optical burst switching, the packet (i.e. burst) control header is sent out along the control path ahead of the actual optical data packet. The idea is that the control header will arrive at the intermediate switching nodes, allowing each switch to perform switching decision computations and install the switch cross-connect setting just-in-time before the actual arrival of the data packet. In this way, the optical data packet can thread its way across the optical network from ingress to egress. The delay between the control header and the data packet increases with the number of hops and the expected processing delay at the intermediate switches. Figure 5.1 depicts the operation of such an optical burst switched WDM node.

Optical burst switching uses one-way reservation according to which a source sends the setup request and then sends the burst without waiting for the setup acknowl-

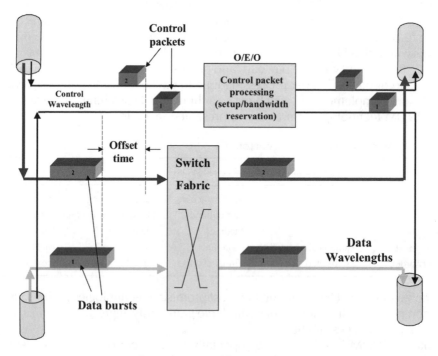

Figure 5.1 Optical burst switching.

edgement. This is due to the fact that burst transmission times can be relatively short. There are three approaches on releasing the bandwidth in burst switching:

- tell-and-go (TAG)
- reserve-a-fixed-duration (RFD)
- in-band-terminator (IBT).

TAG requires an explicit connection tear down step, so it is similar to circuit switching except that it deals with finer granularity traffic and connection request is more dynamic. In RFD, the connection setup request must specify the connection duration. This can be implemented in terms of burst length, for example, bytes, or in terms of burst travelling time, for example, microseconds. IBT carries a header and a trailer in the burst itself.

An important concept of optical burst switching is the offset time, which is the time interval between the setup request and the data burst. The offset time in TAG and RFD is always smaller than or equal to the connection setup time in circuit switching. If the offset time is too long, the intermediate switches need to wait for a while before the data burst arrives. If the offset time is too small, the intermediate switches may need to buffer the burst. When the offset time is zero as in IBT, it works as a packet switch network, whereby the data burst needs to be buffered during setup request processing. When there are more bursts than the switch can buffer, congestion occurs with data burst loss.

5.1.2   What is Optical Packet Switching

In optical packet switching, the packet control header, which can also be treated as a label, is typically sent together with data packet along the same path. To allow for the time needed for switching decision computations and cross-connect setting installation, the data packet is always routed through a local optical delay line upon arrival at an intermediate switch. The delay value is selected so that when the data packet emerges from the delay line, the desired optical cross-connect setting will have been installed. This delay value is local and constant at each intermediate switching node, independent of the particular route paths taken by the packets. Figure 5.2 depicts the operation of such an optical packet switched WDM system.

Packets in the optical network can have either fixed or variable length. A fixed length packet is like an ATM cell and an example of a variable length packet is an IP packet. A variable length packet injects less control information into the network so it is more efficient. However, the packet size in the variable length packet cannot be too large or at least should be smaller than the capacity of the fibre delay line. Therefore, a large size packet may need to be fragmented for transmission. The selection of packet length is based on application and traffic characteristics.

There are two forwarding paradigms in optical packet switching: datagram and virtual circuit. In **datagram forwarding,** the packet header, carried in-band or out-of-band, is examined at each intermediate node, and there is no offset time since the payload and the header are transferred together. This is used in IP. In **virtual circuit forwarding,** virtual circuits are set up before packets are sent over the virtual circuits. This circuit is 'virtual' in the sense that it does not reserve any bandwidth. The virtual

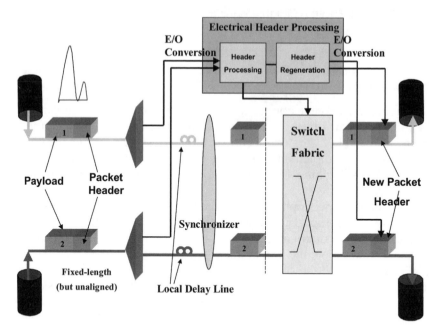

Figure 5.2   Optical packet switching.

circuit has an entry in the switching table. The entry matches an incoming virtual circuit identification number to an output port. In such a way, it separates routing from forwarding. The established virtual circuits are used during forwarding.

5.1.3   Three IP/WDM Networking Architectures

As IP has become the only convergence layer in the computer and communication networks, efficiently and effectively transporting IP traffic in a WDM network is an important task. We describe three IP over WDM networking architectures below.

IP over point-to-point WDM

Under this architecture, WDM-enabled point-to-point optical links are used to provide transport services for IP traffic. WDM devices such as OADM do not form a network themselves. Instead, they provide a physical layer link between IP routers. SONET can be used for transmission framing on the WDM channels. IP packets can be encapsulated in SONET frames using Packet-over-SONET schemes. Many IP router and WDM equipment vendors have products commercially available today that can support IP over point-to-point WDM. Point-to-point WDM systems have seen widespread deployment in long distance networks.

An IP over point-to-point WDM architecture requires the IP routers to be directly connected to each other via multi-wavelength fibre links. Figure 5.3 illustrates such an architecture, in which the neighbouring router for a given router interface is fixed.

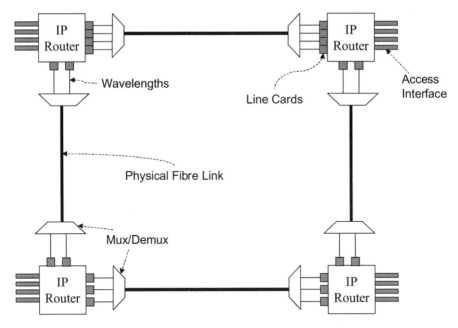

Figure 5.3 IP over point-to-point WDM.

With IP over point-to-point WDM, the network topology is fixed, and the network configurations are all static. Management systems for such networks are typically centralised, with minimal interactions between the IP and WDM layers.

IP over reconfigurable WDM

Under an IP over re-configurable WDM architecture, router interfaces from the IP routers are connected to the client interfaces of the WDM network. Figure 5.4 illustrates such an IP over Re-configurable WDM network. In this architecture, the WDM cross-connects and add/drop interfaces are themselves interconnected into a WDM network with multi-wavelength fibre links. Therefore, the WDM network itself has a physical topology and a lightpath topology. The WDM physical topology is composed of NEs interconnected by fibres; the WDM lightpath topology is formed by wavelength channel connections. Reconfigurable WDM is a circuit switching technology so the wavelength channel setup and tear-down are conducted in preserved phases. It is important to point out that IP traffic switching and wavelength switching never work in the same layer in an IP over reconfigurable networks. This can be translated into an overlay network.

Lightpaths in the WDM network are designed to conform to the IP topology. By appropriately configuring the WDM cross-connects, a given router interface can be connected to any other router interface at any other router. As a result, the neighbouring router for a given router interface is configurable under this architecture. This infers that the physical networks can support a number of virtual topologies subject to the same network resource constraints.

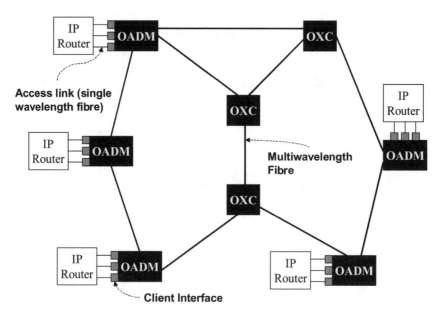

Figure 5.4 IP over reconfigurable WDM.

IP over switched WDM

In an IP over Switched WDM architecture, the WDM infrastructure directly supports a per-packet switching capability, as opposed to simply providing ingress-to-egress lightpaths. As such, it enables a much fine grain sharing than reconfigurable WDM. Various switched WDM approaches have been proposed, including:

- Optical Burst Switching (OBS)
- Optical Label Switching (OLS)
- Optical Packet Routing (OPR).

OBS and OLS use a fat-packet/flow switching paradigm that is different from the conventional IP packet routing. IPv4 itself utilises a best-effort destination-based routing. MPLS is introduced to IP as a value-added service to switch application flows. OLS is similar to MPLS except that it does not support conventional IP destination based packet forwarding. In other words, OBS and OLS (i.e. the core switches) do not understand IP packet headers and therefore cannot forward IP packets. Also, as we discussed before, OBS and OLS prefer medium granularity traffic instead of the fine granularity traffic presented in the IP packets.

OPR represents the optical implementation of conventional IP routing so it supports full IP functionality. Since optical logic processing and optical data buffering technologies are still immature at present, switched WDM systems are typically bufferless. (Although there are optical packet switching projects that attempt the design of optical buffers, those designs remain complex. Therefore most switched WDM system prototype efforts have opted for a bufferless design. Optical delay lines have been used to emulate optical buffers. But these optical delay lines are much less sophisticated than the random access memory.)

Likewise, switched WDM systems rely on electronic processing of the packet header to control the switching actions. This implies that OPR is not as mature as OBS and OLS. In addition to optical buffers, other factors influencing the commercialisation of OPR include fabric switching speed and the reliability and the signal degradation of the switching fabric.

Figure 5.5 shows IP over switched WDM networks. OBS and OLS are represented as OLSR. The key difference between OBS and OLS is that OBS uses fat-packet switching but OLS uses application flow switching. OLS often uses an in-band sub-carrier wavelength to carry the control information, i.e. the flow header. As indicated in the figure, OLSR is usually deployed in a cluster. Within the cluster, only the edge OLSR requires an implementation of the complete IP protocol stack. The edge OLSR also provides electro buffering so incoming IP packets can wait in the queue at the edge in the case of dynamic LSP setup.

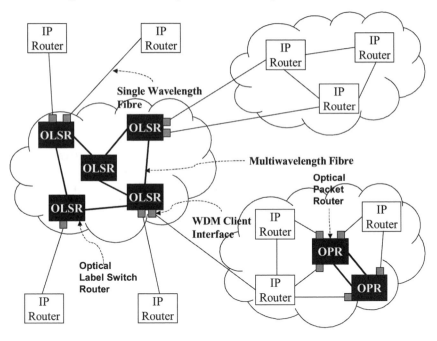

Figure 5.5 IP over switched WDM.

OLSRs are interconnected by fibres supporting multiple wavelength channels. OPR can be deployed just as electrical IP routers except that an OPR has a number of interfaces (i.e. more interfaces than a conventional IP router). In fact, interface savings is one of the main drivers behind OPR over an electrical IP router.

The three architectures presented above are associated with different hardware and control and management software. The IP over Point-to-Point WDM architecture will be gradually replaced by the IP over Reconfigurable WDM architecture since the second architecture can offer much more functionality than that of the first architecture. In addition, the second architecture is more flexible. Through carefully designed network control software and traffic engineering, the second architecture is able to provide much higher network resource utilisation and lower operational cost than the first architecture. Therefore, in the rest of this book, we focus our discussion on the second and the third IP/WDM networking architectures (i.e. IP over reconfigurable WDM and IP over switched WDM).

5.2   IP/WDM Internetworking Models

The last section has presented architectures for constructing IP over WDM networks through connecting conventional IP routers with the WDM network devices. This section describes how the IP network and the WDM network interoperate under these architectures. In particular, we will discuss peering issues in IP/WDM networks in the control plane as well as in the data plane.

5.2.1   IP over Reconfigurable WDM

In the data plane, IP over reconfigurable WDM will always form an overlay network in which IP packets are transported over the WDM lightpath circuits. These WDM lightpath circuits are not virtual paths as is the case in MPLS. When IP packets arrive at an OADM client interface, the corresponding lightpath is already established. Riding on the lightpath will assure that the IP packet crosses the WDM network without any examination in the data plane. In fact, the IP packet is not even aware of the use of any particular transmission technology. It only knows that there is an IP link between the routers. In this case, IP over WDM is the same as IP over any layer 2 technology, for example, IP over ATM or IP over frame relay.

In the control plane, we discuss three interconnection models for IP over reconfigurable WDM:

- overlay
- augmented
- peer model.

Overlay control model

Under the overlay network model, IP networks form the client layer where WDM networks behave as the physical transport network service provider. A WDM network has its own network control and management system, which could be centralised or distributed. It can also introduce its own addressing scheme. To use IP control protocols for WDM networks, a WDM NE must be IP addressed, but the WDM IP address has only local visibility within the WDM network. The routing, topology discovery and distribution, and signalling protocols in the IP network are independent of the routing, topology discovery and distribution, and signalling in the WDM network.

There are two alternatives to interface between IP client and WDM server, described below.

WDM network management system (NMS)

IP clients request services from the WDM NMS, which sits above the WDM transport and control layers. Hence, there is no direct interaction between IP control and WDM control. Once a path request is filed, the NMS connection manager is responsible for path selection and setup. In fact, this is similar to the ATM Soft Permanent Virtual Circuits (SPVC) model. The NMS overlay model is shown in Figure 5.6. In this model, there is a DCN for the WDM network. The DCN provides the WDM control channel, which can be accessed by the IP routers.

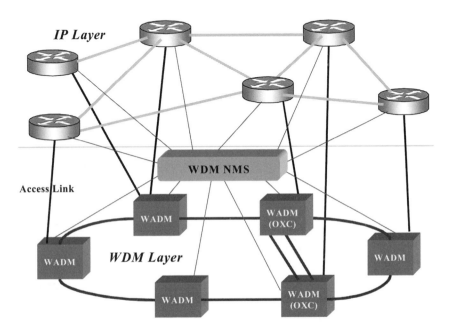

Figure 5.6 NMS overlay control model.

User to network interface (UNI)

Instead of relying on the NMS providing the interface, IP control can directly talk to WDM control through optical UNI. The end-to-end lightpath can be set up dynamically, which requires the WDM edge node to be capable of signalling and reserving bandwidth. The overlay model assumes very limited information sharing between the client and server networks. The UNI only supports simple requests to setup and tear-down lightpaths. This approach is similar to the ATM Switched Virtual Circuits (SVC) model. UNI overlay model is shown in Figure 5.7, where UNI servers are located at WDM network edges. In the figure, there are three sets of interface, UNI, INNI (Internal Network/Network Interface), and ENNI (External Network/Network Interface). UNI represents the boundary between the IP network and the WDM network. The control flows across the UNI depends on the services defined and the manner in which the services are accessed. INNI and ENNI are WDM

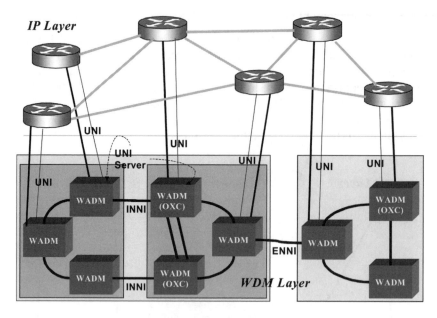

Figure 5.7   UNI overlay control model.

network interconnection interfaces. INNI represents the subnetwork interface within an administrative domain, whereas ENNI represents the inter-administrative domain interface. Hence, INNI and ENNI enforce different policies to restrict control flows between nodes. INNI and ENNI form a NC&M hierarchy for WDM networks. In terms of routing and topological information sharing, these interfaces can be ranked in this order: INNI, ENNI, UNI. The UNI can be categorised as public or private based on the service and context models. Public UNI specification is currently under standardisation. This model requires a control channel to transport the UNI messages between the IP router and the WDM UNI server.

Augmented control model

Under the augmented network model, the reachability information is shared between IP and WDM networks. WDM NEs are IP addressed and the WDM IP address is globally unique. Both IP and WDM networks may employ the same IGP such as Open Shortest Path First Protocol (OSPF), but there are separate routing instances in the IP and WDM domain. Therefore, the augmented model is really an IP inter-domain model. The interaction between IP and WDM can follow an EGP such as Border Gateway Protocol (BGP). OSPF for WDM networks and optical BGP require optical extensions to their counterparts in the conventional IP routing.

Signalling between IP and WDM networks also follows an inter-domain model. Based on the policy defined at the WDM edge, the same signalling protocol is implemented by IP and WDM so a signalling instance can cross IP and WDM networks.

Figure 5.8 shows the augmented IP/WDM model. In the figure, there are three networks, IP network *a* and *b* and WDM network *c*. The two IP networks are controlled by separate IGP running instances, and the WDM network is controlled by an optical version of the IGP. The two IP networks are interconnected directly using EGP. The IP network and WDM network are interconnected using optical EGP.

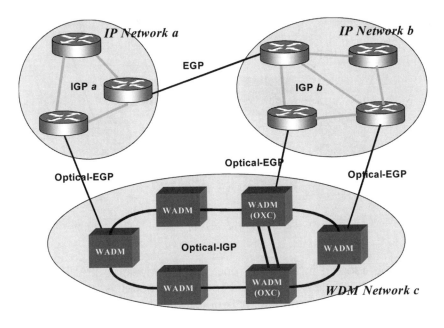

Figure 5.8 Augmented control model.

Peer control model

Under the peer network model, the reachability information is shared between IP and WDM networks and a single routing protocol instance runs over both the IP and WDM networks. In the control plane, WDM switches are just treated as IP routers with a peer-to-peer relationship. So the IP and WDM networks are integrated as a single network, controlled, managed, and traffic engineered in a unified manner. The peer model is shown in Figure 5.9.

Discussion

The three internetworking models presented above differ in the degree of IP/WDM integration. At one hand, the overlay model using NMS provides an indirect interface between IP and WDM networks; on the other extreme, the peer model promises a seamless interconnection between IP and WDM routers in the control plane. The overlay model is likely to be adopted in near-term rapid deployment of relatively static IP/WDM networks. Because of their overall simplified management and control structures, the peering and integrated models

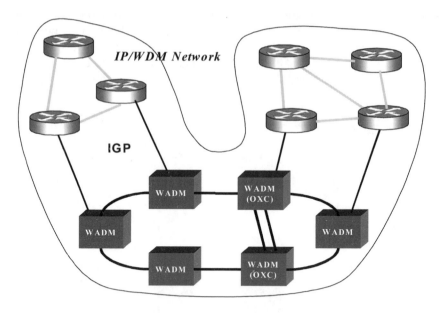

Figure 5.9   Peer control model.

are likely to be adopted in the long run for highly dynamic IP/WDM networks. The selection of the internetworking architectural model is also based on the existing network environment, the network ownership, and the administrative authority.

It is highly likely that these three models will coexist in the future. One may feel that a peer approach is the most efficient. But, optimisation is really to leverage the heterogeneity of the physical network. Thus, for high performance and low latency reasons, an optical network may be intentionally positioned as a high-speed switching network. Such a network will be deployed as a cluster. Quite naturally, it forms an overlay network to any other IP networks.

5.2.2   IP over Switched WDM

OBS and OLS can be implemented using OLSR. We will present IP over OLSR and IP over OPR in this section.

IP over OLSR

OLSR provides a label forwarding mechanism, which is basically a packet switching network. However, since the optical packet is not the IP packet and has its own header, it has to be generated at the edge OLSR. Therefore, OLSR has to be deployed in a cluster in order to take advantage of interface and bandwidth cost savings. The core OLSR is just another layer 2 switch and does not need to implement IP data plane functions.

To support a unified IP-centric control plane, the OLSR can be IP addressed so that it can support IP routing and signalling. In the data plane, IP over OLSR will always form an overlay network, in which IP packets are encapsulated into optical packets at the edge OLSR. However, OLSR is much more flexible than OXC since each optical packet is examined at the intermediate hop. In addition, the label switch path in the forwarding hierarchy is a virtual path, which uses soft state control mechanism to maintain its status. A virtual path does not have to reserve the bandwidth and can cope with finer granularity traffic transport.

Figure 5.10 shows IP over OLSR networks. As indicated in the figure, IP packets are aggregated at the edge of an OLSR network. Within the OLSR network, optical packets are forwarded based on the label (i.e. optical packet header) that they carry.

In the control plane, an OLSR can be implemented using either OBS or OLS. Hereafter, we will use OLS implemented OLSR. Again, the label concept of OLS is similar to that of MPLS. For virtual path setup and label distribution, OLS can reserve a wavelength channel to transport the control information. The OLSR cluster can be controlled by the OSPF protocol with optical extensions. Since OLSR uses an IP address, the IP network and OLSR network can support a unified control plane, i.e. MPLS. A common signalling protocol, for example, RSVP or LDP of MPLS, can be used for path setup and label distribution across the IP and OLSR networks.

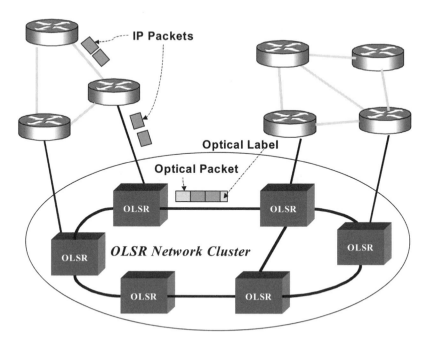

Figure 5.10 IP over OLSR.

In terms of routing interoperability, IP over OLSR can support one of the following configurations:

- *An IGP routing instance is used for the OLSR network,* which will interoperate with other IP networks using EGP. This results in an inter-domain routing model.
- *A single IGP routing instance controls both OLSR and IP network.* This represents a peer-to-peer routing model. However, this requires modification to the existing IGP. For an instance, the OLSR network can be configured as an OPSF area, which means inter-area LSA flooding needs to be updated with OLSR extensions, and intra-area LSA flooding within the OLSR area may also need to be modified to efficiently transport WDM link state information.

To sum up, the first configuration requires extension to the standard IP EGP, whereas the second configuration needs extension to the standard IP IGP.

IP over OPR

IP over OPR is essentially an IP network. OPR-formed optical Internet can be characterised by a number of parallel light channels between adjacent routers. As in conventional IP, there is no clear separation of the traffic-oriented data channel from the control channel. However, using MPLS, traffic can be differentiated and prioritised and cut-through paths can be set up to avoid congestion. These QoS mechanisms have become an essential component of the IP network.

A critical question for an OPR vendor is how much router functionality should be implemented in the optical domain. The answer is related to the cost-efficiency and application characteristics. In the data plane, IP routers and OPRs have a peer-to-peer relationship, in which both can forward raw IP packets.

In the control plane, IP routers and OPRs also form a peer relationship using in-band control. Interconnecting OPRs and IP routers follow the conventional IP solution. That is, for the purposes of routing, a group of networks and routers controlled by a single administrative authority is clustered as an AS. Routers within an AS are free to choose their own mechanism for topology discovery, routing information base construction and maintenance, and route computation. This mechanism within the AS uses IGP. Between a pair of ASs, EGP is used to exchange the network reachability and availability information.

To emphasise optical Internet with WDM technology, the conventional IP protocols need to be modified/extended to address IP/WDM concerns. For example, there are a number of fibre/wavelength ports on the OPR fabric and the question is how to efficiently and effectively use IP addresses. An example of OPR addressing employs link bundling so only one pair of IP addresses is given to the link channels between a pair of OPRs.

Figure 5.11 shows several possible network configurations of IP router and OPR. The top AS in the figure presents a hybrid network using IP router and OPR; the bottom AS consists of optical routers.

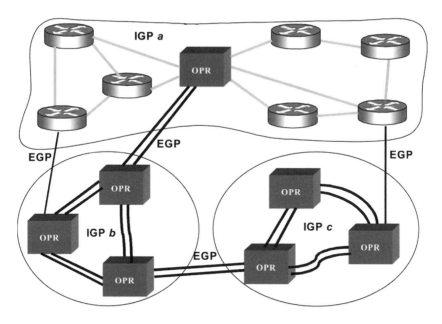

Figure 5.11 IP over OPR.

5.3 IP/WDM Service Models

An IP/WDM network can support two service models: a domain service model and a unified service model. We describe these service models below.

5.3.1 Domain Service Model

Under the domain service model, WDM networks form an optical domain, whereby topological and link state information is transparent from external IP networks. The optical domain has a client-server relationship with the access IP networks in which the optical network provides transport services to the client IP networks. Hence, IP networks and the optical domain operate independently and they need not have any routing information exchange across the interface.

From IP networks, an optical lightpath, possibly over multiple optical switches, is always viewed as a point-to-point link. For TDM clients, the optical lightpath is a large structured, fixed bandwidth pipeline. In the domain service model, the optical domain is able to set up/tear down an optical path requested by the client dynamically. The client network has no knowledge of the optical network path setup mechanism. An IP to optical domain service request is over a well-defined UNI. The current optical UNI has four actions:

- lightpath create: create a lightpath between a pair of termination points in the optical network with specified attributes;
- lightpath delete: delete an existing lightpath;

- lightpath modify: modify certain parameters of the lightpath;
- lightpath status enquiry: enquires about the status of certain parameters of the lightpath (referenced by its ID).

Each of the optical UNI actions is accomplished via a set of messages. Additionally, the following address resolution procedures may be made available over the UNI:

- **Client Registration:** allows a client to register its address(es) and user group identifier(s) with the WDM network. The registered address may be of different types: IP, ATM NSAP, etc. The WDM network associates the client address and user group ID with a WDM-network-administered address.
- **Client De-registration:** allows a client to withdraw its address(es) and user group identifier(s) from the WDM network.
- **Address Resolution:** allows a client to supply another client's native address (e.g., ATM) and user group ID, and get back a WDM-network-administered address that can be used in lightpath create messages.

In addition to static configuration of the interfaces between optical and client devices, a service discovery procedure can be developed to allow a client to automatically determine the parameters of the interconnection with the optical network including the UNI signalling protocol supported. This service discovery routine can also verify local port connectivity between the optical and client devices.

An example of the domain service model is shown in Figure 5.12 (top). The optical domain appears only as an interconnection network 'cloud', where the details within

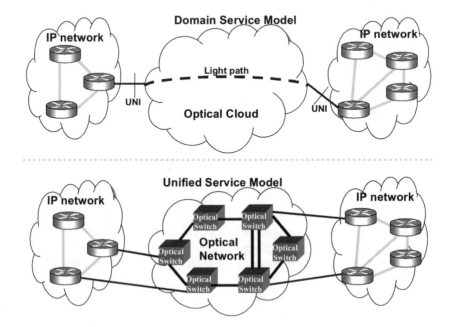

Figure 5.12 IP/WDM service model.

the optical domain are transparent from the client networks. Once a connection is established in the optical domain, the connection appears to be a point-to-point link between the client devices.

5.3.2 Unified Service Model

Under the unified service model, there is a single control plane over client and optical networks, presumably MPLS. From routing and signalling points of view (under the common control plane), there is no distinction between UNI, NNI, and any other router-router interfaces. In this model, services are not specifically defined at an IP-optical interface, but folded into end-to-end MPLS services. For example, an edge switch can create, delete, or modify optical paths as it deals with MPLS LSPs. Although the services (i.e. MPLS over Layer 2 transport) provided are the same as those of the domain service model, they can be invoked in a more seamless manner. Once a lightpath is set up across the optical network, the lightpath becomes a part of the forwarding adjacency and therefore advertises using opaque LSAs. Consequently, the lightpaths provide an overlay of direct connections between optical edge switches across the optical network. Decisions on lightpath setup and end-point selection are similar in both models.

An example of the unified service model is shown in Figure 5.12 (bottom). Optical network reachability and resource availability information are flooded to client IP networks, so an IP client router can compute an end-to-end MPLS LSP across the optical network and then initiate the path setup by invoking the local signalling process.

5.3.3 Services

There are several groups of services that can be offered by the IP/WDM networks.

Dark fibre and lambda channel

The optical WDM network starts as a dark fibre network. Once a connection request comes in, the corresponding wavelength channel or the entire fibre link is reserved. The established connection, i.e. a circuit, appears as a point-to-point pipe for the clients. A client network is interested in this service because it has more control on its routing equipment to support its application needs.

VPNs

This is similar to the dark fibre service except that a virtual channel is set up instead of a reserved circuit of the dark fibre service. A virtual channel gives high levels of bandwidth sharing. It also satisfies finer granularity traffic transport request. However, the virtual channels are maintained using soft state mechanisms, and therefore, they generate a lot of control traffic and also introduce extra complexity into the network control.

Client application services

Optical WDM networks can be designed to support various application services. Examples include:

- *Storage Area Network (SAN):* this is to utilise the abundant bandwidth provided by WDM optical networks to interconnect data storage centres. SAN can be constructed as a private network or an IP storage network.
- *Distributed applications* such as interactive collaborative applications, trading systems, distributed databases.
- Enterprise applications.

5.4   Summary

In this chapter, we have presented the IP over WDM networking architectures based on the available WDM technology. A reconfigurable WDM network supports circuit switching whereas a switched WDM network follows a packet switching paradigm. Examples of the switched WDM are optical burst switching, optical label switching, and optical packet switching.

The idea of switched WDM is to switch optical packets, which are large granularity traffic blocks. The goal of this approach is to have a simple and very low latency network to take advantage of all-optical networking as well as to explore the benefits of packet switching.

Examples of the reconfigurable WDM include IP over point-to-point WDM and IP over OXC. IP over point-to-point WDM can be considered as an early generation and static WDM. Hence, it will be replaced by IP over OXC as OXC becomes mature. To some extent, OXC can be treated as a replacement to the existing Telecom switches and ATM switches especially in the backbone network.

The advantage of the OXC goes to the operational cost savings in a long haul network, for example, fibre deployment, maintenance, and administration. The offset is the cost of the WDM switches, i.e. WDM transmitters, receivers, and amplifiers are still expensive. In fact, that is why the WDM equipment has only been deployed in the backbone networks. However, coarse WDM (CWDM) equipment has appeared and designed to be cost-efficient in the MAN environment.

By comparison, switched WDM networks are still research lab prototypes. Several crucial technologies are not mature yet such as optical buffering, optical packet processing time/optical packet switch time (too long), and the reliability of photonic packet switches. A set of reconfigurable WDM industry has emerged in which there are component vendors supplying WDM-related semiconductors and components, system vendors supplying WDM systems, and software vendors providing network control and management software and tools.

Nearly all data traffic today is IP, so an important issue in WDM optical networking is how WDM optical networks and IP networks interoperate. We presented two sets of IP/WDM internetworking models:

- IP over reconfigurable WDM.
- IP over switched WDM.

In the IP over reconfigurable WDM set, we presented three models:

- overlay control model
- augmented control model
- peer control model.

These models are named as their control plane IP/WDM internetworking models. The overlay control model can be implemented using NMS or UNI. UNI can be considered as a distributed implementation of the interface between IP and WDM. UNI are available at WDM edge switches, which are equipped with signalling protocols. In the NMS overlay model, NMS serves as the sole interface between IP and WDM. The augmented control model is an inter-domain IP control model in which WDM networks are configured as an independent IP domain and controlled by an optical OSPF instance. The optical OSPF instance communicates with other IP OSPF instances through an inter-domain routing protocol such as BGP. The peer control model emulates a pure IP network in the IP/WDM control plane. So an IP router and a WDM control node has a peer-to-peer relationship.

The overlay model forms a client-server relationship between IP and WDM. This approach is simple to implement but complex to manage and control since there are two NMSs: IP NMS and WDM NMS. The peer model integrates the WDM control plane with the IP network but WDM network state information need to be flooded to IP networks. This approach poses challenges on the existing IP control protocols such as OSPF because of the extra WDM-related information having to be dealt with. A compromise between the overlay model and the peer model gives birth to the augmented model.

The selection of IP/WDM internetworking models also depends on network ownership and administration authority. In terms of deployment, only overlay IP/WDM networks have been reported. We discussed two IP over switched WDM models: IP over OLSR and IP over OPR.

A driver behind a commercial technology is the application/service. That is who is going to use the technology, whether users really need the technology, and how much he/she is willing to pay. We presented two service models: domain service model and unified service model. We also presented three groups of optical networking services: dark fibre and lambda channel service, VPN service, and client application service.

6

IP/WDM Network Control

- IP/WDM network addressing
- Topology discovery
- IP/WDM routing
- IP/WDM signalling
- WDM network access control
- GMPLS
- IP/WDM restoration
- Inter-domain network control
- WDM network element control and management protocol
- Summary

Chapters 6 and 7 focus on two functional groups of IP/WDM network transport: network control and traffic engineering. The former refers to the enabling control mechanisms across the network, whereas the latter relates to the optimisation on the usage of IP/WDM networks. Without traffic engineering, network control only can provide the minimal operational functionality under normal circumstances. Traffic engineering can provide the additional driver intelligence for network control software.

Figure 6.1 shows an outline of traffic engineering and network control in IP/WDM networks. IP/WDM traffic engineering deals with issues on efficiently and effectively routing traffic over an established IP lightpath topology, for example, by avoiding congested links and load balancing among multiple equal cost paths. In the case of reconfigurable WDM, it also copes with optimal IP topology design to take advantage of WDM reconfigurability. Reconfiguration also applies to individual light circuits and virtual paths, for example, due to signal degradation, restoration, or rerouting requests. Traffic engineering issues are presented in Chapter 7.

To glue network-wide devices together and support and facilitate traffic engineering, a network control layer is formed on top of individual network elements and interconnecting fibres, i.e. the physical transmission layer. IP network control consists of IGP (e.g. OSPF), EGP (e.g. BGP), ICMP, and RSVP. A WDM network also has a LMP (Link Management Protocol) introduced later in this chapter. IP/WDM network control has the following functionality:

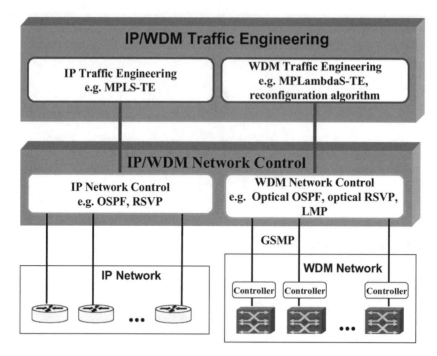

Figure 6.1   IP/WDM network control and traffic engineering.

- IP/WDM network addressing;
- Topology discovery;
- IP/WDM routing, which includes:

  - routing information base design and construction;
  - routing information base maintenance, for example, through periodic flooding and soft state timers;
  - route computation to select fibre path and/or wavelength channels.

- Connection setup and tear-down;
- A signalling mechanism;
- WDM network access control;
- IP/WDM protection and restoration.

Figure 6.2 shows the software architecture for IP/WDM network control. A switch control and management protocol provides the management interface between network control components and the optical switches. It is a master-slave protocol, functioning vertically to connect layered components. Routing, topology discovery, and signalling modules are the main components in the network control, and they function horizontally to link peer instances together. If IP and WDM form an overlay network in the control plane, there is a need for an address resolution server, which is independent from other control modules and can be implemented in a central location or in a distributed fashion. Switch control and management protocol can be co-

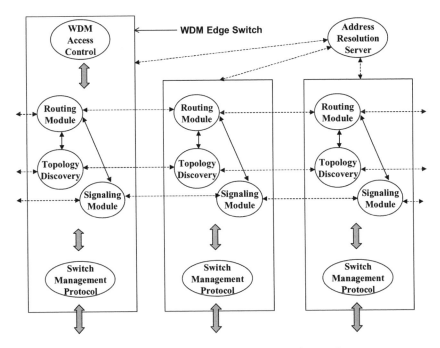

Figure 6.2 Software architecture for network control.

located with the switch controller, whereas an address resolution server can be located anywhere in the network. If IP and WDM form an overlay model in the data plane, one also needs a WDM access control component. This component is always located at the WDM edge switch. We will discuss these components in the rest of this chapter.

6.1 IP/WDM Network Addressing

For interoperability across IP-centric WDM networks, a fundamental issue is addressing. Possible addressing entities in WDM networks include switch interface, wavelength ports, optical links, physical fibres, and optical channels (wavelengths). The issue here is how granular the identification should be as far as network control is concerned. The scheme for identification must accommodate the specification of the termination points in the WDM network with adequate granularity when establishing lightpaths. For instance, an OXC can have many transport interfaces, each of which in turn is related to a number of wavelength channels. There may be a number of parallel fibres between a pair of switches. Hence, it is unreasonable to assume that every wavelength channel or termination port should have a unique IP address. Also, the routing of a lightpath within the WDM network may not depend on the precise termination point information, but rather on the terminating OXC.

Finer granularity identification of termination points is of relevance to the terminating OXC but not to intermediate OXCs. This suggests an identification scheme whereby OXCs are identified by a unique IP address, and a selector identifies further fine-grain information of relevance at an OXC. This, of course, does not preclude the identification of these termination points directly with IP addresses (with a null selector). The selector can be formatted to have an adequate number of bits and a structure that expresses port, channel, and other identifications.

Within the WDM network, the establishment of lightpath segments between adjacent OXCs requires the identification of specific port, channel, or even sub-channel. With a MPLS-based control plane, a label serves this function. The structure of the *optical label* is designed in such a way that it can encode all the required information (including WDM-specific information).

Another entity that must be identified is the Shared-Risk Link Group (SRLG). An SRLG is an identifier assigned to a group of optical links that share a physical resource. For instance, all optical channels routed over the same fibre could belong to the same SRLG. Similarly, all fibres routed over a conduit could belong to the same SRLG. The notable characteristic of SRLG is that a given link could belong to more than one SRLG, and two links belonging to a given SRLG may individually belong to two other SRLGs. While the classification of physical resources into SRLGs is a manual operation, the assignment of unique identifiers to these SRLGs within a WDM network is essential to ensure correct SRLG-disjoint path computation for protection and restoration.

Finally, optical links between adjacent OXCs may be bundled for advertisement in a link state protocol. The component links within the bundle must be identifiable. In concert with SRLG identification, this information is necessary for correct (protection) path computation.

In an overlay IP/WDM network, the WDM layer can have its own addressing scheme, for example, a layer 2 address. To map layer 2 addresses to layer 3 addresses (i.e. IP addresses), an address resolution protocol or mechanism is needed. A peer IP/WDM model supports a unified control plane, according to which IP layer as well as WDM layer uses IP addresses. A unified addressing scheme across IP/WDM networks is certainly more efficient and scalable. Address translation in an overlay IP/WDM addressing is likely to become a performance bottleneck. So far, the main reason for using IP address in the WDM layer is to leverage the control mechanisms such as routing and signalling protocols developed in the IP environment. To support a common, peer control plane, both IP and WDM networks need to have global IP addresses. In this section, we describe both IP and WDM networks with IP address (instead of creating a novel WDM addressing scheme).

6.1.1   Overlay Addressing

An overlay IP/WDM network can use IP addresses in the IP and the WDM layer, but these addresses in different layers are not visible to each other. WDM layer IP addresses can be considered as an example of layer 2 addresses. As such, there is a need for address resolution between IP layer addresses and WDM layer IP addresses. Deploying IP addresses in the WDM layer enables the use of IP control protocols. The IP layer can be controlled by an IGP protocol; the WDM layer can be controlled by the OSPF protocol with WDM extensions. This also implies that although both IP and WDM layers may use OSPF, the IP OSPF and the optical OSPF are separate OSPF instances.

Figure 6.3 shows the IP layer addressing in IP/WDM networks, where there are four IP routers, each with two point-to-point interfaces and one Ethernet interface. In IP working over reconfigurable WDM networks, the old lightpath topology needs to be migrated to the new lightpath topology. However, conventional IP topology is static and network convergence takes time. In addition, dynamic IP topology reconfiguration (such as changing IP interface addresses) may have impact on network stability and cause packet drop. An alternative to dynamically assigned IP addresses is pre-assigning multiple IP addresses to one interface (for example, one primary address and multiple secondary addresses). Different IP addresses for the same interface form different IP subnets, but only one IP subnet is active at a time.

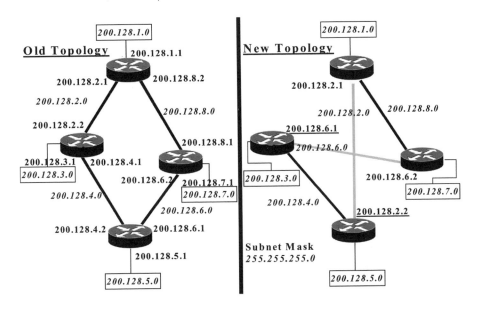

Figure 6.3   IP layer addressing in an IP over reconfigurable WDM network.

In the WDM layer, the control channel is separated from the data channel as shown in Figure 6.4. An out-of-band channel is used to transport the control messages. The data channel represents the data plane WDM network, where the WDM NE interfaces can use physical addresses. The control channel, which is IP addressed, carries the corresponding WDM link information. A control channel is related to one or more data channel links between the two NEs. A link bundle is the collection of all link groups between a pair of neighbouring switches. A link group identified by a link group ID can be formed by links according to the following criteria:

- *Links having the same set of Shared-Risk Link Groups (SRLGs).* A SRLG, used in fault management and protection and restoration, is an identifier of a group of optical links that share a physical resource.
- Links having the same encoding format, for example, OC-192.
- Links having the same protection type, for example, 1 + 1.

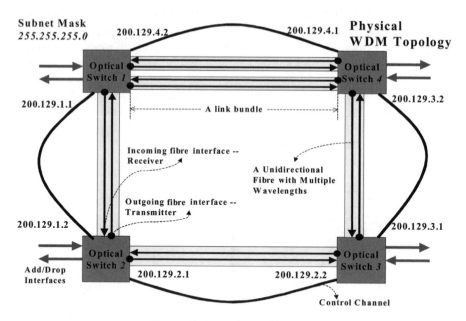

Figure 6.4 WDM layer addressing.

Link bundling saves IP address, which in turn reduces the number of routing adjacencies between neighbouring nodes. The data channel's link state information is flooded in the control channel using OSPF in the format of an opaque LSA. An IP link in the data channel is associated with one bi-directional wavelength channel. The WDM layer IP address can be assigned to the outgoing WDM link to its neighbour switch.

As shown in Figure 6.4, the WDM layer control channel is assigned with IP addresses. There are four optical switches interconnected by WDM links. The data channel information is associated with the control channel and transported over the control channel. This kind of addressing is suitable for IP over a reconfigurable WDM network in which there is no data plane IP function in the WDM layer.

6.1.2 Peer Addressing

IP working over reconfigurable WDM can also foster a peer model in the control plane in which WDM layer IP-addressed switches are peers to IP layer routers. Since reconfigurable WDM can only support circuit switching, there is no need to support IP functions in the WDM data plane. WDM layer IP addressing is the same as the overlay addressing except that the WDM address has a peer-to-peer relationship to an IP network address (i.e. global IP address for WDM network equipment). The WDM layer can employ the link bundling to assign a pair of IP addresses to the link between neighbour switches.

IP over switched WDM always supports peer addressing in the control plane. An IP over an OLSR network has an independent data-forwarding plane that is different from the IP destination-based forwarding. Hence, an OLSR network may not implement the IP data plane functions. This implies, from a data plane point of view, IP working over switched WDM still forms an overlay network. IP over OPR forms a peer network between IP and WDM networks in both the control plane and the data plane. In all cases, link bundling is used to save IP addresses and reduce routing adjacencies.

6.2 Topology Discovery

Routing within the WDM domain relies on knowledge of network topology and resource availability. This information may be gathered and/or used by a centralised system, or by distributed route computation entities. In either case, the first step towards network-wide link state determination is, for each OXC, to discover the status of local links to neighbours. In particular, each OXC must determine the up/down status of each optical link, the bandwidth and related parameters of the link, and the identity of the remote end of the link (for example, remote port number). The last piece of information can be used to specify an appropriate label during signalling for lightpath provisioning.

The determination of these parameters could be based on a combination of manual configuration and an automated protocol running between adjacent OXCs. The characteristics of such a protocol would depend on the type of OXCs that are adjacent (for example, transparent or opaque). In general, this type of protocol can be referred to as a neighbour discovery protocol (NDP), although other management functions such as link management and fault isolation may be performed as part of the protocol. The Link Management Protocol (LMP) is an example of a NDP.

A NDP typically requires in-band communication on the bearer channels to determine local connectivity and link status. In the case of opaque OXCs with SONET termination, one instance of a NDP would run on each OXC port, communicating with the corresponding NDP instance at the neighbouring OXC. The protocol would utilize the SONET overhead bytes to transmit the (configured) local attributes periodically to the neighbour. Thus, two neighbouring switches can automatically determine the identities of each other and the local connectivity, and also keep track of the up/down status of local links.

Topology discovery is the procedure by which the topology and resource state of all links in a network are determined. Topology discovery can be performed using a link state routing protocol (for example, OSPF or IS-IS), or it can be conducted through management interfaces (in the case of centralised path computation). Here, we focus on fully distributed route computation using an IP link state protocol.

Most of the link state routing functionality in IP networks can be applied to WDM networks. However, the representations of optical links and wavelength channels, as well as optical signal QoS parameters, require changes to the link state information. In particular the changes include:

- The link state information may consist of link bundles. Each link bundle is represented as an abstract link in the network topology. Different bundling representations are possible. For an instance, the parameters of the abstract link may include the number, bandwidth and the type of optical links contained in the underlying link bundle.

- The link state information should capture restoration-related parameters for optical links. Example parameters include link protection type and the list of SRLGs for each link.
- A single routing adjacency is maintained between neighbours, which may have multiple optical links or multiple link bundles between them. This reduces the protocol messaging overhead.
- Since link status and channel availability changes dynamically, a flexible policy for triggering link state updates based on availability thresholds can be implemented. For instance, changes in availability of links of a given bandwidth (for example, OC-48) may trigger updates only after the availability changes by a certain percentage.

During topology discovery, neighbourhood information is exchanged through a discovery protocol. This is then consolidated in the network level to produce a topology map. Automatic topology discovery not only reduces network operation costs but also increases the likelihood of a true representation of the network topology. Topology discovery can be regarded as an enabling mechanism for a self-managed network. From an application point of view, up-to-date and dynamic topology knowledge allows the effects of resource sharing on an application's performance to be predicted and nodes and links to be selected with appropriate network connections to match the application needs. The resource representation models and the topology discovery process across WDM domains require inter-domain routing protocols, for example, BGP with optical extensions.

In a reconfigurable WDM network, there is a lightpath virtual topology in addition to a WDM physical topology. The lightpath topology is the IP topology, so once the lightpaths are established, the IP topology (data plane routers) can employ an IGP for topology discovery. For example, OSPF uses the Hello message and the Database Description message for neighbourhood discovery (see Chapter 3). The WDM physical topology can be discovered by the IP control protocol running on the WDM control channel.

6.2.1   OSPF Hello Message

OSPF Hello messages are sent out periodically over all functioning interfaces (including the virtual interfaces). In multicast or broadcast capable networks, the Hello message is also multicast or broadcast over the physical network. Hello messages are also used to elect the designated router and backup designated router. In addition, the Hello message is used to detect and negotiate certain OSPF extensions. The option field in the Hello message (see Figure 6.5) allows a router to reject a neighbour because of a capability mismatch. For example, the E-bit in the options field indicates whether the attached area is capable of processing an AS-external-LSA. When the E-bit is not set, the attached area is basically treated as a stub network. Since IP is a network layer protocol, it runs over different types of networks.

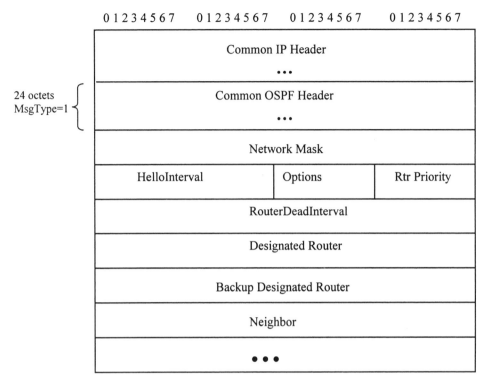

Figure 6.5 OSPF hello message format.

The OSPF Hello protocol performs differently over different subnet types. In an IP subnet, two routers can exchange routing information directly only when they have at least one common subnet. In the Hello message, a router sets the IP source address to the router's address on the subnet and includes the subnet mask. When the other router receives the Hello message, it will accept the Hello if and only if both routers have the same subnet mask and both router interfaces attach to the same subnet. In a broadcast subnet, each router sends a Hello packet to the multicast group address and the receiving router lists the received Hello routers. Therefore, the neighbourhood relationship can be maintained with *n* Hello packets, where *n* is the number of routers in the subnet. During the process of electing designated router and backup designated router, the first OSPF router usually becomes the designated router and the second one to join the subnet becomes the backup designated router. If the designated router fails, the backup designated router is promoted to the designated router. The new backup designated router is selected from the other routers having the highest router priority (which is indicated in the Hello message).

In a point-to-multipoint subnet, the network does not have broadcast capability and does not select a designated router, so the Hello message is sent periodically to directly connected neighbours. If the router and its neighbour locate on the same subnet, the Hello message is accepted. In a NBMA subnet, the network does not have broadcast capability but does select a designated router and a backup designated router. An eligible designated router has a non-zero value for the router priority. During designated router election, eligible routers send each other Hello messages. Based on the router priority, designated and backup designated routers are elected. From then on, the designated and backup designated routers send Hellos to other routers in the subnet. The other routers will only send Hellos to the designated and backup designated routers.

Figure 6.5 shows the OSPF Hello message format. Within the header, the network mask field (four octets) specifies the network mask associated with the interface; the HelloInterval field (two octets) specifies the time interval in seconds that the Hello message should be sent; the options field (one octet) indicates the optional capability of this router. The router priority field (one octet) gives the router priority in numerical values. When this field has the value of zero, this router cannot be selected as designated router, nor backup designated router. The RouterDeadInterval field (four octets) specifies the time interval in seconds that a router will declare its neighbour to be down. The designated router field (four octets) and the backup designated router field (four octet) specify the designated router and the backup designated router for this network. If there is no such router in the network, these fields are assigned to zero. The neighbour field (four octets) gives the neighbour router ID from which the recent Hello has received.

6.2.2   Link Management Protocol (LMP)

In WDM networks, control channels are implemented out-of-band and neighbourhood discovery is implemented using an IGP (assuming the control channels are IP addressed). However, new mechanisms are needed to manage the data-bearing WDM links in terms of link provisioning and fault management. As a result, a link management protocol (LMP) is introduced and is being developed in the IETF. LMP is intended to work together with routing and signalling protocols to provide a common control plane across routers, WDM switches, ATM switches, and SONET network equipment. LMP is especially designed to address these questions:

- How is the control plane coupled to the data plane?
- How are link bundles (or labels) assigned to physical resources?
- How are failures localised, for example, for fast switchover?

In LMP, depending on multiplexing capability, a data-bearing link can be either a 'port' or a 'component link'. Component links are multiplexing capable, whereas ports are multiplexing incapable. For example, an OC-192 SONET cross-connect interface can be divided into four OC-48 channels, i.e. using TDM. In such a case, the link resources must be identified using three levels: TE link ID, component interface ID, and time slot label. Without multiplexing, only two levels of identification are required: TE link ID and port ID. LMP requires that there is at least one active bidirectional control channel between a pair of nodes.

Two main procedures of LMP are control channel management and link property correlation. Control channel management constructs and maintains link connectivity between physically adjacent nodes. This requires a lightweight Hello protocol as a fast keep-alive mechanism. Link property correlation is used to exchange the local and remote property mapping. Two optional procedures offered by LMP are link connectivity verification and fault localisation. Link connectivity verification offers a testing procedure to verify the physical connectivity of the data-bearing links and identify any misconfigurations. Fault localisation localises failures in the WDM network.

LMP messages are IP encoded. In addition to the IP header, LMP has a common header as shown in Figure 6.6. Within the header, the version field (4 bits) gives the version number; the reserved field (12 bits) is reserved for future use. The flags field (1 octet) has defined these values: control channel down (0x01), node reboot (0x02),

Config	TestStatusSuccess
ConfigAck	TestStatusFailure
ConfigNack	TestStatusAck
Hello	LinkSummary
BeginVeify	LinkSummaryAck
BeginVerifyAck	LinkSummaryNack
BeginVerifyNack	ChannelFail
EndVerify	ChannelFailAck
EngVerifyAck	ChannelActive
Test	ChannelActiveAck
ChannelDeactive	ChannelDeactiveACK

link type specification (0x04), LMP-WDM support capability (0x08), and authentication option (0x10). The message type field (1 octet) specifies the type of the message. Currently, there are 22 messages:

Among them, only the Test message (type 10) is delivered over the data channel, and all other messages are sent over the control channel. The length field (2 octets) specifies the total length of the LMP message including this header. The checksum field (2 octets) provides the standard IP checksum for the entire message including the header. The local channel/link ID (4 octets) field represents the control channel ID for Config, ConfigAck, ConfigNack, and Hello messages, and the local TE link ID for all other messages. This ID must have a non-zero value and it is unique node-wide.

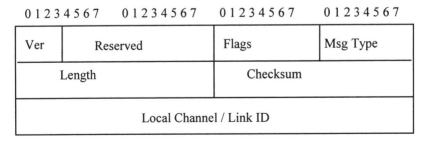

Figure 6.6 LMP header format.

LMP Control channel management

LMP assumes that for each LMP adjacency between a pair of nodes, there is at least one active control channel. LMP does not care for the detail of the implementation of the control channel, for example, it could be an Ethernet link or a wavelength channel. In the case of multiple control channels between two nodes, each control channel needs to be negotiated individually. Two sets of messages are used to maintain the control channel health: Config and Hello messages. The current version of LMP defines five states for the control channel: Down, ConfSnd, ConfRcv, Active, and Up.

Initially, the control channel is in a Down state, where there is no message exchange and all parameters are set to the default values. The Config messages, i.e. Config, ConfigAck, and ConfigNack, implement reliable parameter-negotiation exchange. Once a node starts to send Config message, it enters the ConfSnd state and remains in this state until it receives ConfigAck or ConfigNack. If the timer expires, the node will resend the Config message. When the node starts to negotiate parameters, the control channel enters the ConfRcv state and remains in this state until the parameters are received and acknowledged. Then, the control channel transits into the Active state. This indicates the control channel has been activated between the pair nodes. Hello messages are then employed to synchronise the status of the control channel. These unicast Hello messages are sent periodically and used as a lightweight keep-alive mechanism.

There are two timers for the Hello message, HelloInterval and HelloDeadInterval, respectively. The first timer specifies the interval that the Hello message is sent in milliseconds, whereas the second timer defines the time interval in milliseconds that a node shall wait until it declares its neighbour to be dead. The values for these timers are negotiated using the Config message (i.e. HelloConfig TLV) during control channel activation. The timers and the Hello message are intended to react to control channel failures rapidly so IGP Hello messages are not lost and the link state adjacencies are not dropped yet.

Within a Hello message, there are two 4-octet sequence numbers in addition to the common LMP header as shown in Figure 6.7. The first number, TxSeqNum, specifies the current sequence number for this Hello message; the second number, RcvSeqNum, represents the sequence number of the last Hello message received.

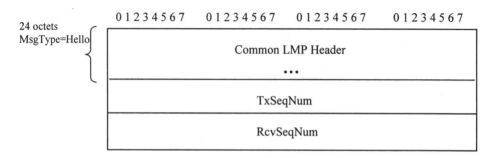

Figure 6.7 LMP hello message format.

When TxSeqNum = 1, this indicates that the control channel has booted or rebooted; when RcvSeqNum = 0, this implies that so far, there are no Hello messages received.

LMP link property correlation

Once the control channel has been established, the next step is to discover the data plane topology. In LMP, this is implemented using the LinkSummary message set including LinkSummary, LinkSummaryAck, and LinkSummaryNack. The LinkSummary message consists of TE Link and Data-Link TLV, used to synchronise the interface ID and correlate the TE link properties such as link multiplexing/demultiplexing capability and the encoding type of the data link.

The data link can be in either active or passive state. The former indicates that the link is used for transmission of a Test message, whereas the latter implies that the link is used for receiving of a Test message. The current version of LMP defines six states for the data link: Down, Test, PasvTest, Up/Free, Up/Allocated, and Degraded. When the link is not in the resource pool, it is in the Down state. During testing, i.e. transmitting or receiving a Test message, the link is in either Test or PasvTest state. Once the link has been successfully tested, the link is put in the resource pool.

Two main states associated with the link in the pool are Up/Free, meaning the link is free, and Up/Allocated, meaning the link is in use. However, for an Up/Allocated link, its control channel may go down. If that happens, the link enters the Degraded state, where the data-bearing link is in fact in use by clients.

In terms of topology discovery, the design of LMP is to co-exist with IGP such as OSPF. Dynamic WDM link state information can be flooded using opaque LSA and the link state and TE databases are maintained by OSPF. As such, functions already provided in OSPF are not repeated in LMP. The LMP specification can be found in [Lang01].

6.3   IP/WDM Routing

Routing is a technique by which traffic, across the network, can reach its destination from the source node. We describe several routing-related issues in the context of IP/WDM networks. We present an optical extension to the widely deployed OSPF protocol for WDM networks. We describe routing protocols and mechanisms first and then discuss routing behaviours.

6.3.1   Routing Information Base Construction and Maintenance

The quality of the routing decision is very closely related to the availability, the accuracy, and the detail of the routing information. Routing information can be available at one central location, several sites, or every node. Nevertheless, routing information in a network needs to be collected and properly maintained. A conventional IP network uses a fully distributed routing information base. For simplicity, RIP keeps only a distance vector that uses the hop count as the only metric. A more scalable and fast-converging approach is a link state protocol such as OSPF, which maintains a link state database. To synchronise the copies of the link state databases, a reliable flooding mechanism is developed in OSPF using these messages: *link state update*, and *link state acknowledgement*.

Figure 6.8 shows the OSPF reliable flooding in an example network with six nodes. In Figure 6.8(a), node A has observed a link state change and starts to flood the link state change using link state update packets. Node A sends the LSU (Link State Update) to all neighbours: B, C, and F. In Figure 6.8(b), nodes B, C, and F start to relay the flooded message to all their neighbours except the one that the LSU arrived at, i.e. node A. Figure 6.8(c) shows the next round when node D and E floods the LSU. To acknowledge the LSU has arrived safely, the receiving node sends a link state acknowledgement to the sending node. Note if the receiving node receives from the sending node the same LSU that it has sent, it assumes that the sending node has received a LSU. Hence, it will not explicitly send a link state acknowledgement. This is shown in Figure 6.8(d).

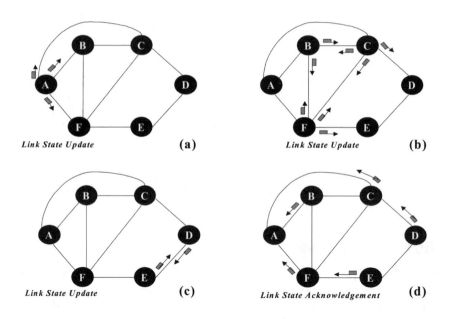

Figure 6.8   OSPF reliable flooding.

In the WDM NMS of an overlay model, the configuration and connection manager maintains the routing information base. The database is initially populated through defined management interfaces and then updated by NE using notifications. To increase the WDM routing information availability, WDM networks can also implement an optical link state database (i.e. wavelength MIB), which can be distributed to every switch. An example of the implementation extends the IP OSPF to flood opaque type **link state advertisements** (LSA). The distributed optical link state database can be separated from the standard IP link state database, but can be maintained and synchronised using the same flooding mechanism (see Figure 6.8). OSPF extensions for WDM networks are discussed later in this chapter. Instead of having separate routing information bases and NMS for IP and WDM, a more efficient approach is to design an integrated link state database for IP/WDM networks, which is also desirable in the peer model.

In IP inter-domain routing, BGP routers maintain a BGP routing table in addition to the IGP routing table. For optical inter-domain routing, a lightpath MIB is needed at the WDM edge in addition to the wavelength MIB. BGP with extension can be used for the control channel to exchange inter-AS WDM reachability and availability information. If various vendors supply NEs in the optical subnets or domains, network interoperability may require special attention.

6.3.2   Route Computation and WDM Switching Constraints

Once the routing database is constructed, route computation can be conducted through a routing algorithm. The IP protocol set uses the shortest path first (SPF) algorithm, although a better term is really the optimum path algorithm. After a metric is selected, for example, the least costly or the shortest distance, the source SPF tree is computed according to one of the single-source shortest path algorithms.

SPF algorithms

The most well known single-source SPF algorithm is **Dijkstra's algorithm**. Given a graph $G(V, E)$, where V is the set of vertices and E is the set of edges, and given a source vertex s, $s \in V$, a shortest path tree to every vertex in G can be obtained. The algorithm maintains a set, S, of vertices whose final shortest path weights from s have already been determined, and defines:

$d(s_j) =$ shortest distance from s to s_j
$w(s_i, s_j) =$ weight of (s_i, s_j), $s_i \in V, s_j \in V$.

The algorithm assumes $w(s_i, s_j) > 0$ for each edge $(i, j) \in E$:

- Dijkstra algorithm starts with, $S_0 = \{s\}$, $s \in V$.
- Given $S_k = \{s, s_1, \ldots, s_k\}$ at kth step, Dijkstra algorithm conducts the following computation:

  - for each $s_j \in S_k$, find $b_j \notin S_k$, $\min(w(s_j, b_j))$;
  - find $s_q \in S_k$ such that $\min(d(s_q) + w(s_q, b_q))$;
  - set $s_{k+1} = b_q$, $S_{k+1} = S_k \cup \{s_{k+1}\}$, $d(s_{k+1}) = d(s_q) + w(s_q, s_{k+1})$.

Dijkstra's SPF algorithm presents results in terms of minimal cost for each destination from the given vertex s. It does not automatically store the SPF paths. Also, it only deals with non-negative weight edges. The **Bellman-Ford algorithm** solves the single-source shortest-paths problem in the more general case in which edge weights can be negative. The algorithm detects the 'negative-weight cycle' from the source vertex s and reports a Boolean value to indicate no solution exists. If there is no such cycle in G, the algorithm generates the shortest path tree together with the weights. The Bellman-Ford algorithm uses the notation of $D_i^n =$ cost of minimum path from the vertex i to the source vertex s using n edges. Hence, we have $D_s^n = 0, \forall n, D_i^s = \infty, i \neq 0$. If there is no edge between vertex i and j, $w(i, j) = \infty$. The Bellman-Ford algorithm conducts these operations for each n:

- $D_i^{n+1} = \min_j(w(i, j) + D_j^n), \forall i \neq 0$.
- terminates after at most m iterations, where m = the number of nodes. No negative-weight cycles imply $D_i^m = D_i^{m-1}$.

Observe that Dijkstra's algorithm computes the shortest path with respect to link weights for a single connection at a time. This can be very different to the paths that would be selected when a batch of connections between a set of endpoints is requested for a given optimising objective. Due to the complexity of some of the routing algorithms (high dimensionality, and non-linear integer programming problems, for example) and various criteria by which one may optimise the network, it may not be possible or efficient to run a full set of these versatile routing algorithms in a distributed fashion on every network node. However, it may still be desirable to have a basic form of path computation capability running on the network nodes, particularly in restoration situations, where quick recovery is required.

Dynamic routing and wavelength assignment

In WDM networks, for example, one can install a full set of routing algorithms in nodes that are lightpath-termination capable (lightpaths are explicitly routed), and equip the rest of the nodes with only basic routing algorithms. Such an approach is in line with the use of MPLS for traffic engineering, but is quite different to the standard OSPF or IS-IS usage in IP networks where all nodes must run the same route computation algorithm. Comparing with static routing and wavelength assignment presented in Chapter 4, dynamic routing and wavelength assignment must deal with dynamic and short-lived lightpaths and must be able to compute the route for a lightpath request in real time. Due to the complexity involved in the routing and wavelength assignment problem (a problem formulation can be found in Chapter 4), dynamic routing and wavelength assignment can employ heuristic-based algorithms for lightpath determination.

Figure 6.9 shows the possible alternatives to form a dynamic routing and wavelength assignment algorithm. Based on the wavelength interchange capability in the physical network, the algorithm can be set to achieve certain objectives. Example objective functions including minimising the number of hops in a full wavelength interchangeable WDM network, minimising the number of wavelength conversions in a WDM network with limited number of wavelength converters, or minimising the lightpath request blocking probability in a WDM network with wavelength continuity constraints.

To save time, path computation for each node pair can be conducted beforehand and the source-orientated routing tree is stored at the source node. The primary stored path may not be connected in a dynamic environment for a number of reasons such as no wavelength available or a link cut. To overcome this, alternate paths can also be computed and stored together with the primary path. Once a lightpath request is received, the algorithm tries to allocate wavelengths along the primary

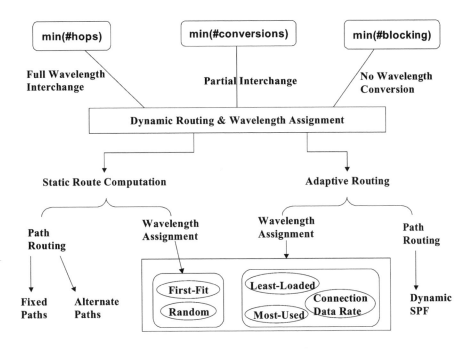

Figure 6.9 Dynamic Routing and Wavelength Assignment.

path. If this fails, the algorithm searches the alternate paths and attempts to find an alternate path that can be set up based on the network wavelengths availability. In the presence of a distributed topology discovery and routing protocol (for example, OSPF for IP networks), link state information can be collected and link state updates can be disseminated within the convergence time. As such, adaptive routing can be performed using the dynamic link state and topology information. When a link path is determined through either stored static routes or adaptive routing, the next step is to allocate wavelength channels to the link path.

We enumerate five methods in Figure 6.9. Wavelength selection based on these methods differs in respect to the current path and future lightpath blocking probability. Two simple methods are random wavelength selection and wavelength first-fit selection. The random method selects a wavelength from the available wavelength set randomly. The first-fit method selects the smallest wavelength index from the available wavelength set, where the wavelength indexes (integer value) can be set according to certain criteria. For example, the wavelengths are ranked in ascending order according to their lambda and the ranks are assigned to the wavelength indexes. Dynamic wavelength assignment methods such as least-loaded, most-used, and connection data rate require the knowledge of a dynamic network environment, so they are applicable to adaptive routing. The least-loaded method selects the least loaded wavelength route over the alternate paths between the given node pair. The most-used method allocates the most utilized wavelength to the lightpath. The connection data rate method assigns the best-fit wavelength to the lightpath request according to the connection data rate.

Lightpath protection

Another issue specific to WDM lightpath routing is lightpath protection. Because of the circuit nature of the reconfigurable WDM network, it is sometimes required that a backup lightpath can take over, should a fault or failure occur with the primary lightpath. In this way, traffic between the two endpoints can keep flowing regardless of the failure. When exclusive protection is used, a backup path, which does not traverse any link that is part of the same SRLG (shared risk link group) as links in the primary path, must be computed. Thus, it is essential that the SRLGs in the primary path be known during alternate path computation, along with the availability of resources in links that belong to other SRLGs.

The backup path routing algorithms are also known as disjoint path algorithms. There are two groups of disjoint path routing algorithms: shortest pair of disjoint paths and maximally disjoint paths. The shortest pair disjoint path algorithms are able to compute node or link disjoint paths. In the node-disjoint pair algorithm, for a given node pair, the algorithm first finds the shortest path for a given network graph, then deletes the links incident on the (found) shortest path nodes (except the endpoint nodes) from the given graph, and finally finds the shortest path (i.e. the node disjoint path) on the residual graph. In the link-disjoint pair algorithm, for a given node pair, the algorithm first finds the shortest path for a given graph, then removes the (found) shortest path links from the given graph, and finally finds the shortest path (i.e. the link disjoint path) on the residual network graph. The maximally disjoint path algorithms aim for searching the shortest pair of paths that are maximally disjointed, i.e. the pair of paths with a minimum number of common nodes and links.

WDM switching constraints

The computation of a primary route for a lightpath within a WDM network is essentially a constraint-based routing problem. The constraint is typically the availability of the wavelength required for the lightpath along with administrative and policy constraints. Although there are various dynamic routing and wavelength assignment algorithms as we discussed in the previous section, the most deployed routing algorithm is still the constraint SPF (CSPF). Possible constraints in WDM optical networks for dynamic routing and wavelength assignment include:

- wavelength availability
- wavelength interchange capability
- fibre bandwidth
- number of wavelengths per transport interface
- switching fabric type: opaque or transparent
- client signal format requirement or line rate
- fabric port multiplexing capability
- multiple fabric boundary constraint
- quality of signal constraints, such as optical signal to noise ratio (OSNR), wavelength dispersion, preferred wavelength with less crosstalk, path stability, for example, avoiding the fast switching fibre.

Since current IGPs (Interior Gateway Protocols) such as IS-IS (Intermediate System-Intermediate System) and OSPF (Open Shortest Path First) are topology-driven, route selection is based on the shortest path computations using simple additive link metrics. This means that the routing protocols do not consider the characteristics of offered traffic and network capacity constraints. Recent advances in MPLS open new possibilities to overcome some of these limitations by allowing users to establish explicit route LSPs (Label Switched Paths) at the edge devices. However, in order to achieve this, IGPs have to be modified so that they can distribute extra information for path computations in addition to the topology information. In optical WDM networks where there are more capacity constraints, for example, limited transparent wavelength interchange and limited label swapping, several WDM extensions to these IGPs have to be made so that the edge switch can establish appropriate optical paths.

6.3.3   OSPF Extensions

OSPF or OSPF like protocol can be used for the construction and the maintenance of a highly available global routing information base. The topological information is used by the CSPF module to compute the constraint-based optical paths. Most of the optical extensions to OSPF are the modifications (or definitions) of the message formats in the form of opaque LSAs. The routing module, for example, OSPF, can interact with an external Traffic Engineering (TE) module through a TE database and the optical switch through the Optical Switch Control Protocol (OSCP) as shown in Figure 6.10. OSCP is introduced later in this chapter. The routing module periodically collects the values of the various optical parameters through OSCP messages and floods the information to other switches/routers so that each switch can maintain its own view of the TE database. An external TE module can retrieve the TE database information that has been built by the routing protocol via a resident TE agent. SNMP can be used for accessing the TE database if the database is constructed according to standard MIB formats.

Opaque LSA

Every optical switch/router in an OSPF routing area (an AS may consist of multiple areas) maintains a distributed database that describes the routing topology. In order to build the link state database, every router is responsible for describing its local piece of the routing topology in link state advertisements and floods it to all routers in the routing domain. To support traffic engineer, OSPF defines enhancements to the standard LSAs, known as **opaque LSAs**. Opaque LSAs provide a generic mechanism to encapsulate application-specific information in a specific opaque type in order to send and receive application-specific information.

The opaque LSAs, like any other LSAs, use the link state database distribution mechanism to flood the information throughout the topology. Traffic engineering in conventional IP network utilizes this capability to distribute traffic-related parameters. In an optical network, the same mechanism is used to distribute the WDM and optical-specific parameters. The format of the opaque LSA packets is presented in this section. Currently, two types of opaque LSAs have been defined: **capacity-**

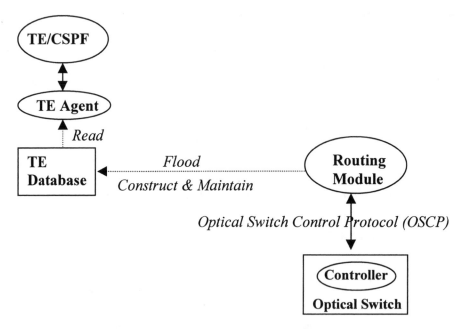

Figure 6.10   Network control process interactions.

related and **traffic-related LSAs**. Opaque LSAs consist of a standard LSA header followed by a 32-bit aligned application-specific information field. Figure 6.11 shows the format of opaque LSA.

The first 16 bits store the advertisement's age. The value starts from zero when the LSA is first issued, and it increments by one each second from that point forward. When an advertisement's age reaches 3,600 (one hour), it is considered out of date. The options byte identifies the capabilities of the router that generated the LSA. It has the same format as the options field of the *Hello* packet. The Link State Type byte distinguishes the various types of link state advertisements. Opaque LSAs use 9, 10, and 11 in this field based on the flooding scope associated with it. For example, the scope of flooding may be link-local (type 9), area-local (type 10), or the entire OSPF

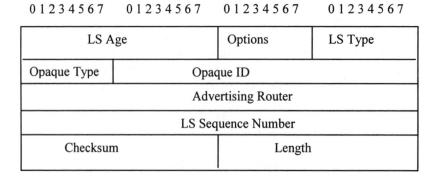

Figure 6.11   Opaque LSA header format.

routing domain (type 11). The link state ID field in the standard LSA header is divided into an opaque type field (8 bits) and an opaque ID field (the remaining 24 bits). Opaque type values in the range of 0–127 are allocated by IETF and opaque type values in the range of 128–255 are reserved for private and experimental use. The opaque ID uniquely identifies each link. To distinguish a reissued advertisement, routers use the link state sequence number. Each time the advertising router reissues an advertisement, it increments the sequence number. The link state Checksum field is a special error-detection code known as Fletcher's checksum. It covers the entire advertisement except the link stage age field. The length field indicates the size, in bytes, of the advertisement. The final opaque information field is the place where we define application-specific information. The WDM-specific information can be encoded in this field. In what follows, we present the format of several example LSAs. Other information may be possibly defined using the same format.

Optical switch capacity-opaque LSA

WDM optical switches have more capacity constraints than the electrical routers. For example, it is possible that some of the intermediate optical switches do not support the wavelength interchange capability or label swapping capability. Furthermore, the number of wavelengths within a fibre, or the number of parallel fibres between two adjacent optical switches can be one of the parameters to consider when we compute a better routing path. In this opaque LSA, we define three parameters related to the capacity of each optical switch:

- wavelength interchange capability;
- label swapping capability;
- the total number of wavelengths between two adjacent optical switches (including the parallel fibres).

Figure 6.12 shows the format of this opaque LSA. Only the payload information is presented since every opaque LSA uses the same LSA header format. If the port # field is –1, the following capacity items represent the capabilities of the entire system (for example, the total number of ports installed).

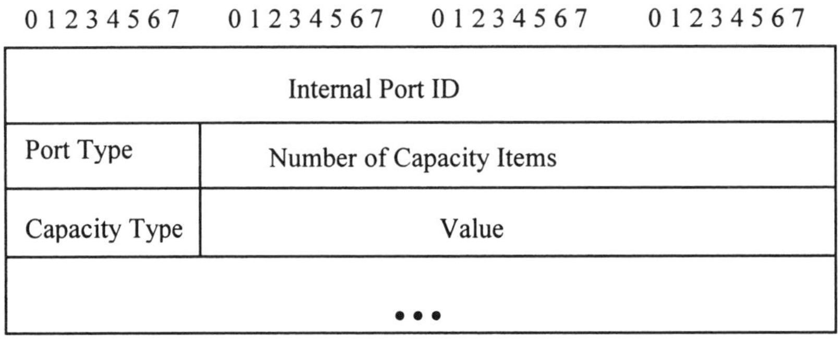

0 1 2 3 4 5 6 7	0 1 2 3 4 5 6 7	0 1 2 3 4 5 6 7	0 1 2 3 4 5 6 7
Internal Port ID			
Port Type	Number of Capacity Items		
Capacity Type	Value		
•••			

Figure 6.12 Capacity opaque LSA payload.

Message fields

- *Opaque type* = 128; this type of opaque LSA has the value of 128.
- *Internal port ID* – 4 octets; this field specifies the internal port ID.
- *Port type* – 1 octet, internal port type. The following types can be defined:
    - Type 1: Input Port of a Transport Interface
    - Type 2: Output Port of a Transport Interface
    - Type 3: Input Port of a Client Interface
    - Type 4: Output Port of a Client Interface.
- *Number of capacity items* – 3 octets, the number of capacity items enclosed in this packet.
- *Capacity type* – 1 octets; value – 3 octets. The following capacity types have been defined:
    - Type 1: Wavelength Interchange Capability,; Value: 1 (Yes), 0 (No)
    - Type 2: Label Swapping Capability; Value: 1 (Yes), 0 (No)
    - Type 3: Number of wavelengths; Value: Total number of wavelengths
    - Type 4: Number of ports in the system; Port #: –1, Value: Total number of ports.

Optical switch connection-opaque LSA

A reconfigurable optical switch distinguishes the client interface from the physical transport interface. The capacity-opaque LSA floods the fabric configuration. In circuit switched WDM networks, there are cross-connects at local fabric that reserve the ports and wavelengths. In addition, optical connections are usually associated with protection/restoration schemes. In such a case, the local cross-connect information needs to be flooded to inform connection managers on local fabric usage and connectivity. Figure 6.13 shows the format for this opaque LSA. As in capacity opaque LSA, this LSA is also preceded with the common opaque LSA header.

The message fields in an opaque LSA for optical switch connection are defined as follows:

- *Opaque type* = 129; this type of opaque LSA has the value of 129.
- *# Cross-connects* – 4 octets; this field specifies the total number of cross-connects at NE encoded in this packet. A NE may have more than one fabric. A cross-connect can be distinguished by its source and sink port ID.
- *Trail ID* – 4 octets; this field represents the trail ID. The trail ID and the source NE of the trail together provide a unique identification of the end-to-end lightpath.
- *Upstream/downstream NE ID* – 4 octets; this field specifies upstream/downstream NE ID of the cross-connect.
- *In/out port ID* – 4 octets; this is the source/sink Port ID of the cross-connect.
- *In/out wavelength ID* – 2 octets; this field represents the wavelength ID of the source/sink port of the cross-connect. This will also indicate the upstream/downstream link wavelength ID.

01234567   01234567   01234567   01234567

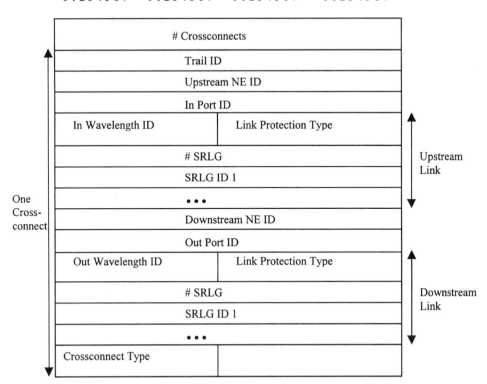

Figure 6.13   Connection opaque LSA payload.

- *Link protection type* – 2 octets; this field specifies the link protection type of the upstream/downstream wavelength link. The following link protection types have been defined:
    - Type 1: Unprotected
    - Type 2: Shared
    - Type 3: Dedicated 1 + 1
    - Type 4: Dedicated 1:1.
- *# SRLG* – 4 octets; this is the total number of SRLG (Shared Risk Link Group) associated with the upstream/downstream wavelength link.
- *SRLG ID* – 4 octets; this field specifies the SRLG ID. A set of links or wavelength channels constitutes a SRLG if they share a resource whose failure may affect all channels in the set. SRLG can be defined differently from system to system. The defined SRLG is then configured accordingly. For example, all wavelength channels over a fibre share the same SRLG ID. A link or channel may belong to multiple SRLGs.

Traffic-related opaque LSAs

In addition to the capacity and connection-related parameters, traffic-related parameters are also needed for WDM network traffic engineering. For example, the total
number of dropped packets between two adjacent OLSRs should be distributed to
edge OLSRs so that they can use this information to compute a better optical path.
Figure 6.14 shows the format of the traffic-related opaque LSA. This LSA is designed
to disseminate NE-related traffic engineering information. If the port # field is −1, the
following traffic parameters will be the items related to the entire system.

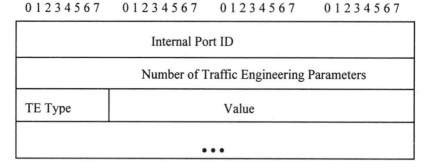

Figure 6.14 Optical traffic engineering NE opaque LSA payload.

The message fields of a traffic-related opaque LSA are defined as follows:

- *Opaque type* = 130; this type of opaque LSA has the value of 130.
- *Number of traffic engineering parameters* – 4 octets, the number of traffic engi-
  neering parameters enclosed in this packet.
- *TE type* – 1 octet; value – 3 octets. The following TE types can be defined:

  - Type 1: Total number of dropped packets; Value: the total number of dropped
    packets per minute
  - Type 2: Average packet duration; Value: the average duration
  - Type 3: Signal quality; Value: the highest data rate supported by the port
    (used in optical packet switching).

Figure 6.15 shows the link opaque LSA for optical traffic engineering. This LSA is
mainly used to disseminate link-related traffic engineering information. It assumes
that a link can be formed by parallel fibres. Again, the LSA is preceded by the
common opaque LSA header.

The message fields of a link opaque LSA for optical traffic engineering are defined
as follows:

- *Opaque type* = 131; this type of opaque LSA has the value of 131.
- *Link ID* – 4 octets; the link identification. In the case of point-to-point WDM
  connection the link ID is the neighbour WDM NE IP address.
- *Fibre ID* – 4 octets; the fibre identifier. When there are parallel bi-directional
  fibres, the fibres are uniquely identified by their fibre IDs. A common naming
  scheme can be adopted for parallel fibres between two NEs.

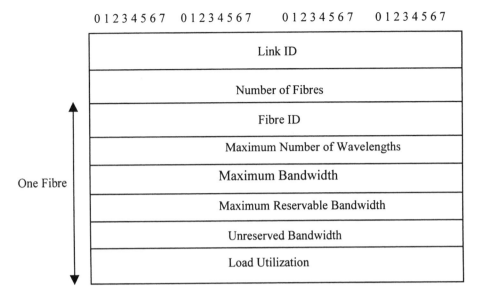

Figure 6.15 Optical traffic engineering link opaque LSA payload.

- *Maximum number of wavelengths* – 4 octets; this represents the maximum number of wavelength channels that can be supported on this fibre.
- *Maximum bandwidth* – 4 octets; the maximum bandwidth can be used on this fibre.
- *Maximum reservable bandwidth* – 4 octets; the maximum bandwidth can be reserved on this fibre.
- *Unreserved bandwidth* – 4 octets; the amount of bandwidth not yet reserved on the fibre.
- *Load utilization* – 4 octets; the current fibre link load utilization.

6.3.4 Routing Behaviour

A routing protocol implements the mechanisms for disseminating routing information within a network, and based on this routing information the route computation algorithm determines how the traffic is routed across the network. How the routing protocol behaves and how the routing algorithm performs in the network is another interesting and important area, also referred to as routing behaviour. The detailed performance study of Internet routing is beyond the scope of this book. We discuss several common routing behavioural issues in this section.

Routing loops

A well-known problem of distributed routing is routing loops, which exist due to the nature of parallel decision making in a distributed environment. A routing loop can be either an information loop or a forwarding loop according to its cause. Figure 6.16

shows examples of information and forwarding loops. An information loop forms when the up-to-date information is derived from the information it provided earlier. In the example of Figure 6.16(a), each node maintains a distance vector, which describes the distance to all other nodes of the network in number of hops. When the edges AB and BC are down, nodes A and C assume the other has a path to node B, but in fact their decision is based on the information they provided. RIP has this problem, so it defines the maximum number of hops to reach a destination, for example, 16. When this number is reached, a node assumes that the destination node is unreachable. A forwarding loop forms when the traffic forwarded by a router is eventually returned to the router. Figure 6.16(b) shows an example of forwarding loop among nodes C, D, and E.

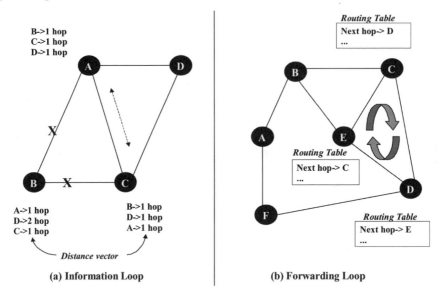

(a) Information Loop   **(b) Forwarding Loop**

Figure 6.16   Routing loops.

A routing loop can be either a persistent or a temporary loop according to the duration of the loop. Temporary loops are more common and cause less damage or performance degradation than persistent loops. Temporary looping can be caused by failures and it can rapidly lead to congestion on network components. Persistent loops are more serious and may require shutdown of part of the network. Persistent loops not only cause congestion and waste network resources but also potentially break down operational services.

Routing oscillation

Routing oscillation refers to irregular routing patterns or the fluttering of the route paths. Routing oscillation is desirable in certain circumstances. For example, it may be the effect of load balanced routing, which splits traffic onto different routing paths. OSPF is equipped with ECMP that employs a fair sharing allocation scheme among equal cost paths to the destination. Also, deflection routing may cause routing oscillation. In a bufferless WDM network, deflection routing is particularly attractive since it can be used as a contention resolution scheme. In conventional IP networks, deflection routing is employed to avoid congestion and packet dropping.

However, routing oscillation can cause the following problems. First, it results in unstable network paths. The applications using the paths suffer the most. The round trip time and available bandwidth are no longer traceable. Hence, the application QoS or the SLA may not be guaranteed. Second, when TCP traffic splits over different paths, the traffic may arrive out-of-order at the destination. Out-of-order reception often causes timeouts and adds to delay at the receiver. In the case of timeout, a retransmission wastes network resources and may face the same challenge of fragments serialisation at the receiver. Third, routing oscillation usually increases the degree of traffic asymmetry, which makes traffic engineering and network optimisation difficult. In the example network of Figure 6.17, node G performs load balance routing, where it splits traffic of the same destination onto different paths. When congestion is detected over the edge CF, node F employs deflection routing to avoid the congestion.

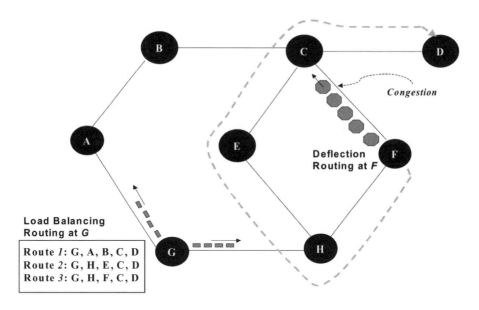

Figure 6.17   Routing oscillation.

Routing failures

Routing in a highly distributed system may fail for several reasons;

- Routing failures due to control software failures. For example, a protocol daemon crashing. Another example is different versions of the same protocol existing in the network.

- Routing failures due to hardware failures.
- Routing failures due to temporary outages. Temporary outages may be caused by heavy congestion or by temporarily losing connectivity. WDM network reconfiguration alters connectivity that might also cause temporary outage.
- Unreachable destination due to too many hops. For instance, an IP datagram has a TTL field in its header, which times its existence in terms of the number of hops that can be traversed in a network.

Routing stability

Routing stability refers to how often the routes change and how stable they are over time. We introduce two definitions of routing stability below:

- **Prevalence:** routing stability can be defined in terms of prevalence that is the probability of observing a given route during a given time frame. When we observe a route, prevalence indicates how likely we will see the same route again in the future. Higher levels of prevalence mean a more stabilised routing environment.
- **Persistence:** routing stability can be also defined in terms of persistence, that is the time frame that a given route persists before it changes. When we observe a route at one particular time, persistence tells us how long after that time that the route is likely to change. Higher levels of persistence means a more stable routing environment.

Prevalence and persistence can be coupled to describe the routing stability. High levels of prevalence and persistence represent a very stable network; high levels of prevalence but low levels of persistence indicate an unstable network possibly with routing loops; low levels of prevalence but high levels of persistence represent a stable network; low levels of prevalence and persistence indicate a very unstable network. An unstable routing environment is likely to introduce these problems:

- The properties of the network paths are unpredictable.
- The experience and the performance study based on previous measurements are unreliable.

6.3.5   Routing Scalability

A desirable routing solution or architecture should be scalable, i.e. it can deliver required performance even when the network scales to a certain large size. For example, OSPF is known to be scalable to at least 200 nodes. Scalability should be given in terms of

- type of network which could be, e.g. an IP network or optical circuit switched network,
- network topology, which is a combination of links and nodes.

Routing scalability can be measured in

- response time: once the routing server receives a request, how fast can it compute and provide the routing path.
- availability: what is the waiting time of a request before it is processed by a routing server.

Path optimality refers to the quality and accuracy of a routing path. Although path optimality is always desirable, routing scalability has different objectives as listed above. Figure 6.18 shows how to design a scalable routing solution.

First, is to employ routing hierarchies. One level of routing hierarchy is also known as a flat routing system in which every node peers with any others. In a hierarchical routing system, certain nodes form a routing backbone. Traffic from non-backbone nodes travel to their designated backbone nodes; then they are sent through the backbone to reach the general area of the destination. They continually travel from the last backbone node through one or more non-backbone nodes to the final destination. Hierarchical routing systems name logical groups of nodes as domains, areas, or autonomous systems. Hence, hierarchical routing means that certain nodes in a domain can communicate with nodes in other domains, while others can communicate only with nodes within their domain. Multiple routing hierarchical levels may exist, with nodes at the highest hierarchical level forming the routing backbone. With hierarchical routing, intra domain nodes need to know only about other nodes within their domain, their routing algorithms can be simplified, routing related computations or tasks can be reduced, and routing update traffic can be condensed accordingly.

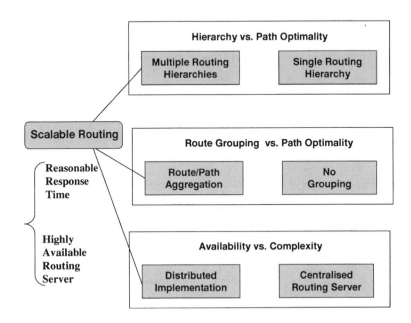

Figure 6.18 Routing scalability.

For example, OSPF version 2 supports two levels of hierarchy and groups routers into areas (see Chapter 3 for details). Private Network to Network Interface (PNNI) is another link state routing protocol standardised by the ATM Forum. PNNI implements hierarchic partitioning of networks into Peer Groups (up to 104 levels). The PNNI hierarchy is configured automatically based on switch addresses using dynamic election protocols.

Second, is to use route/path aggregation. Routing information aggregation within each peer group can significantly reduce routing traffic. Both BGP and PNNI support path aggregation. OSPF supports summary LSAs in addition to router LSA and link LSA. An Ethernet network as a group only generates and advertises one summary LSA (by its designated router). Route/path aggregation and routing hierarchies simplify object entities and reduce related information flow, but the abstraction may hinder path optimality.

Third, is to support distributed implementation of routing servers or daemon processes. The distributed approach provides high availability through multiple presences of servers. However, these parallel servers require coordination and synchronization, and distributed routing information bases require consistency. Hence, it is much more complex than its centralised counterpart.

6.4 IP/WDM Signalling

At the physical layer, signalling refers to the transmission of an optical signal over a fibre; at the network control layer, signalling involves a series of distributed processes to accomplish certain tasks, for example, establishing a circuit across the network. We use RSVP (Resource Reservation Protocol) and its optical extension as an example to illustrate the signalling process and operation in IP/WDM networks.

6.4.1 RSVP Overview

RSVP is an IP-based signalling protocol that enables IP applications to obtain certain QoS for their data flows in IP networks. RSVP reserves resources with QoS requirements through flow specification, which guarantees how inter-networking handles the application traffic. The flow specification includes the QoS requirement (RSpec), a description of the traffic flow (TSpec), and the service class of an application. Through reservation, certain parameters in the node packet scheduler and classifier are set to specific values so that application flows are provided with the corresponding services when they arrive. There are different styles of reservation based on the functionality provided by the resource manager (the RSVP daemon). Reservation can be either distinct in which each reservation of a session from a sender is processed separately, or shared in which all reservations are processed as one reservation (since they are known not to interfere with each other). A distinct reservation can also be specified with an explicit scope through fixed-filter, for example, by explicitly specifying a sender.

Once the routing path is determined through a routing protocol or an external traffic engineering application, the host RSVP daemon starts a session and sends a

RSVP path message along the routing path to the destination. When the path message reaches the destination, the receiver initiates the reservation and sends a reservation request message along the reverse routing path to the sender. During reservation, each node including the receiver and the sender is responsible for choosing its own level of reserved resources, a process known as **admission control**, to determine whether it can supply the desired QoS. If the admission control succeeds, the corresponding parameters at the node's packet classifier and scheduler are set and the reservation request message relays towards the data source. If the admission control fails, an error message is sent to the source. RSVP defines four groups of messages:

- *Path messages,* containing the state of the path in each node, are sent hop-by-hop along the route path from the source towards the destination.
- *Reservation request messages,* detailing the reservation information, are sent hop-by-hop along the route path from the destination towards the source.
- *Tear-down messages,* which are used to remove the path and the reservation without waiting for the cleanup timer to expire. The two tear-down messages are:
  - path tear-down messages;
  - reservation request tear-down messages. The sender, the receiver, or any router can initiate the tear-down messages.

- Error and confirmation messages, which include:
  - path error messages are used to report errors for path messages, and sent hop-by-hop along the route towards the source host;
  - reservation request error messages are used to report errors for reservation request messages, and sent hop-by-hop along the route towards the destination;
  - reservation request acknowledgement messages are used to acknowledge the reservation request message, and sent hop-by-hop along the route towards the receiver.

Since RSVP is used for application flows, it is unidirectional. A bi-directional path has to be set up by running RSVP signalling twice, reversing the position of the sender and receiver. RSVP channels are soft state maintained, in which channel states are maintained at each node and applied with timers. These channels should be periodically refreshed through path and reservation request message updates. Otherwise, when the timer expires, the channel states will be deleted and the resource will be released.

In IP networks, not every router supports RSVP. For those do not support RSVP, the QoS orientated resource reservation simply cannot be guaranteed. Assuming the non-RSVP routers do not present any QoS problems, RSVP has a tunnelling operation to traverse the non-RSVP network clouds. Tunnelling requires RSVP and non-RSVP router forward path and reservation request messages based on the local routing table. Therefore, the path and reservation request messages are relayed by both RSVP and non-RSVP routers hop-by-hop between the source and the destination. For example, when the path message is transported across a non-RSVP cloud, the

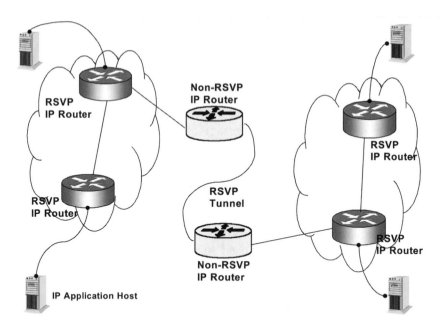

Figure 6.19   RSVP tunneling.

copy of the path message containing the last RSVP router address is passed along based on its destination address. Figure 6.19 shows an example of RSVP tunnelling.

6.4.2   RSVP Extension for Optical Networks

Optical packet switching possesses similarities with a conventional IP network since both of them use in-band control, a store and forward paradigm, and switch fine granularity traffic. However, pure optical packet switching is not mature yet as there is no optical buffer available. Fibre delay lines can be regarded as a primitive alternative to an optical buffer but they are not random access memories and certainly cannot implement sophisticated packet scheduling and classification functions. Optical burst switching provides similar functionality to optical label switching but the latter uses subcarrier wavelengths to carry the control information, reducing the synchronisation complexity between the header and the payload. In the rest of the section, we focus on reconfigurable WDM networks and optical label switching networks.

Figure 6.20 shows an example optical WDM network for optical RSVP implementation. As shown in the figure, the control plane is separated from the data plane. Control signalling and information are propagated in the control plane and actual data travels in the data plane. In the figure, the network data plane has six WDM switches – four edge and two core nodes. Likewise in the control plane there are six control nodes.

Each WDM switch has a corresponding control node and the control nodes are

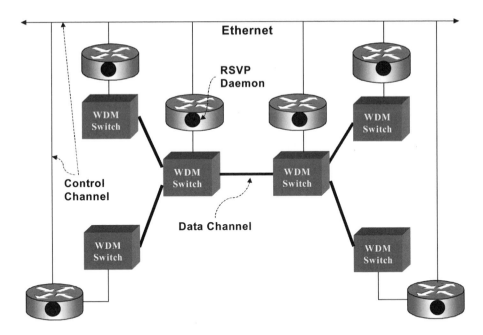

Figure 6.20   An example WDM network for optical RSVP.

physically connected using point-to-point Ethernet links. Data channel topology and other control-orientated information are transported over the control channel. As we discussed before, LMP is responsible for data plane to control plane mapping and property correlation. The signalling protocol such as RSVP is used to set up or tear down end-to-end optical label switch paths. The node at the edge that initiates the RSVP signalling for LSP setup is called the ingress node; the node where the LSP terminates is known as the egress node.

In Figure 6.20, the connectivity among the control nodes are shown in thin lines to indicate that these are the control channels and are not meant for data plane communication. RSVP daemons are installed at each control node. These RSVP daemons communicate with each other using RSVP PATH and RSVP RESV messages. The PATH message is initiated by the RSVP daemon at the ingress node, whereas the RESV message is initiated by the RSVP daemon at the egress nodes. As we recall, a router as well as a computer host can initiate RSVP.

6.4.3   RSVP Extension Implementation Architecture

Figure 6.21 illustrates the software architecture of RSVP at each control node. As we discussed earlier, each control node hosts one RSVP daemon among other software modules such as routing protocol daemon and optical switch control protocol (OSCP). RSVP at the ingress node provides an API, known as RAPI (RSVP API), to allow interaction between the RSVP daemon and the user application. Using this RAPI, a user application can initiate RSVP signalling to set up LSP between a pair of

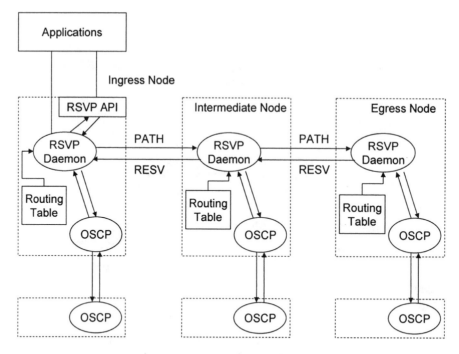

Figure 6.21   RSVP software architecture.

ingress and egress nodes with the help of MPLS-based label allocation and distribution technique.

When signalling is initiated at the ingress node by the user application, the RSVP daemon sends a PATH message to the RSVP daemon at the next control node. In order to obtain the route to the next control node, it consults the routing table that is prepared by the routing daemon running the OSPF protocol. RSVP can also support explicit routing in which path details are carried in the RSVP packets (as in source routing). Once the PATH message reaches the egress node, it generates a RESV message, which travels back to the ingress node along the route recorded by the PATH message. Labels are allocated and distributed according to the scheme explained above. The labels are then sent to the optical switches from each control node through OSCP. RSVP also queries the optical switches for hardware-related information (for example, whether the optical label switch node is label-swappable or not) through OSCP. This information is used for determining the label allocation scheme to be used at a particular optical label switch node.

6.4.4   RSVP Message Extensions

This section defines the message extensions to the RSVP protocol for optical networks. Each RSVP message has a 16-byte RSVP header. The header contains 4-bit version field, 4-bit flags field, 1-octet RSVP type, 2-octet checksum field, 2-octet message length, 2-octet reserved, 4-octet message identifier, 1-octet reserved, and 3-

octet fragment offset field. The RSVP header is appended with RSVP objects. The PATH and RESV messages in RSVP can be used to exchange the WDM specific information and set up/tear down lightpaths. RSVP object has a common 4-octet header. There are 3 fields in the object header: Length (2 octets), Class-Number (1 octet), and Class-Type (1 octet).

PATH label message

In optical label switched networks, based on the label swapping capability of the outgoing port, two types of **label request objects** will be used: a label request with a suggested value and a label request with a possible label range. Figure 6.22(a) and (b) illustrate the formats of the two label request objects.For the two types of the label request objects, the first field indicates the size of the object in bytes. The next two fields, class number and class type, distinguish different objects. RSVP defines 15 different types of objects. For an instance, one can use 21 as the class number for both types. The class type of the label request object with a suggested label value is type 1 and it is followed by a real label value. For the label request object with a possible label range, the class type 2 can be assigned and a list of both minimum and maximum label values follows.

A representative feature of peering model is explicit routing through which an outside peering entity can request an explicit routed path to/through a network cloud. RSVP can be used as the signaling protocol for explicit routing. RSVP PATH message in MPLS defines an explicit route object (ERO) that carries a set of routing hops after the common object header. Each routing hop consists of a route type (strict

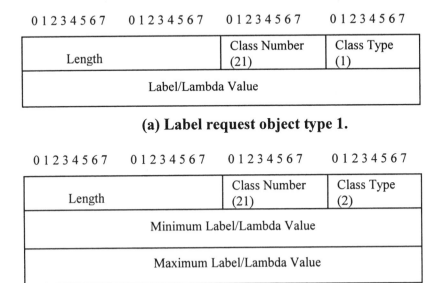

(a) Label request object type 1.

(b) Label request object type 2.

Figure 6.22 PATH message label request object format.

or loose), an IP address, and a route prefix. For optical networks or GMPLS, Figure 6.23(a) shows a format of type 3 of class 21 to encode explicit route information.

Finally, a PATH label request object used for fibre path and local wavelength assignment is defined. Without external routing entities, the routing path from default routing protocol details the fibre path. If the signalling protocol is implemented with certain intelligence, it can make decisions on wavelength assignment. This message extension may need the corresponding signalling process functional extension to accomplish wavelength assignments. The idea is to collect the wavelength availability information in the PATH message and the egress makes a decision on wavelength selection. Figure 6.23(b) shows the format of class type 4. This is useful only in the presence of a wavelength continuity constraint, i.e. the assumption is that no switch in the optical network supports wavelength interchange. During signalling, each node maintains a list of in-use wavelengths and conducts a union operation on the incoming in-use wavelength set and the local in-use wavelength set. The resulting

```
0 1 2 3 4 5 6 7   0 1 2 3 4 5 6 7    0 1 2 3 4 5 6 7    0 1 2 3 4 5 6 7
```

Length		Class Number (21)	Class Type (3)
ER Type	Address Type	Reserved	
Port ID			
Wavelength (Channel/Label/Time Slot) ID			
• • •			

(a) Label request object type 3.

```
0 1 2 3 4 5 6 7   0 1 2 3 4 5 6 7    0 1 2 3 4 5 6 7    0 1 2 3 4 5 6 7
```

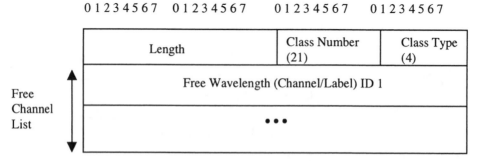

(b) Label request object type 4.

Figure 6.23 PATH message label request object format for explicit path setup and local wavelength assignment.

wavelength set is passed on towards the egress. In such a way, the egress can select an available wavelength channel from end-to-end.

The PATH message fields are encoded as follows:

- *RSVP_LABEL object* has these values: Class-Number = 21, Class Type = 3 or 4.
- *ER Type* – 1 octet; the explicit routing type, "0" for strict explicit route and "1" for loose explicit route. The former specifies each hop one after another for the entire request path. During strict explicit routing, the local routing table is not used. The latter selectively specifies some hops for the request path. RSVP consults the local routing table for route information between hops.
- *Address Type* – 1 octet; this field represents the route-hop address type. The following values can be defined: 0 for domain, 1 for cloud, 2 for network element, 3 for port. It indicates the level of peering. It has a similar functionality as network prefix.
- *Port ID* – 4 octets; this is the port identifier on the network element (i.e. switch). A port ID usually encodes the following information: network, cloud, shelf, circuit pack, port.
- *Wavelength ID* – 4 octets; this field specifies the wavelength ID used on the switching fabric port.
- *Free Wavelength ID* – 4 octets; this field specifies an available wavelength ID/channel on the switch. Each switch provides an update on the free wavelength/channel list so that the destination switch can make a decision on wavelength/channel selection.

RESV message

As mentioned in the previous section, the RESV message has to be modified so that it contains the **label object**. The modified RESV message will contain the label object in the RSVP object field. Figure 6.24 shows the format of the label object. Like the label request object, the first field indicates the size of the object in octets. The next two fields are the class number and class type. For example, 22 can be assigned as the class number. For the class type 1, 2, 3 and 4 label request objects (class number 21), we can use a simple label object whose format is shown in Figure 6.24. The last field in the RESV message label object is the real label or lambda value that has been reserved at the switch. For the explicit routed lightpath (with the label request object class number = 21 and class type = 3), the lambda

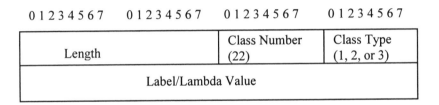

Figure 6.24   RESV message label object format.

value of the label object in fact is the in-wavelength ID of the corresponding label request object. For local assignment lightpath (with the label request object class number = 21 and class type = 4), due to the wavelength continuity constraint, the label request object has the same lambda value from egress to ingress.

6.4.5   Hybrid Label Allocation Scheme for Optical Networks

In theory, label allocation in MPLS can be conducted in downstream orientated or upstream orientated fashion. Downstream label allocation means that a request to bind labels to a specific LSP (Label Switched Path) is initiated by an ingress node through the RSVP PATH message. Then, labels are allocated downstream and distributed (propagated upstream) by means of the RSVP RESV message. In (electronic) MPLS where the label swapping capability is implicitly assumed, the allocation of a label value is trivial since the significance of the label value is only local between two adjacent OLSRs and we need to choose any label value that is not currently used. However, in optical networks, for example, OLS, some optical switches such as OLSR may not support the label swapping capability and each switch has to maintain the same label value up to the switch that can handle label swapping. By comparison, upstream label allocation means labels are in fact allocated at the ingress first, and the allocation propagates towards the egress. Once the egress allocates its label, it can send a confirmation message to its previous hop to trigger a hop-by-hop confirmation towards the source. Or the egress can simply send a notification to the ingress directly to inform that the path has been set up.

A hybrid label allocation scheme adaptively using both downstream and upstream label allocation is introduced. It is presented in the context of OLS networks in particular to address the problem of lacking optical label swapping capability. Optical label swapping can be implemented in the optical domain but it is very expensive. So an operational OLS network usually involves both label-swapping and non-label-swapping nodes. Using the conventional upstream or downstream label allocation in such a network, the signalling messages become very complex and the signalling process itself is also complicated. Instead, without complicating the signal message, we define a simple allocation rule that adaptively uses upstream and downstream allocation.

The hybrid label allocation scheme works as follows:

- When the downstream node is swappable and the upstream node is non-swappable, downstream label allocation is performed.
- When the downstream node is non-swappable and upstream node is swappable, upstream label allocation is performed.
- When the downstream node is non-swappable and upstream node is also non-swappable, upstream label allocation is performed.
- When the downstream node is swappable and upstream node is also swappable, downstream label allocation is performed.

Figure 6.25 illustrates an example scenario of OLS label allocation and distribution under this scheme. As shown in this figure, the PATH message is originated from the ingress node and the RESV message is originated from the egress node.

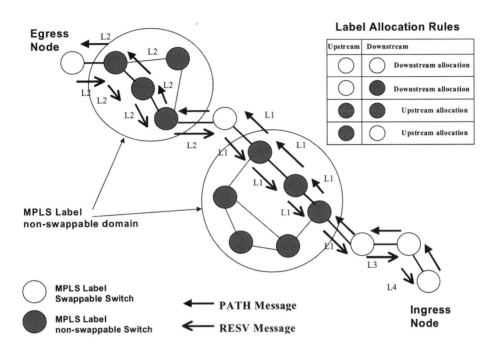

Figure 6.25 Hybrid label allocation scheme.

The label allocation is performed according to the scheme described above. Label L1 is allocated by the first label non-swappable node. This label is propagated until the PATH message reaches a label swappable node. Another label non-swappable upstream node allocates another label L2, which is propagated by the PATH message until it reaches the egress node. Since the egress node has already been assigned the input label of L2, this label is carried by the RESV message to the upstream nodes until a label swappable node is reached. At the label swappable node, another label L1 is allocated and propagated by the RESV message until it reaches another label swappable node. From then on, all the way to the ingress node, all nodes are label swappable. So, for each such node, the RESV message generates a new input label at the downstream node that is used by the upstream node as the output label. These labels are L3 and L4 as shown in the above illustration.

When the size of the label space is large, i.e. a large number of wavelength channels per fibre, the above hybrid label allocation scheme can be improved through label space partitioning in MPLS non-swappable domains. A simple static partitioning scheme is to divide the entire label space evenly into segments and assign each segment to each edge (i.e. ingress node). As such, during label allocation, the ingress node determines the lambda/label from its label space segment. This will guarantee that other switches in the non-swappable domain can support the same lambda/label. This is desirable when the label space is large and there is a high level of concurrency on path setup. However, this approach fragments the label space so the label space may not be fully utilized.

6.4.6 Discussion

IP/WDM networks need to decide the NC&M functions that should be provided by control protocols or by optical network management systems. IP encodes most of the NC&M functionality into network control, which by nature is a distributed process. On the other hand, the Telecom optical networks implement most of the NC&M functionality using a management system, which by nature follows a centralised, hierarchical structure. An interesting question is how much NC&M functionality should be provided using distributed control processes. A signalling protocol as a control mechanism can be relatively simple such as the case in explicit lambda path setup; it can be quite intelligent or complex such as the case in local wavelength assignment.

At the current stage, optical networks are more complex than conventional IP networks. Part of the reason is that the optical network development lacks inter-operability and standardisation among vendors. In such an environment, fully distributed control faces enormous implementation complexity. For example, a hybrid optical network may consist of different types of optical switches and various constraints such as wavelength continuity. Using an IP-centric control plane, the optical-specific information can be disseminated. However, a purely distributed, scalable route computation solution seems difficult as circuit switched lightpaths also require resource reservation. A compromise solution is to share the function implementations between distributed network control and centralised management systems. For example, in circuit switched optical networks, a signal protocol is only used for signalling path setup and the lightpath route computation is conducted by an external (centralised) entity.

A scalable and high performance network control should integrate signaling with routing in which routing (through highly available routing tables) injects intelligence into the network and then signaling performs intelligent travelling and triggers resource reservation or release. A good example of a scalable control system is SS7.

6.5 WDM Network Access Control

WDM access control is responsible for mapping IP packets to wavelength channels. Figure 6.26 shows an example of WDM metropolitan area network (MAN). A WDM MAN can be formed using a ring topology with one point of presence (PoP) to the long haul network and several wavelength access points for the access networks. An important issue in WDM access control at the access point is the packet-to-wavelength mapping function (MF). The design of the MF can be based on the IP side, i.e. examining the IP packet header in particular on these attributes:

- destination address
- source address
- type of service (TOS).

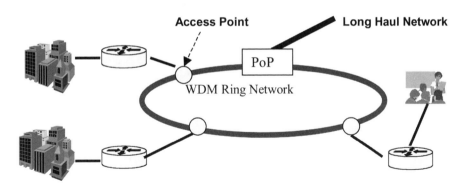

Figure 6.26 A WDM metropolitan area network.

The MF can also be designed to address the WDM characteristics. There may exist multiple paths with different line rate or signal quality between the same source and sink NE. For example, within each fibre, the channels in the middle of the spectrum produce better signal quality than that of channels using the spectrum extremes. The quality of an optical signal can be measured in terms of wavelength power, dispersion, and OSNR. Therefore, the MF can be constructed based on the following components:

- connection rates
- quality of the optical channel
- load on links.

The anatomy of a MF is shown in Figure 6.27, where the source population represents IP packets and the target population represents wavelengths. Assuming there are n wavelengths in the target, $f_1(n)$ packets can be mapped into wavelengths in one step. However, when the mapping is complex and needs to be performed in many points along a packet delivery path, an intermediate population may be desirable. The first step mapping,

$$f_{21}_{n1>n} (n_1),$$

is designed to reflect the common components shared by mapping function along the path, and the second step mapping, $f_{22}(n)$, is customised to the particular needs of each mapping point. Thus, one has a choice to repeat the entire one-step mapping

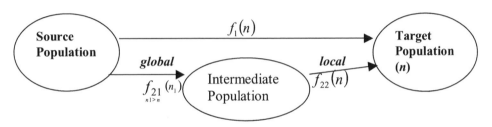

Figure 6.27 Packet to wavelength mapping function.

everywhere, or to perform the first mapping only once at the head end of the path and have the result passed to all other mapping points on the path. An example of using an intermediate population is for aggregating short IP packets. When two-step MF is used, a Wavelength Equivalence Forwarding Class (WEFC) is generated after the first step. WEFC is useful when the classification is needed elsewhere, for example, in the process of forwarding.

6.6   GMPLS

GMPLS is introduced to generalise the MPLS architecture to also consider non-packet-based control planes in addition to the conventional packet networks. GMPLS defines four types of interfaces as shown in Figure 6.28:

- packet switch capable (PSC)
- time division multiplex (TDM)
- lambda switch capable (LSC)
- fibre switch capable (FSC).

These interfaces forward data based on the content of the packet/cell header, the data's time slot, the receiving wavelength, and the receiving fibre, respectively.

Based on the IP/WDM inter-networking architecture, GMPLS can be applied in different models. For packet switched optical networks, an edge LSR is equipped

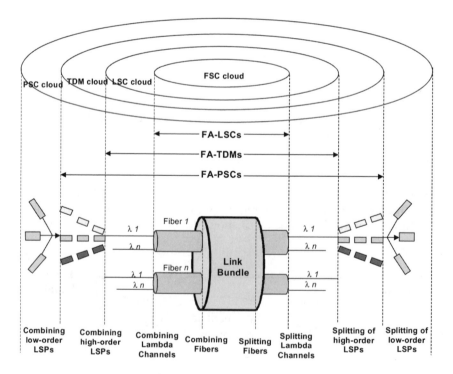

Figure 6.28   GMPLS hierarchy.

with the control and forwarding components of both label switching and conventional routing. When an edge LSR receives a packet without a label, the LSR uses the conventional forwarding component to determine the FEC that this packet belongs to and the next hop. When an edge LSR receives a labelled packet, the LSR uses the label-forwarding component to locate the FEC and the next hop. For traffic engineering, GMPLS inherits multiservice and explicit routing capability from MPLS. GMPLS can be used to support either independent or integrated IP/WDM traffic engineering.

For circuit switched optical networks such as OXC, GMPLS includes the LMP for neighbourhood discovery in optical networks. In terms of signalling, GMPLS extends the base functions of RSVP-TE and CR-LDP to enhance the basic LSP properties. In particular, GMPLS defines a generic Generalised Label Request (GLR) and specialised GLRs. When there are specific characteristics that cannot be carried by the generic GLR, the specialised GLR is used. GLR can be implemented as a new object or TLV in an RSVP-TE path message instead of using the regular Label Request, or in a CR-LDP Request message in addition to the existing TLVs. The conventional MPLS label is extended to a generalised label to allow the representation of not only data packets labels, but also labels that identify timeslots, wavelengths, and fibres. Waveband switching is introduced to switch a set of contiguous wavelengths together to a new waveband. Waveband switching can reduce the distortion on the individual wavelengths and thus allow tighter separation of the individual wavelengths.

For signalling efficiency, upstream label suggestion and restriction are allowed. In addition to the MPLS unidirectional LSP, GMPLS can set up bi-directional LSP, where each of the unidirectional paths share the same traffic engineering requirements including protection and restoration, label switch routers, and resource requirements such as latency. To continue to support explicit routing, the explicit route object is extended to include interface numbers as abstract nodes to support unnumbered interfaces. Embedding labels in an explicit route to control the placement of an LSP is more likely to be used for non-PSC links. To improve routing scalability, GMPLS supports link bundling by which an IP link can consist of a number of bi-directional WDM fibres. IP links without IP addresses are unnumbered links. The GMPLS framework supports unnumbered links.

6.6.1 Discussion

The spirit of GMPLS is to propose a common IP-based control plane for all types of networks including optical circuit switched networks. GMPLS is different from OIF UNI approach in the sense that it calls on network peering instead of the circuit request/delete interface in OIF UNI (overlay). GMPLS peering can be applied to addressing, signaling, and routing. A global addressing scheme is the foundation of network peering; GMPLS proposes IP addressing for all networks. Signaling represents the agreed message formats for automatic circuit setup and teardown; GMPLS suggests the use of RSVP and LDP as the standard signaling mechanism. Routing provides the intelligence into the network through routing tables (so that signaling can provide its best performance as a dynamic, distributed process); GMPLS proposes the use of OSPF and BGP for routing reachability and availability information sharing.

6.7   IP/WDM Restoration

As IP/WDM advances, an important network operational issue is restoration that aims at a resilient network providing network survivability. Survivability is the ability of a network to recover traffic affected by failures. Restoration can be implemented in two fashions, provisioned and non-provisioned. Provisioned restoration is also known as protection. Protection copes with predetermined failure recovery since protection resources are reserved beforehand and can be used by low priority traffic if pre-emption is allowed. Non-provisioned restoration deals with dynamic discovery of alternate routes from the spare network resources for disrupted traffic once the failure is detected. Dynamic restoration has better resource utilization than protection but it requires a longer restoration time and involves extra complexity in signalling and routing.

A restoration facility or mechanism can be provided in respect to the layering in the IP/WDM network and the IP/WDM inter-networking architecture. An IP/WDM network can transport IP/MPLS directly over WDM, or IP over ATM over SONET over WDM. In a multi-layered IP/WDM network, restoration can be provided either in a single layer or in a co-ordinated fashion among different layers. Lower layer protec-

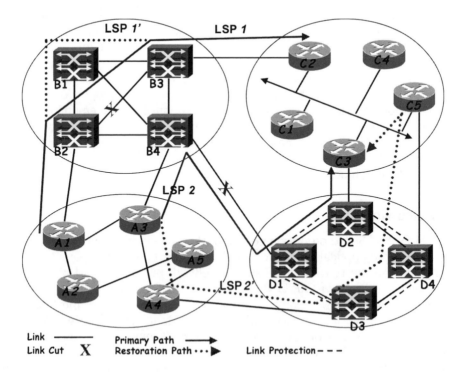

Figure 6.29   IP/WDM restoration.

tion uses primitive protection schemes such as automatic fibre switching in SONET, whereas higher layer restoration aims at service level restoration.

Figure 6.29 shows several types of IP/WDM restoration. In the WDM-D network (bottom-right corner) in the figure, each fibre link is protected by a backup fibre (shown as dotted line in the figure) in the ring, i.e. automatic protecting switching (APS). This type of fibre protection needs only a simple network control and management mechanism and provides fast restoration, for example, less than 50 ms. However, this approach has low utilization as only 50% of the network capacity is used for the traffic. Another type of IP/WDM restoration is path protection/restoration, which is more bandwidth-efficient but slower in terms of restoration time. Path restoration is enforced by the NMS either as user requested or as QoS mechanisms provided by the network. In a WDM network, path protection can be offered in a finer granularity, i.e. the lambda channel. In IP/MPLS/WDM networks, path protection can be designed for virtual channels.

In principle, dynamic restoration can be applied to sub-wavelength channels or very fine-granularity virtual channels, but that level of restoration usually generates a significant amount of control traffic and therefore hinders its 'real time' performance and subsequently scalability. Path protection/restoration has three variants:

- *Link protection/restoration:* In link protection, a dedicated link disjoint backup route is set up in advance during the lightpath or LSP creation process for the primary link. In case of link failure, once its adjacent nodes detect it, the upstream adjacent node switches to the backup path if there is a protection path or dynamically computes a route to replace the damaged link. LSP 1 in Figure 6.29 shows an example of link restoration.
- *Lightpath or LSP protection/restoration:* When a link failure is detected, its adjacent nodes notify the ingress and egress nodes of the lightpaths of LSPs traversing the failed link. The ingress and egress nodes either use the provisioned end-to-end backup path or compute a new end-to-end path based on the current network capacity and condition. Once the backup path is established, the ingress and egress nodes switch from the failed primary path to the backup path. A special kind of lightpath restoration is disjoint-link path restoration, which is designed for multiple simultaneous failures on the primary path. LSP 2 in Figure 6.29 shows an example of disjoint-link LSP restoration.
- *Partial protection/restoration:* Partial protection/restoration provides restoration for a segment of the entire path or several links. In terms of protection scope, it is in between link and lightpath restoration.

Path protection can use dedicated paths or shared paths. Both 1 + 1 and 1:1 have dedicated protection paths, whereas 1:N and M:N employ shared protection paths. In 1 + 1 path protection, a signal is transmitted simultaneously over two disjoint paths and a selector is used at the receiving node to choose the better signal. In 1:1 path protection, a dedicated backup path is reserved and pre-established for the primary path. In M:N path protection, N primary signals are transmitted along disjoint paths, and M backup paths are set up for shared protection switching among the N primary paths. For 1:N protection, N primary paths share the same backup path.

Restoration can be provided in IP/MPLS. Lower layer restoration emphasises network survivability whereas higher layers such as MPLS restoration focus on traffic. Since MPLS binds packets to a path via the labels, it is imperative the MPLS be able to provide protection and restoration of traffic. MPLS supports fast signalling and introduces protection/restoration priority to provide a differentiating mechanism for applications requiring high reliability. IP/MPLS allows more efficient use of network capacity in terms of restoration. IP/WDM restoration involves four steps, fault detection, fault localisation, fault notification, and restoration.

The computation of a primary route for a lightpath within a WDM network is essentially a constraint-based routing problem. The constraint is typically the availability of the wavelength required for the lightpath, maybe along with administrative and policy constraints. The computation of a backup path is essentially a disjoint (diverse) path problem. In terms of graph theory, there are two levels of disjoint: *link disjoint* and *node disjoint*. If there is no single link common to two paths, these two paths are link disjoint. A path traverses a set of nodes. If there is no single node (except the two endpoints) common to the two node sets each of which is traversed by a path, these two paths are node disjoint. Obviously, node disjoint is more diverse than link disjoint. It is solely a design choice of which level of diversity is sought. A reasonable strategy, however, is searching for a node disjoint path first; if not found, look for a link disjoint path. This way may assure maximal level of protection.

It is critical that the SRLGs in the primary path be known during alternate path computation, along with the availability of resources in links that belong to other SRLGs. Assuming the above information is available for each bundle at every node, there are several approaches possible for path computation.

The primary path can be computed first, and the (exclusive or shared) backup is computed next based on the SRLGs chosen for the primary path. The primary path computation procedure can output a series of link groups that the path is routed over. Since a link group is uniquely identified with a set of SRLGs, the alternate path can be computed right away based on this knowledge. However, if the primary path setup does not succeed for lack of resources in a chosen link group, the primary and backup paths must be recomputed. In another case, if the backup path setup does not succeed for lack of resources under the SRLGs constraint, the primary, and consequently the backup path, may be recomputed.

It might be desirable to compute primary paths using bundle-level information (i.e., resource availability in all link groups in a bundle) rather than specific link group level information. In this case, the primary path computation procedure would output a series of bundles the path traverses. Each OXC in the path would have the freedom to choose the particular link group to route that segment of the primary path. This procedure would increase the chances of successfully setting up the primary path when link state information is not up-to-date everywhere. But the specific link group chosen, and hence the SRLGs in the primary path, must be captured during primary path setup, for example, using the RSVP-TE Route Record Object. This SRLG information is then used for computing the backup path. The backup path may also be established specifying only which SRLGs to avoid in a given segment, rather than which link groups to use. This would maximise the chances of establishing the backup. Alternatively, the primary path and the backup path can be computed together in one step. In this case, the paths must be computed using specific link group information.

To summarise, it is essential to capture sufficient information in link bundle LSAs to accommodate different path computation procedures and to maximise the chances of successful path establishment. Depending on the path computation procedure used, the type of support needed during path establishment (for example, the recording of link group or SRLG information during path establishment) may differ.

6.7.1   Provisioning Case Study

Figure 6.30 shows an example of lightpath provisioning and link provisioning. The figure uses an IP-centric control plane in which each switch is IP addressed and controlled by OSPF and RSVP. With respect to routing, the switches are controlled by one OSPF instance. With respect to management, the switches are grouped into sub-networks, for example, according to geographical locations. Each subnet has an EMS (Element Management System). The crown network is managed by a NMS (Network Management System). Although network control can be purely distributed, it is reasonable to assume that network management is hierarchical. There are 8 wavelengths on each unidirectional fibre. Each link on the figure represents a bi-directional fibre. Let us also assume that all the wavelength channels on a fibre link have a same SRLG.

In the case of lightpath protection, NMS receives a connection request from a client, decides the protection scheme (for example, 1 + 1, 1:1, 1–1), and notifies the

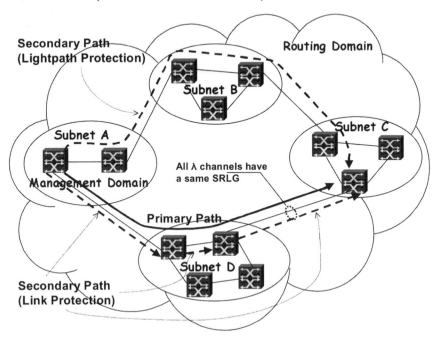

Figure 6.30   Lightpath protection vs. link protection.

edge switch for path setup. The notation '1–1' is used to refer to a secondary path which is computed but the secondary path is not signalled. The edge switch computes the end-to-end primary and secondary path based on its local routing table maintained by OSPF, and signals the primary (and secondary if 1 + 1 or 1:1 protection) lightpath setup using RSVP. If failure occurs on the primary path during 1:1 transmission, the edge switch will switch over to the secondary path. If the 1–1 protection scheme is implemented, the secondary path needs to be signalled before data can be transmitted.

In the case of link protection, NMS receives a connection request, decides the protection scheme, and notifies the edge switch for path setup. The edge switch computes the end-to-end primary path and the first hop secondary path, and signals the primary path setup using RSVP. Hop by hop, the secondary path is computed. If the 1:1 protection scheme is requested, each hop is also responsible for its secondary path segment setup using RSVP. If failure occurs on the primary path during transmission, the intermediate switch to the failure spot will switch over to the secondary path segment. If the 1–1 protection scheme is implemented, the secondary path segment needs to be signalled before data can be transmitted. Note in our example in the figure the link protection secondary path computation does not take into account SRLG.

6.7.2 Restoration Case Study

Figure 6.31 shows an example of subnet segment restoration and network restoration. For subnet segment restoration, the primary path is computed and set up by the

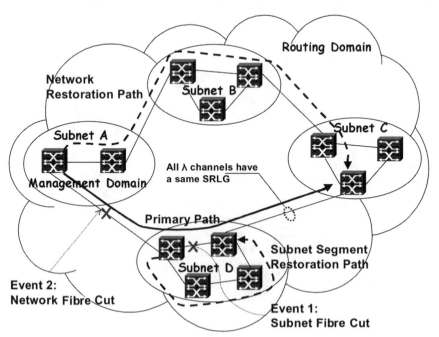

Figure 6.31 Subnet segment restoration vs. network restoration.

edge switch. There is no secondary path signalled or computed. Once failure occurs on a subnet segment of the primary path, the intermediate switch to the failure link is responsible for computing and signalling a secondary path dynamically. In the absence of active fault management mechanisms in IP/WDM networks, RSVP will detect the failure. This is shown in the figure as event 1 – subnet fibre cut. The intermediate switch (in the subnet) consults its local routing table, computes a disjoint path based on SRLG, and signals the disjoint path using RSVP. Then, it performs a switchover to bypass the affected link segment. For co-ordination and consistency, it sends an alarm to NMS once the fibre cut is detected and sends a notification to NMS once the switchover is performed.

For network restoration, the primary path is computed and signalled first. If failure occurs on a segment of the primary path, using RSVP, the failure information (including its affected SRLG) is relayed to the edge switch. Then, the edge switch decides a secondary path, signals the path, and performs switchover to the path. The edge switch should also send an alarm and a notification to NMS for courtesy. This is shown in the figure as event 2 – network fibre cut. Once the edge switch receives the RSVP failure message, it computes and sets up a secondary end-to-end lightpath.

6.8   Inter-domain Network Control

Chapter 5 summarised the architectural models for IP/WDM networks. In an overlay model, an IP network and a WDM network can interact through UNI while a WDM network and another WDM network can interact through NNI. In an augmented model, WDM networks are deployed as WDM-domain IP networks so a WDM domain can interface to other WDM domains and IP domains through BGP (see Figure 6.32). In a peer model, inter-domain IP routing issues are handled by BGP. Note that the conventional BGP (for example, BGP version 4) needs to be modified for inter-domain IP/WDM networks. An augmented model requires an optical BGP located at the WDM domain boundary but other IP networks with conventional BGP are intact. A peer model demands an integrated IP/Optical BGP that has to be implemented by each IP and/or WDM networks.

The primary requirements for IP/WDM inter-domain routing can be summarised as follows:

- Distribute reachability information throughout an inter-networking. An inter-networking consists of an interconnected set of networks. These networks may be under different routing and/or administrative domains.
- Maintain a clear separation between distinct administrative or routing domains.
- Hide information on the internal structure of the distinct administrative or routing domains.
- Limit the scope of IGP protocols. This is for security, scalability and policy reasons.
- Provide for address/route aggregation.
- For control and management of multiple IP/WDM domains a routing scheme must also distribute availability information throughout an inter-network. An inter-network may consist of interconnected networks using different technologies (i.e. IP networks and WDM networks).

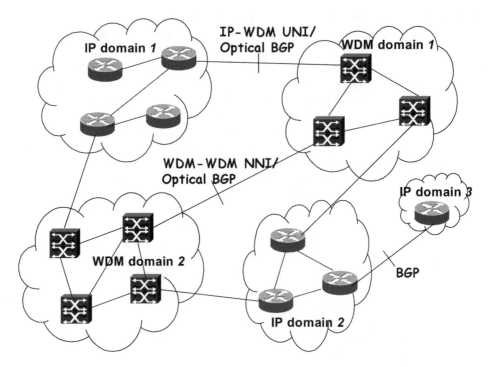

Figure 6.32 IP/WDM inter-domain network control.

The IP-centric control architecture for WDM networks can be extended to satisfy the functional requirements of WDM inter-networking. Routing and signalling inter-action between WDM domains can be standardised across the domain boundary. For the joint control and management of the network, an integration of the intrado-main management system and its inter-domain counterpart is required. BGP is currently deployed across different autonomous systems (AS), i.e. domains, on the Internet. So it is a natural starting point to study multidomain IP/WDM networks routing solutions.

A set of interconnected WDM networks should be functionally similar to a single WDM network. It should be capable of dynamically provisioning and restoring light-paths across WDM domains. To achieve this, one needs an addressing scheme to uniquely identify lightpath end-points in different WDM networks, a protocol to determine reachability of end-points across WDM domains, a signalling protocol to provision lightpaths spanning multiple WDM networks, and a procedure for the restoration of multidomain lightpaths. BGP is developed for IP networks. The conventional BGP cannot fulfil all these requirements. We discuss what are the required additional functions from a routing point of view for inter-domain IP/WDM networks, and extensions to support lightpath provision across ASs. We illus-trate the mechanism by use of an application, i.e. support of optical virtual private networks.

6.8.1   IP/WDM Network Reachability vs. Availability

The semantics of reachability and availability in packet switched networks such as IP networks can be quite different from those of circuit switched networks such as in OXC networks. In the context of packet switched networks, the reachability more or less implies the availability and vice versa. In IP datagram networks, through best-effort routing, availability is exploited along the way as a packet being forwarded across the network. Routing protocols do not carry specific information on availability. This is desirable because network resources are shared, i.e. no reservation. In circuit switched networks, however, resources are dedicated to specific connections. Thus, if reachability is defined as whether or not a physical connection path is present between two points in the network, availability can be defined as whether resources along this path are currently available to accommodate a new connection. A link state routing protocol can be used to obtain network topology (reachability) and resource (availability) information in WDM networks. However, a WDM optical network imposes its own constraints. For example, the OXCs along the path must also be capable of transporting the desired signal type with or without wavelength conversion. This results in additional constraints upon WDM network routing, not typically found in packet networks routing such as OSPF. Example constraints are signal format compatibility, maximum hop count, total delay (mostly propagation), and link protection type. In addition, the optimisation objectives can be more diverse than simply 'the shortest path' upon making routing decisions. For instance, it may be desirable to route traffic in such a way as to optimise overall network capacity, or reliability, or some weighted combination of selected objectives.

IP datagram packet forwarding is used on a hop-by-hop basis (no connection established ahead of data flowing). While with circuit switched optical networks end-to-end connections must be explicitly set up subject to network topology and resource status constraints. While the topology and resource status information can be obtained via routing protocols in both cases, the way that the information is being used differs from one case to another. The routing protocols in the circuit switched case are not involved with data forwarding. For a given stream of data, the topology and resource information is used in circuit setup before the data starts flowing through the network. On the contrary, the IP routing protocols are precisely involved with data plane forwarding decisions for every packet in the IP datagram case. The topology and resource information is normally being used throughout the duration of the data stream. Therefore, it is possible that packets belonging to the same data stream may flow through the network along different paths.

However, this does not imply routing is unimportant in the WDM data plane functionality, but that its service impacting effect is secondary. For example, topology and resource status inaccuracy will affect whether a new connection can be established (or a restoration connection can be established). But, it will not cause changes to the way that an existing connection works. Another property of circuit switched WDM networks is the possible separation of control channel and data channel. In this case, an established data channel may continue to operate even when the corresponding control channel is down. The flip side of this is that not all data channel problems would be caught by the control channel, leading to potential holes on the health of the data channel.

This observation tends to lead to a general guideline for extending the protocol suite developed for IP networks to support control and management of WDM networks. Due to the hop-by-hop forwarding nature of IP datagram networks, data plane forwarding decisions at different nodes must be consistent along the data path. This implies that topology and resource information required by one node for routing purposes is identical to that by another node. Hence, information conveyed by IP routing protocols is adequate to meet the needs. When incorporating information fields into WDM routing protocols, unlike for IP routing protocols, one should be concerned less about how the information will be later utilized. For WDM networks, any information that is potentially useful for network control and management can be incorporated into the routing protocol. Whether a route computation algorithm fully utilizes this information or whether two route computation algorithms use this information in the same/different manner is a different problem. Route computation across optical networks can be explicitly computed. The optical route computation problem is fundamentally a constraint-based routing problem that, for example, can be solved by each ingress edge node independently.

6.8.2   Inter-domain Routing Information Exchange

We organise our discussion on inter-domain routing information exchange into two groups, IP and WDM inter-domain and WDM and WDM inter-domain.

Routing information exchange between IP and WDM domain

Exchanging routing information over an IP domain and WDM domain (IP controlled) boundary would allow the border routers to advertise IP address prefixes within their domain to the WDM domain and to receive external IP address prefixes from the WDM domain. The WDM network transports the reachability information from one IP domain to others. The propagation of the address prefixes from one IP domain to another through the WDM domain assumes the exterior BGP (EBGP) is running between IP border routers and WDM border OXCs. Within the WDM domain, the interior BGP (IBGP) is used between the border/edge OXCs. The case where there are multiple WDM domains present between two IP domains will be discussed in the next section. Like conventional IP inter-domain routing using BGP, the IP address prefixes within the WDM domain are not advertised to routers in the IP domain.

A border OXC receives external IP prefixes from a router. It includes its own IP address as the egress point before propagating these prefixes to its BGP peers, which may be border routers in other IP domains or border OXCs in other WDM domains. The recipients of this information need not propagate the OXC addresses further, but they must keep the association between external IP prefixes and egress OXC addresses. When a specific external IP address is to be reached, the border router can determine if an optical path has already been established to the appropriate egress OXC or a path must be dynamically established.

Optical virtual private network (OVPN)

When optical virtual private networks are implemented, the address prefixes advertised by the border OXCs must be accompanied by some OVPN identification. Border OXCs can then filter external addresses based on OVPN identifiers before propagating them to routers. Once a router has determined reachability to external destinations, the dynamic provisioning of optical paths to reach these destinations may be based on traffic engineering mechanisms implemented in the router. The simplified BGP approach is also referred to as the **partial peer routing model**. It works as follows:

- Each border router belonging to an OVPN registers a set of < IP address, OVPN identifier > pairs to the border OXC to which the border router connects.
- The IP addresses of all border routers belonging to an OVPN are propagated across the WDM network.
- These addresses are conveyed to each router that needs to register as a border router in the OVPN. Within the WDM network, a border OXC is assumed to originate routing advertisements for external IP addresses registered with it. This would allow interior OXCs to route lightpaths destined to external IP addresses to the correct destination OXCs.

Once border routers in an OVPN receive the address of other border routers within their own OVPN, they may construct an OVPN topology dynamically through signalling. Assuming that each router has at least two interfaces to the WDM network, a linear topology may be built automatically. Over this topology, the border routers may run their own IP routing protocol, for example, OSPF. In this case, the lightpaths between the border routers will be represented as virtual links in the OSPF link state database. The initial topology may be modified dynamically, based on traffic engineering algorithms that are implemented in the OVPN. Thus, the simple reachability protocol described above provides a mechanism for bootstrapping end-to-end IP routing within the OVPNs across the WDM network.

Border OXCs filtering external addresses based on OVPN identifiers before propagating them to routers would involve all border OXCs to change their configuration as changes occurred in the OVPNs. Therefore, this approach may be acceptable for a single WDM transport service provider serving a small number of VPN customers.

When a WDM service provider is supporting a large number of VPN customers, it is desirable for the provider to be capable of what is known as single-end provisioning, where changing of a connection to a given OVPN would involve configuration/ provisioning changes only on the OXC adjacent to that connection.

Within a given OVPN each router port has an identifier that is unique within that OVPN, but may not be unique across multiple OVPNs. One way to overcome this is to assign each port an IP address that is unique within a given OVPN and use this address as a port identifier. Another way to overcome this is to assign each router port an interface index that is unique within a given router, assign each router an IP address that is unique within a given OVPN, and then use a tuple < interface index, router IP address > as a port identifier.

Within the WDM network, each access port on an OXC has an identifier that is only unique within that network. One way to overcome this is to assign each port on an OXC an interface index, assign each OXC an IP address that is unique within the service provider network, and then use a tuple < interface index, OXC IP address > as a port identifier within the provider network. As a result, each customer port identifier (CPI) is unique within a given OVPN, and each provider port identifier (PPI) is unique within the service provider network.

Each OXC maintains a port information table (PIT) for each OVPN that has at least one port on that OXC. A PIT contains a list of < CPI, PPI > tuples for all the ports within its OVPN. A PIT on a given OXC is populated from two sources: the local information received from the routers attached to the ports on those OXCs, and the remote information received from other OXCs. The local information is propagated to other OXCs by using BGP with multi-protocol extensions. To restrict the flow of this information to only the PITs within a given OVPN, BGP route filtering based on the Route Target Extended Community can be used as follows.

When adding a new OVPN, the service provider allocates a new BGP Route Target Extended Community that will be used for the purpose of this OVPN. Each PIT on an OXC is configured with the Route Target associated with the OVPN of that PIT, and that creates an association between a PIT and an OVPN. When exporting local information to other OXCs using BGP, this information is tagged with the Route Target Community. When importing remote information into a particular PIT, an OXC can add the information to the PIT if and only if the information with the Route Target Community is equal to the one configured for the PIT. When a service provider adds a new OVPN port to a particular OXC, this port is associated at provisioning time with a PIT on that OXC, and this PIT is associated (again at provisioning time) with that OVPN.

Once a port is configured on the OXC, the customer router that is attached via this port to the OXC passes to the OXC the CPI information of that port. This can be accomplished by using GMPLS signalling. This information, combined with the PPI information available to the OXC, enables the OXC to create a tuple < CPI, PPI > for such port, and then use this tuple to populate the PIT of the OVPN associated with that port.

An OXC uses the information in its PITs to provide customer routers connected to that OXC with the information about CPIs of other ports within the same OVPN. These remote ports are referred to as target ports. An OXC can pass target ports information to customer routers using GMPLS signalling. Having acquired this information, a customer router uses GMPLS signalling to request the provider network to establish a lightpath connection to a target port. The request originated by the customer router contains the CPI of two ports. One is the router's own port that the router decides to use for the lightpath connection, and the other is the target port. When the OXC attached to the customer router that originated the request receives the request, the OXC identifies the appropriate PIT, and then uses the information in that PIT to find out the PPI associated with the CPI of the target port carried in the request. The PPI should be sufficient for the OXC to establish a lightpath connection. Ultimately the request reaches the customer router associated with the target CPI. If the customer router associated with the target CPI accepts the request, the lightpath connection is established. Note that a customer router need not establish a lightpath connection to every target port that the router knows about – it can select a subset of target ports to which the customer router will try to establish lightpath connections.

In addition to its CPI and PPI, a port may also have other information associated with it, which may describe characteristics of the channels within that port, such as encoding supported by the channels, bandwidth of a channel, total unreserved resources within the port, etc. This information is necessary to ensure that ports at each end of a lightpath connection have compatible characteristics, and that there are sufficient unallocated resources to establish a lightpath connection. Distribution of this information is identical to the distribution of the CPI information. Updating this information due to establishing/terminating of lightpath connections should use a threshold-control mechanism to contain the volume of control traffic caused by such distribution (see Topology Discovery Section 6.2).

It may happen that for a given pair of ports within an OVPN, each of the customer routers connected to these ports would concurrently try to establish a lightpath connection to the other customer router. If having multiple lightpaths between a pair of ports is not supported or not allowed by the policy in use, a way to resolve this is arbitration based on the value of CPI (for example, the router with the lower CPI value is required to terminate the lightpath connection originated by this router). This option could be controlled by configuration on the customer routers.

Routing information exchange between WDM domains

Provisioning an end-to-end lightpath across multiple WDM domains involves the establishment of path segments in each WDM domain sequentially or in parallel. The first path segment is established from a source OXC to its border OXC in the source WDM domain. From the border OXC a path segment is established to a border OXC in the next WDM domain. Provisioning then continues in the next WDM domain and so on until the destination OXC is reached. To automate this process, a source OXC must determine the route to the destination OXC from within the source WDM domain. An inter-domain routing protocol must therefore run across the multiple WDM domains. It is desirable that such a protocol allows the separation of routing between domains. This allows proprietary provisioning schemes to be implemented within each domain while end-to-end provisioning is performed.

These objectives may be satisfied by running BGP with extensions between border OXCs. For this purpose, the OXC IP addresses need to be globally unique. Using EBGP, adjacent border OXCs in different WDM domains can exchange reachability of OXCs and other external IP endpoints (border routers). Using IBGP, the same information is propagated from one border OXC to other OXCs in the same WDM domain. Thus, every border OXC eventually learns of all IP addresses reachable across different neighbouring WDM domains. These addresses may be propagated to other OXCs within the WDM domain thereby allowing them to select appropriate border OXCs as exit points for external destinations.

It is clear that border OXCs must keep track of many IP addresses corresponding to different remote OXCs and IP border routers. The overhead for storage and propagating these addresses can be reduced if OXC addresses within a WDM domain can be aggregated to a relatively few IP network prefixes. This is indeed possible if OXC addresses within a WDM domain are derived from a small set of IP network prefixes.

Once border OXCs acquire reachability information regarding remote destinations, this information may be shared with other OXCs within the WDM domain to enable end-to-end path provisioning. To accomplish this, a source OXC within a WDM domain must determine the border OXC through which the ultimate destination can be reached. Also, if there is more than one such border OXC, a procedure must be available to select one of them. Furthermore, policy decisions may be involved in selecting a particular route. These issues are similar to inter-domain routing in the Internet.

Dynamic provisioning model

The dynamic provisioning model is used to provision a lightpath across multiple WDM domains. The model illustrates the primary information exchange and key steps in setting up lightpaths. It allows either an IP border router or a third party like a management system to initiate a provisioning request. In either case, the source and destination end-points must be specified. While the source end-point is implied if a border router initiates such a request, identifiers for both end-points must be specific enough to provision a lightpath in between. In general, an end-point identifier may take the form of a tuple $<$ OXC IP address, Port Identifier $>$. With the specified end-points, the routing of a lightpath is done as follows:

- The source OXC looks up its routing information corresponding to the specified destination IP address:

  - If the destination is an OXC in the source WDM domain, a path may be directly computed to it.
  - If the destination is an external address, the routing information will indicate a border OXC that would terminate the path in the source WDM domain. A path is computed to the border OXC.

- The computed path is signalled from the source to the destination OXC within the source WDM domain. The complete destination endpoint address specified in the provisioning request (either $<$ OXC IP address, Index $>$ or $<$ IP address $>$) is carried in the signalling message.
- The destination OXC in the source WDM domain determines if it is the ultimate destination of the path.

  - If so, it checks if the destination endpoint identifier specified in the message includes a port identifier. In this case, it completes the optical path set-up using the port identifier. If the port identifier is not included, the address corresponds to a border router. In this case, the port through which the border router performed registration is used to complete the path set-up.
  - If the OXC is not the ultimate destination, it determines the address of a border OXC in an adjacent WDM domain that leads to the final destination. The path setup is signalled to this OXC using inter-domain signalling. The next OXC then acts as the source for the path.

OSPF for information exchange

When some trust relationship presents among WDM transport domains and client IP domains, the routing information exchanged across the domain boundaries could be summarised using a hierarchical routing protocol such as OSPF with each domain configured as an area. OSPF supports a two-level hierarchical routing scheme through the notion of areas (see Chapter 3). Routing within each area is flat, while detailed knowledge of an area's topology is hidden from all other areas. Routers attached to two or more areas are called area border routers (ABRs). ABRs propagate aggregated IP prefixes from one area to another using summary LSAs. Within an OSPF routing domain, all areas are attached directly to a special area called the OSPF backbone area. The exchange of information between areas in some way is similar to the BGP method of propagating reachability. The use of a single OSPF routing domain with multiple areas is beneficial from the point of view of ease of migration, as providers migrate to optically switched backbones.

One should note that OSPF areas are collections of network segments, not collections of routers. An ABR is a router that has operational interfaces belonging to two or more areas. The hierarchical routing structure is applicable to IP and WDM interworking. In this case, the physical backbone is replaced with a WDM network, which is simply achieved by replacing each ABR with an OXC. While the data plane characteristics of the WDM network are completely different from those of the OSPF backbone area, the control plane remains essentially the same. As long as OXCs participate in the OSPF routing, the WDM network can serve as the OSPF backbone area, flooding summary LSAs between different areas. The WDM network advertises external addresses into each area, along with the address of the OXC corresponding to each address and a cost metric associated with it. The cost information may be used to select a delivery path across client networks among alternatives.

BGP for Information Exchange

BGP is designed for dissemination of reachability information across domains. BGP is path-vector based routing protocol, which means a network element through running of BGP can construct a connected graph that is a subgraph of the domain level topology (i.e. each network node is a domain or an abstract node). With this condensed (or abstract) topology information, there is less guarantee on path selection optimality. Capability of traffic engineering may also be limited because of the lack of complete topology information. However, the application-orientated policy routing capability inherent in BGP is undeniably a significant advantage for inter-domain control and management.

IP domains interconnected by WDM domain(s) should have three types of routes in their routing database: fixed routes, forwarding adjacencies, and potential forwarding adjacencies. The fixed routes are generally associated with fixed connections between routers. The forwarding adjacencies are defined by established lightpaths so they exist only on IP border routers that directly connect to a WDM domain. The potential forwarding adjacencies are present between any two IP domain border routers that are interreachable through one or more WDM domains (but no lightpath has been created to connect the two yet). After setting up a lightpath through signal-ling, a potential forwarding adjacency becomes a forwarding adjacency; after tear-down a lightpath, its corresponding forwarding adjacency, becomes a potential forwarding adjacency. When GMPLS is used as the signalling scheme, in IP networks, label stacking can be used for packet forwarding over multiple subnet-works or across non-GMPLS domains (or unknown domains). There is no such equivalent label stacking in a WDM optical network.

Explicit routed hops or path segments can be stacked for LSP creation through multiple domains. An explicit hop can be either node ID, domain ID or existing LSP ID as defined in MPLS signalling. In the case of explicit routing, a source node (which may be a source router or an egress edge OXC of the WDM domain adjacent to the IP domain that the source router belongs to) may only include domain IDs in its explicit routed hop list. When a path-create request enters a transit WDM domain, the ingress OXC is responsible for computing an explicit routed hop list for its domain and pushing this list to the top of the original explicit routed hop stack. When it leaves the WDM domain, the egress OXC needs to pop the domain-explicit routed hop list off the stack.

When the path operation message crosses domain boundaries, they should be authenticated. Each domain needs to validate this information in order to guarantee network operation integrity.

6.9   WDM Network Element Control and Management Protocol

Routing and signalling provide network-level functions and mechanisms to control a network. For interoperability and standardising operations, network elements can be modelled into a NE-level MIB. To synchronise the activities between network-level and NE-level control modules and provide updated NE information to network-level control entities, a network element management protocol is needed. We review several network element management protocols in this section.

6.9.1   Simple Network Management Protocol (SNMP)

IP networks use the Simple Network Management Protocol to allow both for management data to be collected remotely from devices and for management devices to be configured remotely by managers. All SNMP messages are transported using UDP. SNMPv1 only supports four operations:

- *get* and *get-next* operations for retrieving data;
- *set* operation for writing data;
- *trap* operation for a device to send an asynchronous notification.

SNMPv2 extends the protocol operations by defining an SNMPv2 trap PDU, an inform message, and a get-bulk request message:

- The *inform* operation is an extension to the trap operation with reliable message delivery. The inform message allows the management entity to send events to its manager and then get explicit acknowledges on them.

- The *get-bulk* operation is used to retrieve portions of a table. It is equivalent to repeated *get-next* operations.

Each managed entity (i.e. a device) maintains a structured collection of objects in its MIB. The MIB definition is specified in the Structure of Management Information (SMI) format that represents the set of rules for object naming and definition. SNMPv2 also defines additional error values. Under SMI, the MIB-II has defined these groups: *system, interfaces, at, ip, icmp, tcp, udp, egp, transmission,* and *SNMP.* Each group defines managed objects. For example, the system group has these managed objects:

sysDescr	sysName
sysObjectID	sysLocation
sysUpTime	sysServices
sysContact	

SNMPv3, supporting a modular architecture, extends the protocol in two major areas: administration and security. SNMPv3 no longer keeps separate concepts of SNMP agent and SNMP manager. Instead, it introduces a generalised SNMP entity concept in its architecture. A SNMP entity is made up of two pieces: a SNMP engine and SNMP applications. The engine provides the fundamental components:

- dispatcher
- message processing subsystems
- security subsystems
- access control subsystems.

The SNMP internal applications are built on top of the SNMP engine. In particular, they play the role of command generators, command responders, notification originators, notification receivers, and proxy forwarders. SNMPv3 has defined a command header that consists of these fields:

- message version
- message ID
- maximum message size
- message flags
- message security model.

SNMP is popular mainly because of its simplicity. However, it does not define any further detailed network management services and applications. Since SNMP originates from the IP world, it does not address the rich and complex network management functionality (such as configuration, connection, fault, and performance management) required in the telecom world.

6.9.2   General Switch Management Protocol (GSMP)

GSMP is an IETF effort to provide a general-purpose protocol to control and manage an ATM switch. GSMP, a master-slave protocol functioning asymmetrically, separates the control mechanisms for the forms of switching from the actual switch fabric that allows for independent upgrade of control planes as well as switching planes. GSMP allows a controller to establish and release connections across the switch, add and delete leaf nodes on a multicast connection, manage switch ports, request configuration information, request and delete reservation of switch resources, and request statistics. It also allows the switch to inform the controller of asynchronous events such as a link down. GSMP packets are variable length and are encapsulated directly into AAL5 with an LLC/SNAP header 0x00-00-00-88-0C to indicate GSMP messages. GSMP can operate across an ATM link connecting the controller and the switch. The control channel or virtual channel must be established during the initialisation.

In GSMP, a switch can be considered as a group of logical ports, each of which consists of an input fabric port and an output fabric port. GSMP requests can refer to the logical port as a whole as well as to the specific input or output fabric port. ATM traffic carried in ATM cells is switched according to its virtual paths and virtual channels. Virtual paths on a port or link are referenced by their virtual path identifiers (VPI), whereas virtual channels on a port or link are referenced by their VPI and virtual channel identifiers (VCI). A virtual path connection of a switch connects an incoming virtual path to one or more outgoing virtual path(s). The virtual path connection is identified by the input port and its VPI. A virtual channel connection of a switch connects an incoming virtual channel to one or more outgoing virtual channels. The virtual channel connection is identified by the input port it arrives at and its VPI and VCI. As defined in ATM, a group of VCIs can shared the same VPI so that the switch can allow fine granularity traffic and support flexible multiplexing schemes. GSMP does not explicitly support multipoint-to-point and multipoint-to-multipoint connection but they can be emulated using multiple point-to-point connections and multiple point-to-multipoint connections. The concept of virtual channels over the same virtual path also allows differentiated QoS to be associated with each virtual channel. The QoS parameters of a virtual channel connection can be configured using GSMP QoS messages. As such, when contention occurs for an input port, the ATM cell belonging to higher priority VPI/VCI will get through the switch first.

GSMP defines two types of messages:

- adjacency protocol messages
- request-response messages.

Adjacency protocol message

The adjacency protocol message is used to maintain the link status between the switch and its controller and further to synchronise state and parameters between the two. The adjacency protocol message format is shown in Figure 6.33.

The message fields in GSMP adjacency protocol are defined as follows:

- *Version #* – 1 octet; the GSMP protocol version number. GSMP is capable of automatic version negotiation, by which the switch and the controller agree to the highest protocol version both sides understand.
- *Message type* – 1 octet; the adjacency protocol message type. This field has the value of 10.

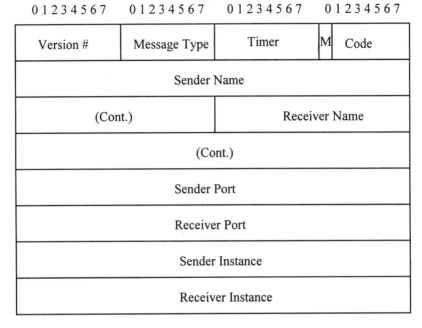

Figure 6.33 GSMP adjacency protocol message format.

- *Timer ID* – 1 octet,; the time interval that the sender sends the periodic adjacency protocol message in the unit of 100 ms.
- *M-flag* – 1 bit; the master/slave flag. When the bit is set, the sender is the master. Otherwise the sender is the slave.
- *Code* – 7 bits; this field specifies the function of the message. The following functions have been defined:

  - Type 1: SYN
  - Type 2: SYNACK
  - Type 3: ACK
  - Type 4: RSTACK

- *Sender Name* – 6 octets: in SYN, SYNACK, and ACK functions, this field represents this message sender; in RSTACK function, this field represents the receiver that caused the RSTACK to be generated.
- *Receiver Name* – 6 octets; in SYN, SYNACK, and ACK functions, this field represents this message receiver; in RSTACK function, this field represents the sender that caused the RSTACK to be generated.
- *Sender Port* – 4 octets: in SYN, SYNACK, and ACK functions, this field represents the local port number where the message is sent; in RSTACK function, this field represents the receiver port number that caused the RSTACK to be generated.
- *Receiver Port* – 4 octets: in SYN, SYNACK, and ACK functions, this field represents the port number (of the other side of the link) where the message is received; in

RSTACK function, this field represents the sender port number that caused the RSTACK to be generated.

- *Sender Instance* – 4 octets: in SYN, SYNACK, and ACK functions, this field represents the sender's instance number; in RSTACK function, this field represents the receiver's instance number that caused the RSTACK to be generated. The instance number is used to detect when the link comes back up after going down and when the entity at the other side of the link changes its identity.
- *Receiver Instance* – 4 octets: in SYN, SYNACK, and ACK functions, this field represents the receiver's instance number; in RSTACK function, this field represents the sender's instance number that caused the RSTACK to be generated.

The GSMP adjacency protocol defines three times the value carried in the Timer field as the loss of synchronisation interval. Once the switch passes the loss synchronisation interval, it assumes its controller is lost and tries to establish synchronisation with its backup controller (that could be more than one).

Request-Response Messages

The GSMP request-response messages have the message format shown in Figure 6.34. Six classes of messages are defined: connection management, port management, state and statistics, configuration, quality of service, and events.

```
0 1 2 3 4 5 6 7    0 1 2 3 4 5 6 7    0 1 2 3 4 5 6 7    0 1 2 3 4 5 6 7
```

Version #	Message Type	Result	Code
Transaction Identifier			
Message Body			

Figure 6.34 GSMP Request-Response message format.

The message fields of GSMP request-response messages are defined as follows:

- *Result* – 1 octet; this field indicates the result for the request/response message. The request message has the following result types:
  - Type 1: NoSuccessAck. Value: the request message does not expect a response message if the request outcome is successful.
  - Type 2: AckAll. Value: the request message expects a response message in spite of the outcome. The response message has the following result types.
  - Type 3: Success. Value: OK. In the case of multiple response messages, this type is applied to the final message only.
  - Type 4: Failure. Value: the outcome of the request message is failure.
  - Type 5: More. Value: this is used to indicate there are multiple response

messages. For example, the response message has to be fragmented for delivery. All response messages except the final message has this type to indicate there is more response messages. The fragmented response messages have the same transaction identifier.

- *Code* – 1 octet; this field is reserved to transport further information regarding the result in the response message. In case of failure, this field can be encoded with a detail error code.
- *Transaction Identifier* – 4 octets; a request message sets a transaction identifier to differentiate request messages and its response message(s) using the same transaction identifier. As such, the sender can associate the request message and all corresponding response message(s). In the event message, the switch sets this field to zero.

The Connection Management messages are used by the controller to add, remove, update, and verify virtual channel connections and virtual path connections across the switch. Five request messages have been defined: Add Branch, Delete Tree, Verify Tree, Delete All, Delete Branches, and Move Branch messages:

- *Add Branch*: This message type is used to establish a new virtual channel connection or a new virtual path connection or add an additional branch to an existing virtual channel connection or virtual path connection.
- *Delete Tree*: This message type is used to remove an entire virtual channel connection or virtual path connection.
- *Verify Tree*: This message type is used to retrieve a virtual channel connection or a virtual path connection. GSMPv2 no longer uses this message type.
- *Delete All*: This message type is used to remove all connections on a switch input port.
- *Delete Branches*: This message type is used to remove one or more branches of a virtual channel connection or a virtual path connection.
- *Move Branch*: This message type is used to move a branch of an existing connection from its current output port VPI/VCI to a new output port VPI/VCI in a single transaction. This message has the same effect as deleting a branch and adding a new branch to a connection.

The Port Management messages are used to enable, disable, loop back, reset, and configure the label range for a port. Two messages have been defined as below:

- *Port Management*: This message type is used to set a port to be in-service, out-of-service, or loop-back. It also allows updates to the transmit cell rate. The message type has a reset operation that can be applied to a specific input port or event flags.
- *Label Range*: This message type is used to set the range of VCIs and VPIs for a specified port.

The Configuration Management messages allow the controller to discover the capabilities supported by the switch. These messages are usually issued at initialisation time. Three types of Configuration Management messages have been defined: Switch Configuration Message, Port Configuration Message, and All Ports Message:

- *Switch Configuration Messages* are used to query the global configuration information of the switch.
- *Port Configuration Messages* request the switch for the configuration information of a single switch port.
- *All Ports Configuration Messages* request the switch for the configuration information on all of its ports.

The State and Statistics messages are used to request the statistics related to the switch input ports and outputs ports, virtual channel connections, virtual path connections, and QoS classes. They also allow the controller to retrieve the connection state of a switch input port. The following message types have been defined:

- *Connection Connectivity*: this message type is used to verify whether one or more specified virtual channel connections or virtual path connections have been carrying traffic.
- *Port Statistics*: this message type requests statistics related to a specified port.
- *Connection Statistics*: this message type requests statistics related to a specified virtual channel connection or virtual path connection of a given port.
- *QoS Class Statistics*: this message type is used to request the statistics related to a specified QoS class of a given port.
- *Report Connection State*: this message type is used to query the connection state of a specified port for a single virtual channel connection, a single virtual path connection, or the entire input port.

The event messages allow the switch to inform the controller of certain asynchronous events that have occurred in the switch port specified in the Port Number field. Event messages are not acknowledged. Event messages are not sent during initialisation. The following types of event messages have been defined: Port Up, Port Down, Invalid VPI/VCI, New Port, and Dead Port message:

- *Port Up* message is used to inform the controller that the port status has changed to the Up state.
- *Port Down* message is used to inform the controller that the port status has changed to the Down state.
- *Invalid VPI/VCI* message is used to inform the controller the ATM cells have arrived at an input port with a VPI/VCI that is currently not used by any connections.
- *New Port* message informs the controllers that a new port has been added to the switch.
- *Dead Port* message informs the controller that a port has been removed from the switch.

The QoS messages are used by the controller to group virtual path connections (VPC) and virtual channel connections (VCC) into QoS classes and allocate QoS resources to QoS classes and individual connections. GSMP supports an abstract switch model for QoS. The model has four main functions: a policer, a classifier, a regulator, and a scheduler. The policer is a single input and output device that can be used to filter or tag cells by attaching to VPC/VCC or aggregated traffic. The classifier groups VPC/VCC into QoS classes. The regulator offers either a policing function or a shaping function for each QoS class. The scheduler located on the output port distributes the available bandwidth to the QoS classes and individual connections. Figure 6.35 shows the process flow of the GSMP QoS model.

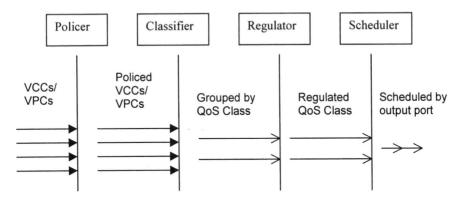

Figure 6.35 Process flow for GSMP QoS model.

GSMP has defined the following QoS message types:

- *QoS Configuration*: this message type allows the controller to discover the QoS capabilities of each switch port in respect to the above QoS model (see Figure 6.35).
- *Scheduler Establishment*: this message type is used to configure a scheduler for a specific output port.
- *QoS Class Establishment*: this message type is used to configure a QoS class on a specific port or modify parameters for an existing QoS class.
- *QoS Release*: this message type is used to remove a QoS class or a scheduler. The resources associated with it will also be freed.
- *VCC QoS Connection Management*: this message type is used to request VCCs across the switch with the specified QoS parameters.
- *VPC QoS Connection Management*: this message type is used to request VPCs across the switch with the specified QoS parameters.

6.9.3 Optical Switch Control Protocol (OSCP)

GSMP was originally developed for ATM networks although recently there are efforts on extending the protocol to control WDM optical switches. In this section, we present an optical switch control protocol (OSCP). Our purpose is to highlight WDM switch-specific information. We adopt the same request-response message header as defined by GSMP (see Figure 6.34), so OSCP can be considered as an extension to GSMP.

In the IP-centric optical networks, OSCP can be implemented as a daemon that accepts connection requests from RSVP and OSPF. Using the specified messages, the OSCP daemon controls the switch fabric setting. OSCP is a simple master-slave, request-response protocol. The controller implements the master portion of the OSCP protocol, and the optical switch implements the slave portion of the protocol. The master sends a request and the slave issues a positive or negative response when the operation is completed. The OSCP messages have the format shown in Figure 6.34. Six classes of messages can be defined: Connection management messages, Logical port management messages, Configuration management messages, Statistics messages, Event messages, and Edge node management messages.

Reliable delivery of OSCP messages is accomplished by using a TCP connection between the master and the slave protocol modules.

The Connection Management messages are used by the controller to initialise, add, remove, update, and check the forwarding table entries located at the WDM switch. In OXC networks, they allow the controller to set up, tear down, add a leg, and query cross-connects on the switch fabric. Nine request messages can be defined: Add_Cross-connect, Add_Leg, Delete_Connection, Query_Connection, Add_Label_Branch, Remove_Label_Branch, Verify_Label_Branch, Get_Table, and Put_Table. The first four messages are used in the OXC networks, where there is a pre-reserved connection setup time. The connection refers to the wavelength channel that requires resource reservation. The last five messages are used for OLS networks, where traffic is forwarded along virtual channels. The virtual channels identified by labels (local visibility) do not require the same level of physical resource reservation. Hence, a virtual channel provides a finer granularity of bandwidth sharing than a wavelength channel.

- *Add_Cross-connect*: this message type is used to set up one or more cross-connects on the switch.
- *Add_Leg*: this message type is used to add leg(s) to existing connections. Some switching fabrics support point-to-multipoint signal splitting, which can be used to emulate physical layer multicasting. By adding a leg to an existing connection, the cross-connect connects an incoming wavelength of an input port to the same wavelength of two output ports. If the switch port supports wavelength conversion, the output wavelength channel can use a different optical frequency from the incoming wavelength.
- *Delete_Connection*: this message type is used to remove a switch connection that can be either a cross-connect or a leg. If the cross-connect is deleted, its related leg(s) will be automatically deleted.
- *Query_Connection*: this message type is used to retrieve connection information. If the port number is specified, all the connections on the port are replied back. If the port number and the wavelength are specified, only that connection (on that wavelength) is sent back. If the port number is unspecified, all connections on the switch are enclosed in the response message.
- *Add_Label_Branch*: this message type is used to add a forwarding table entry into the forwarding table.
- *Remove_Label_Branch*: this message type is used to remove a forwarding table entry, part of the forwarding table entries or the entire table entries from the forwarding table.

- *Verify_Label_Branch*: this message type is used to retrieve a forwarding table entry associated with the corresponding incoming port number and incoming label pair.
- *Get_Table*: this message type is used to retrieve the entire content of the forwarding table associated with the incoming port number.
- *Put_Table*: this message type is used to transfer the entire content of the forwarding table associated with the incoming port number into the WDM switch.

The Logical Port Management messages are used to enable, disable, and check the label-switching capabilities associated with the corresponding port. Two messages, *Read_Logical_Port_Status* and *Write_Logical_Port_Status*, are defined as below:

- *Read_Logical_Port_Status*: This message type is used to read the current switching mode of the specified logical port. If the label-switching mode is disabled in the port, no activities related to the header processing are conducted. This allows a port to support the provisioned connection between the two WDM switches.
- *Write_Logical_Port_Status*: This message type is used to update the switching mode of the specified logical port. Two switching modes are currently supported: label switching enabled and disabled.

The Configuration Management messages allow the controller to discover the capabilities supported by the WDM switch. These messages are usually issued at initialisation time. Three types of Configuration Management messages can be defined: Switch Configuration Message (*Get_Switch_Configuration*), Port Configuration Message (*Get_Port_Configuration*), and Label Configuration Message (*Get_Label_Configuration*):

- The Switch Configuration message (*Get_Switch_Configuration*) is used to query the global configuration information of the WDM switch.
- The Port Configuration message (*Get_Port_Configuration*) requests the WDM switch for the configuration information of a single switch port. In WDM networks possibly with wavelength continuity constraints, a switch port has to indicate its wavelength interchange capability.
- The Label Configuration message (*Get_Label_Configuration*) allows the controller to discover the label-switching capabilities supported by the WDM switch such as forwarding table granularity, maximum forwarding table size, and minimum or maximum label value.

The Statistics messages are used to query the statistics occurred within the WDM switch. One type of Statistics message in OLS networks is defined: the Packet Statistics Message (*Get_Packet_Statistics*). The Packet Statistics message requests the packet statistics for the switch port (output port) specified in the Port Number field.

The Event messages allow the WDM switch to inform the controller of certain asynchronous events occurred in the switch port specified in the Port Number field. Event messages are not acknowledged. Event messages are not sent during initialisation. Three events, *Port_Up*, *Port_Down*, and *QoS_Fault*, are defined. The *QoS_Fault* message type is used to report signal QoS related faults.

The Performance Monitor messages allow the controller to set thresholds for certain signal QoS parameters. Once these thresholds are broken, the switch generates QoS Fault events. Optical signal QoS parameters include Optical Signal-to-Noise Ratio (OSNR), Wavelength Power, Fibre Power, and Wavelength Registration.

The Edge Node Management messages are used by the controller to set various parameters related to the edge LSR. Three types of Edge Node Management messages can be defined: FEC (Forwarding Equivalence Class) Label Mapping Table Management Messages, Router Id Management Message, and Traffic Aggregation Management Messages.

The FEC Label Mapping Table Management messages are used by the controller to initialise, add, remove, update, and check the FEC label mapping table entries located at the edge WDM switch. Three request messages can be defined: *Add_FEC_Label_Branch*, *Remove_FEC_Label_Branch*, and *Verify_FEC_Label_Branch*:

- *Add_FEC_Label_Branch*: This message type is used to add a FEC label table entry into the FEC label mapping table. The label branch to be added is specified by a FEC type, length of address prefix, a FEC element, priority, and the corresponding label value. Each classified FEC will have a *corresponding priority* associated with it. Since multiple branches can be specified within a single message, the entire table entries can be set by this request message.
- *Remove_FEC_Label_Branch*: This message type is used to remove a FEC label table entry, part of table entries, or the entire entries from the FEC label mapping table. If the FEC type is –1, the entire table entries will be removed. The label branch to be removed is specified by a FEC type, length of address prefix, and a FEC element.
- *Verify_FEC_Label_Branch*: This message type is used to retrieve a FEC label table entry associated with the corresponding FEC type, length of address prefix, and FEC element pair. If this message contains multiple subentries, they are also retrieved in response to this request message. If the FEC type is –1, the entire table entries will be retrieved.

We have presented three network element management protocols in this section. SNMP has been widely used and supported in IP networks. It is the only IP-industry-accepted standard for network element management. There are a number of third-party software vendors who supply SNMP applications and management tools. GSMP was originally developed for ATM networks. Comparing to SNMP, GSMP as a standard is less accepted in the ATM industry. With the emergence of the WDM industry, a WDM network element management protocol is needed. Recently, GSMP has been extended to control WDM optical switches. Although there are similarities between ATM switches and WDM optical switches, the WDM optical switches have unique properties such as wavelength channels, wavelength ports, all-optical features, and optical signal QoS issues. In addition, given the lessons learned from IP and ATM integration, one expects a closer integration between IP and WDM. The different architectures of inter-networking IP and WDM are presented in Chapter 5. As the WDM industry grows and matures, a standard WDM network element management protocol and a WDM NE MIB will emerge. Until then, OSCP can be used to control and manage reconfigurable WDM NEs and OLSRs.

6.10   Summary

IP/WDM inter-networking models and network architectures are introduced in Chapter 5. However, there are common network control issues in spite of various network architectures and inter-networking models. In this chapter, we have presented a detailed discussion on network control for IP/WDM networks.

We discussed the IP/WDM network addressing, in particular, overlay addressing and peer addressing. We presented a section on topology discovery. We examined the IP topology discovery scheme in OSPF, i.e. OSPF Hello protocol. We introduced the LMP protocol, which is an IETF effort to couple the WDM data plane to the WDM control plane, to model the physical resources such as link bundling, and to localise failures for fast switchover. We gave an in-depth discussion on IP/WDM routing. We discussed the routing information base construction and maintenance by examining the flooding routines in the OSPF protocol. With respect to route computation, we presented the SPF algorithms, the dynamic routing and wavelength assignment, and the disjoint path/node routing for protection. With respect to WDM switching constraints, we enumerated a list of possible constraints in the WDM optical networks. To control WDM optical networks, we discussed the extension to the standard IP OSPF protocol in particular the opaque LSA formats. We discussed the classical issues associated with routing behaviour such as routing loops, routing oscillation, routing failures, and routing stability.

In terms of IP/WDM signalling, we studied RSVP and proposed RSVP extension including extension implementation architecture and RSVP message extensions such as PATH and RESV messages. We discussed WDM access control to map IP packets to WDM wavelength channels. We reviewed the GMPLS hierarchy. A relatively new concept to IP networks and engineers is protection and restoration. We gave the definition of protection and dynamic restoration, and discussed three variants of path protection/restoration: link protection/restoration, lightpath or LSP protection/ restoration, and partial protection/restoration. To clarify the concepts of path and link protection and subnet segment and network restoration, we presented two case study sections.

We also introduced inter-domain network control by discussing issues on IP/WDM reachability and availability and inter-domain routing information exchange. We organised our inter-domain routing discussion into routing information exchange between IP and WDM and routing information exchange between WDM domains. Finally, we discussed the network element management protocol. We reviewed the SNMP protocol used in the IP networks, the GSMP protocol used in the ATM networks, and the OSCP protocol proposed for the WDM networks.

6.10.1   Network Control vs. Network Management

Network control can be separated from network management. Typically, network control functions include addressing, neighbourhood discovery, routing, signalling, and network element control protocol. In a network environment, resources are distributed so control mechanisms are usually distributed. IP networks implement distributed control mechanisms such as OSPF and RSVP. Distributed control makes the distributed entities peer-to-peer. A distributed networking system is more scalable than its centralised counterpart. It is also more robust and available. With the appropriate mechanisms and protocols in place, a distributed system is more reliable and efficient. However, network management (as in any type of management) is usually centralised or hierarchical. Its tasks include resources abstraction, performance monitoring and analysing, and group decision-making. IP network management follows the SNMP approach, where a manager communicates directly with the managed entities (their software representation are known as agents) using SNMP protocol. The router functionalities and configurations are standardised and modelled into the router SNMP MIB.

In conventional telecom networks, network management has been used extensively. In fact, in some telecom networks, there is no distributed control at all. Instead, a management hierarchy is constructed, where the resources are modelled and grouped into subnets. All decisions have to be made by the network level managers as EMSs or NEs have little knowledge of the entire network. For example, the TMN framework groups network management functions into configuration, performance, fault, security, and accounting. Then these management functionalities are implemented in a hierarchical fashion, i.e. each EMS has its own managers, for example, configuration, performance, fault managers, but each EMS is independent from other EMSs.

As communication and computer networks converge, IP and telecom network management and control will integrate inevitably. Given the current industry practice and network deployment, IP network control is certainly more scalable. However, to address QoS and to provide a carrier grade of services, management functions and related mechanisms (such as those defined in Telecom networks) need to be implemented.

7

IP/WDM Traffic Engineering

- What is IP over WDM traffic engineering?
- Modelling of IP over WDM traffic engineering
- IP over WDM traffic engineering functional framework
- Teletraffic Modelling
- MPLS traffic engineering
- Lightpath virtual topology reconfiguration
- Reconfiguration for packet switched WDM networks
- Simulation study of IP over WDM reconfiguration
- IP over WDM traffic engineering software design
- Feedback-base closed-loop traffic engineering
- Summary

7.1 What is IP over WDM Traffic Engineering?

IP over WDM traffic engineering aims at utilizing IP/WDM resources (for example, IP routers and electrical buffers, WDM switches, fibres and wavelengths) efficiently and effectively, to transport IP flows and packets. IP/WDM traffic engineering includes IP/MPLS traffic engineering and WDM traffic engineering as shown in Figure 7.1.

MPLS traffic engineering deals with issues of flow allocation and label path design. Using MPLS-explicit path control, MPLS traffic engineering provides load balancing over an existing IP topology. The MPLS LSPs work as virtual paths sharing the established IP topology. WDM traffic engineering releases the assumption of static IP topology in a WDM network. WDM traffic engineering copes with issues of lightpath topology design and IP topology migration. In reconfigurable WDM networks, MPLS and WDM traffic engineering work in different layers, i.e. one for the IP layer and another for the WDM layer. In optical packet switched networks, MPLS and WDM traffic engineering can be implemented either in an overlay approach or with an integrated fashion. The former is similar to the overlay IP over reconfigurable WDM networks (the data plane), where MPLS LSPs (virtual paths) are allocated to the established WDM light circuits. The latter constructs lightpaths, allocates flows

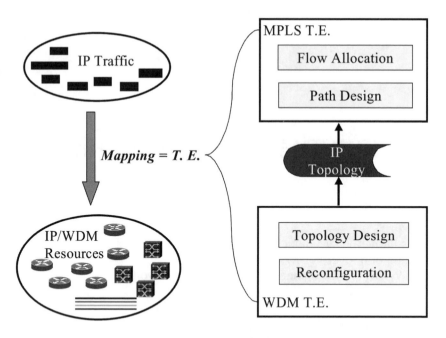

Figure 7.1 IP/WDM traffic engineering (TE).

onto the lightpaths, and forwards traffic in an integrated fashion. We describe MPLS and WDM traffic engineering in this chapter.

7.2   Modelling of IP over WDM Traffic Engineering

Traffic engineering in IP over WDM networks can be pursued in two fashions: overlay and integrated traffic engineering. With *overlay IP/WDM traffic engineering*, there is a traffic engineering module for each of the IP and WDM layers. Operations in one network can be independent from those in the other network. Traffic engineering solutions developed for either IP networks or WDM networks can be directly applied to each layer, respectively. The overlay client-server network is a natural match to the notion of overlay traffic engineering. With *integrated traffic engineering*, performance optimisation with respect to selected objectives is pursued co-ordinately across both IP and WDM network elements. As emergence occurs of more sophisticated hardware that integrates functionality of both IP and WDM at each NE, integrated traffic engineering can be performed more efficiently.

7.2.1   Overlay Traffic Engineering

The principle of overlay traffic engineering is that optimisation is pursued for one layer at a time. This means that an optimal solution in a multi-dimensional space is sought by sequentially searching different dimensions. Obviously, the optimal solution is search-sequence-dependent, and not guaranteed to be a global optimal. The earlier presence of a dimension in the search sequence favours better optimality towards that dimension. An advantage with overlay traffic engineering is that mechanisms can be tailored to best meet the needs of a particular layer (either IP or WDM) for selected objectives. Figure 7.2 shows overlay traffic engineering.

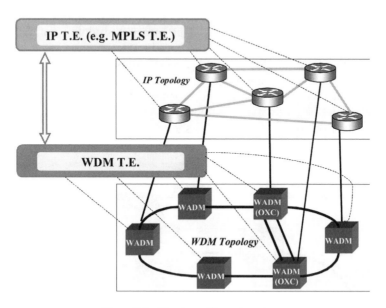

Figure 7.2   Overlay traffic engineering.

Overlay traffic engineering can be constructed by connecting IP routers to an OXC-based WDM network through an OADM. IP/WDM networks constructed this way render the OXC-based WDM network, the server layer that is facilitated by the physical network composed of optical NEs and fibres. Each fibre carries multiple wavelengths whose routing is flexibly reconfigurable. The client layer (i.e. the virtual network) is formed by IP routers connected by lightpaths embedded in the physical network. Topology of a virtual network is reconfigurable due to the reconfigurability of lightpaths in the server layer. Interfaces of an IP router connected to OADM are reconfigurable interfaces. That means IP neighbours connected through such reconfigurable interfaces can be changed by updating the underlying lightpath configuration. In IP/WDM networks, congestion control can be realised not only at the flow level using the same topology but also at the topology level using lightpath reconfiguration. Therefore, not only can a traffic source regulate its stream of packets before sending them to the network, but the network also can adapt itself to the traffic pattern in selectable timescales. In the IP layer, congestion control provides the basis for traffic engineering; i.e. how to transfer bit streams along their routes quickly to their destination. In the WDM layer, allocation control is used to manage network resources (such as wavelength) and allocate them to virtual IP links. The WDM layer allocation control can be static, i.e. fixed at the beginning of the connection request, or it can be dynamic and changed over the duration of the connection. It is also this flexibility that permits the WDM layer to provide connections to the upper layer with different quality of service.

7.2.2   Integrated Traffic Engineering

The principle of integrated traffic engineering is that optimisation is pursued at both IP and WDM networks simultaneously. This means the global optimal solution is sought in a multi-dimensional space. Integrated traffic engineering is applicable to networks in which functionality of both IP and WDM is integrated at each NE. When IP and WDM functionality is integrated, an integrated control plane for the networks becomes feasible, which in turn provides a natural match for an integrated traffic engineering model. IP traffic management and WDM resource control and management are considered together. Figure 7.3 shows integrated traffic engineering.

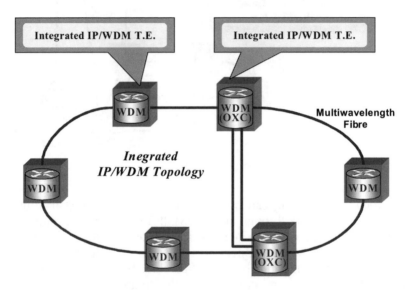

Figure 7.3   Integrated traffic engineering.

7.2.3   Comparison of the Two Models

The relationship between performance optimisation and resource allocation distinguishes the overlay approach and integrated approach. With overlay traffic engineering, performance optimisation such as load balancing and traffic routing can be done at the IP layer, which is essentially separated from WDM physical resource allocation that is done in the WDM layer. Because of this, performance optimisation at the IP layer can use reconfiguration as well as traditional mechanisms without involving reconfiguration. When reconfiguration is not involved, it means the performance optimisation is sought within a fixed set of resources (for a fixed IP topology). When reconfiguration is employed, dynamic resource allocation for a virtual topology is involved. Then the performance optimisation at the IP layer has a choice regarding how much it wants to consider the resource states of the WDM layer, where the actual physical resource allocation is taking place. On the other hand, performance optimisation and network resource allocation are combined in integrated traffic engineering. If the performance optimisation involves a variable set of network resources, resource allocation is automatically encompassed in the optimisation.

The traffic engineering models can be implemented in either centralised or distributed fashion. Table 7.1 shows 4 choices on TE models implementation. The overlay approach intuitively favours a centralised or hierarchical implementation, in which there is one IP layer TE and one WDM layer TE, where the two TEs communicate through either WDM edge UNI or interfaces between IP NMS and WDM NMS. In a centralised overlay implementation, the central IP layer NC&M manager and the central WDM layer NC&M manager gather state information about its own layer, respectively. However, this approach may not scale because clearly there are bottlenecks at NC&M managers of both IP and WDM. The integrated approach maps naturally to a distributed implementation for traffic engineering. That is each site is capable of exploiting congestion control, and makes resource allocation decisions based on the locally maintained IP/WDM network state information. The distributed implementation of traffic engineering improves availability and flexibility, but it faces complex synchronisation problems that exist due to the nature of parallel decision making in distributed locations.

Table 7.1   T.E. models implementation.

	Overlay model	Integrated model
Centralised implementation	Centralised overlay TE	Centralised integrated TE
Distributed implementation	Distributed overlay TE	Distributed integrated TE

In summary, the overlay approach may not perform efficiently as network size increases, because both IP and WDM NMS servers become potential bottlenecks. The integrated approach faces enormous implementation complexity. Synchronisation among a large number of IP/WDM nodes regarding network state and configuration information takes considerable time to converge. The selection of overlay vs. integrated traffic engineering and their respective implementation choice is operational network and application traffic dependent. Nevertheless, the functional TE framework presented covers the two models of two implementation approaches. The components within the framework are general for traffic engineering applications in IP/WDM networks.

7.3   IP over WDM Traffic Engineering Functional Framework

The fundamental enabling mechanism in the traffic engineering framework is lightpath and virtual path provisioning on demand. A unique property in a WDM network is the lightpath and virtual topology reconfigurability. That is, for a physical fibre topology, the physical WDM network can support a number of virtual topologies formed by lightpaths. As shown in Figure 7.4, the main functional components in a traffic engineering triggered reconfiguration framework include the following components:

- *Traffic monitor*: this component is responsible for collecting traffic statistics from the routers/switches or links. To support this traffic engineering framework, IP/WDM networks monitor IP traffic.
- *Traffic analysis*: this component is applied to the collected statistics for decision-making. In case of an update, the analysis report is produced.
- *Bandwidth projection*: this component is used to predict the bandwidth requirements in the near future based on past and present measurements and traffic characteristics.
- *Signal performance monitor*: this component is responsible for observing the optical signal QoS for wavelength channels. Signal QoS presents a complex yet dynamic factor to wavelength routing and fault management. WDM fault management is not the focus of this TE framework. So, signal QoS is only used by lightpath reconfiguration.
- *Traffic-engineering reconfiguration trigger*: this component consists of the set of policies that decide when a network-level reconfiguration should be performed. This can be based on traffic condition, bandwidth projection and other operational issues, for example, suppressing the influence of transitional factors and reserving adequate time for network convergence.

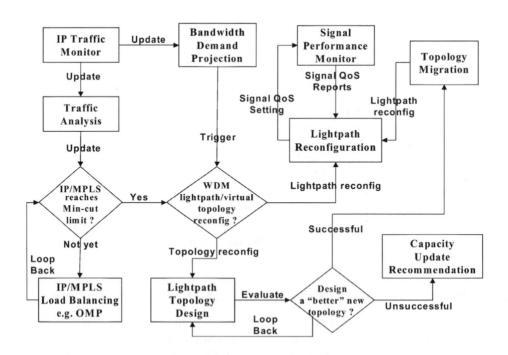

Figure 7.4   IP/WDM traffic engineering functional framework.

- *Lightpath topology design*: this component computes a network topology based on the traffic measurements and predictions. This can be considered as optimising a graph (i.e. IP routers connected by lightpaths in the WDM layer) for specific objectives (for example, maximising throughput), subject to certain constraints (for example, nodal degree, interface capacity), for a given demand matrix (i.e. traffic load applied to the network), which in general is a NP-complete problem. Searching for the optimal graph can be computationally expensive. Since changing traffic patterns trigger reconfiguration, an optimised graph may no longer be optimal when its configuration is actually completed. A practical approach uses heuristic algorithms that focus on specific objectives such as cost-effective, fast convergence, and/or minimal impacts on ongoing traffic instead of on global optimality.
- *Topology migration*: this component consists of algorithms to schedule the network migration from an old topology to a new topology. Even if WDM layer resources are sufficient to support any migration sequence (a most demanding example can be adding all new connections before tearing down any unwanted ones), there are still other issues concerning the migration. For example, as WDM reconfiguration deals with large-capacity channels (for example, up to OC-192 per wavelength), changing allocation of resources in this coarse granularity has significant effects on a large number of end-user traffic. A migration procedure consists of a sequence of establishing and taking down individual WDM lightpaths. Traffic flows have to adapt to the lightpath changes during and after each migration step.
- *Lightpath reconfiguration*: this component is used to reconfigure individual lightpath, i.e. path tear-down and setup. This in turn requires the following modules:

  - *Lightpath routing algorithm* is needed for lightpath computation. When the route of the lightpath is not specified, this component computes the explicit routing path. If a routing protocol is available (for example, OSPF with optical extensions) the routing path can also be obtained from the local routing table.
  - *Path setup/tear-down mechanism* is needed for path setup or tear-down, for example, a signalling protocol.
  - *Interface management* is responsible for an interface and its related information update. Lightpath reconfiguration may reassign WDM client interfaces to a different lightpath, which will affect the WDM interface to IP network. IP routing requires IP addressing and allows packet forwarding only within one IP subnet. The new IP topology may require IP interface address changes.

7.3.1 IP/WDM Network State Information Database

A network state information database is required to control and manage an IP/WDM network. Based on the traffic engineering model and its implementation approach, the network state information database (including TE) is constructed and maintained accordingly. For example, in the integrated approach an entire integrated IP/WDM database is maintained at each site and the synchronisation among the database is taken care of by a distributed protocol; in the overlay approach the IP database is stored separately from the WDM database.

Network state information for traffic engineering considers two aspects: resources and their usage. Conventional representation of network resources for the purpose of packet routing can be simply the topological information. However, traffic engineering requires more information, for example, total bandwidth and current usage on each link. Two tiers of routing exist in an overlay IP/WDM network. One is routing lightpaths across the physical network, and another is routing data through these lightpaths. Traffic engineering can be practised in both tiers. WDM traffic engineering is interested in not only the utilization status of the network resources but also the optical characteristics of the WDM optical connections and the signal quality. When overlay traffic engineering is attempted, the objective functions at different layers may even be different. In the case of integrated traffic engineering, traffic control and resource allocation are considered together so that optimisation objectives must be co-ordinated.

Although different traffic engineering models require different design and implementation of the network state information database, many common attributes are shared in both cases as discussed below. In an overlay approach, the IP layer's network state information database consists of the following information:

- *IP virtual topology* is a directed graph, where vertices represent IP routers and edges represent lightpaths. This is identical to what is required by a standard link state routing protocol. Also included are the data rates and signal formats that each IP interface can support. This is useful for dynamic configuration.
- *IP link status* includes link capacity and its utilization (as a percentage). Other measurements (for example, the number of packets dropped at a router interface) required by traffic engineering algorithms, should also be included.

At the WDM layer, the managed object is the physical network, where the load is represented by lightpath trails. Traffic engineering practice at this layer is undertaken through embedding the IP virtual topology into the physical network. The network management operations are wavelength assignment and routing. If wavelength continuity is required along a trail, a single wavelength must be assigned; if a WDM NE is capable of wavelength interchange, different fibre hops may use different wavelengths. A network state information database at the WDM layer, therefore, includes these components:

- *Physical topology* is a directed graph, on which vertices represent WDM NEs and edges represent fibres.
- *NE properties* indicate switching capability and port availability. A NE can perform fibre switching (i.e. connecting all wavelengths on an incoming fibre to an outgoing fibre using the same wavelengths), or wavelength switching (i.e. connecting a specific wavelength on an incoming fibre to the same wavelength on one or more ongoing fibres). Furthermore, a signal can be converted into a different frequency through wavelength conversion. A NE has a limited number of add/drop ports, so there may be contention on inserting/dropping signals at the switch fabric.
- *Fibre state* includes the number of wavelengths, directionality, link protection type, and optical signal quality such as total wavelength power for the fibre,

wavelength registration, individual wavelength power, optical SNR of each wavelength.
- *Light path state* contains the source NE ID, add port ID, sink NE ID, drop port ID, wavelength ID (of each fibre hop) and its directionality, fibre ID, bit rate, end-to-end optical SNR, SRLG (Shared Risk Link Group) IDs. Optionally, lightpath priority, lightpath pre-emption level.

In integrated traffic engineering, the wavelength topology and the fibre topology are combined; the optimisation is on wavelength routing subject to fibre constraints. Therefore, the contents of the IP layer database and the WDM layer database in the overlay model presented above will be merged into one single IP/WDM network state information database.

7.3.2   IP to WDM Interface Management

How to efficiently and effectively control interfaces between IP and WDM is critical to proper traffic engineering in IP/WDM networks. The approach should take advantage of hardware features yet be flexible and extendable since novel network hardware is emerging quickly. Presently, a single IP interface can couple with only one lightpath, which renders an IP over WDM network not much more special than an IP over any virtual circuit transport network. Traffic engineering, or network control and management in general, for this type of IP/WDM networks is open to adopt any existing techniques developed.

In the software domain, IP/WDM requires corresponding software to manage the IP to WDM hardware interface and translate between IP addressing and WDM addressing schemes (when necessary). In an overlay approach, an address resolution scheme is needed to maintain the mapping between the two layers. Note the WDM layer may also use IP addresses instead of physical addresses, but the address resolution scheme is still required because the IP layer and the WDM layer use two different addressing spaces and different routing instances. An advantage of the overlay approach is leveraging on the existing control mechanisms for the IP network and the WDM network. In an integrated approach, each IP/WDM NE interface is IP addressed. Therefore, there is only one addressing scheme used, i.e. the IP addressing, and possibly only one routing instance. However, the conventional IP protocols such as OSPF need to be extended to cover IP/WDM networking concerns.

7.3.3   Examples of Reconfiguration Triggers

As shown in Figure 7.5, reconfiguration can be triggered by several factors. Example triggers include:

- traffic engineering
- fault
- protection/restoration
- network maintenance.

The fault trigger has fault detection, root cause analysis, and a fault manager. Once

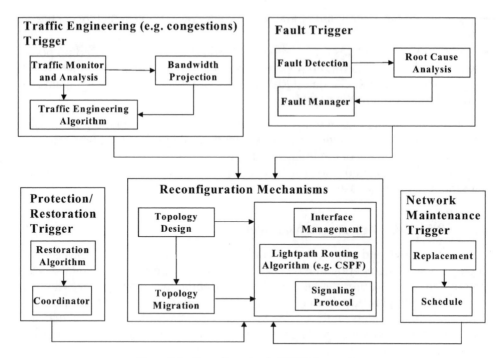

Figure 7.5   Reconfiguration in IP/WDM networks.

a fault is detected and its root cause is determined, the fault manager decides what to do. The fault network component can be isolated and connections affected are rerouted. The protection/restoration trigger can be used to support dynamic restoration. Once the primary path is down, the restoration algorithm can compute (or precompute) a disjoint backup path. The co-ordinator is responsible for path switch over from the primary to the backup path. The reconfiguration mechanisms provide facilities for path setup and tear-down, i.e. a signalling protocol. The distributed signalling protocol plays an important role in protection/restoration. Network maintenance involves replacement and operations scheduling, which can be facilitated using reconfiguration.

There are three main components in the reconfiguration mechanisms: topology design, topology migration, and lightpath reconfiguration. The lightpath reconfiguration component consists of three modules: interface management, lightpath routing algorithm, and signalling protocol. The reconfiguration mechanisms can be invoked when certain trigger conditions are met.

7.3.4   Traffic Monitoring and Measurements

IP over WDM networks are used to deliver IP traffic. Network traffic measurement obtained through IP monitoring and collection mechanisms is critical to the traffic engineering framework. The accuracy of the measurement directly affects the framework performance since the measurement provides the framework input describing the dynamic network condition. In a continuous, closed-loop process, network traffic measurement can trigger the reconfiguration algorithm as well as evaluate the effectiveness of the reconfiguration.

What statistics need to be collected depends on the optimisation objectives and the heuristic algorithm being used. For example, traffic volume is important when the optimisation objective is global throughput; one-way delay between a selected pair of nodes is important when latency between the node pair is to be minimised.

Depending upon traffic engineering objectives, traffic measurement and load statistics collection must be able to flexibly select granularities of aggregation, sampling and measurement. Aggregation granularity is related to the property of endpoints of the monitored flow. Sampling granularity compromises the sheer volume of the traffic being monitored and the processing power of monitoring devices. Measurement granularity determines the time period over which traffic data are averaged. Proper selection of these granularities can help control the operational cost of traffic engineering. In addition to statistics available in SNMP MIBs, long-term aggregation of statistics and their analyses are especially important in determining a new topology. Key elements of these analyses are aggregating traffic data at the IP layer into traffic matrix statistics and predicting the traffic demands of the near future.

Traffic monitoring and statistics collection is not an easy task to implement due to the sheer volume of the traffic and the high capacity of modern Internet trunks. Without careful consideration, it may significantly degrade network performance. In addition, traffic matrix estimates may not be available directly with most contemporary router equipment; link loads are available, but they may not hold enough information to estimate a reasonable matrix.

According to [Kali99], traffic monitoring should follow these guidelines:

- uses standard metrics;
- measures unidirectional properties;
- uses dedicated machines for accurate measurements;
- measures continuously;
- provides long-term performance data;
- provides real-time access to performance data;
- provides end-to-end measurements.

Traffic monitoring methods and tools

Traffic monitoring capability can be implemented at either packet level or network level. We review several software tools available for traffic monitoring.

Packet level monitoring: This method requires monitoring every packet, for example, at the packet source. In particular, the packet header can be examined and the related information can be extracted from the header. The two popular software tools for packet level monitoring are as follows.

- **'tcpdump'**

  This tool can print out the headers of packets on a network interface that match the boolean expression. It can also be run with the '-w' flag, which saves the packet data to a file for later analysis, and/or with the '-b' flag, which causes it to read from a saved packet file rather than to read packets from a network interface. In all cases, only packets that match *expression* will be processed.

  'tcpdump' also has a '-c' flag. Without the '-c' flag, it continues capturing packets until it is interrupted by a SIGINT signal or a SIGTERM signal; with the '-c' flag, it will capture packets until it is interrupted by a SIGINT or SIGTERM signal or the specified number of packets have been processed. Finally, it will also report the number of packets dropped by the kernel, for example, due to buffer overflow, and the number of packets related to the filter, i.e. the packets that match the filter expression.

- **'libpcap'**

  This is the library that one can use to capture packets from the network card directly. It provides an implementation-independent access to the underlying packet capture facility provided by the operating system [Stev98]. The 'libpcap' application has the following general format:

  - Specify which interface one wants to sniff on, e.g. eth0.
  - Initialise 'libpcap', which can be operated multiple devices. Each sniff is identified by a file descriptor (or a session identifier).
  - Define the rule set to detail what type of traffic one wants to sniff, compile the rule set, and apply the rule set to the session or the file.
  - Execute primary loop for sniffing.
  - Close the session or the file descriptor.

Network level monitoring: This method requires monitoring network level behaviour, for example, network congestion or performance bottlenecks. Network level monitoring can be conducted in three fashions: active measurement, passive measurement, and control monitoring. The active measurement approach sends data through the network and observes the results; the passive measurement approach inserts a probe to a link between network nodes, and summarises and records information about the traffic flowing on that link; the control monitoring approach captures and analyses network control information such as routing and network management information.

In particular, network level monitoring can be performed using Ping-based tools, traceroute, SNMP, and network monitoring devices. For example, one can use Ping to measure the round trip response time, the packet loss percentages, the variability of the response time, and the lack of reachability (i.e. no response for a succession of Pings):

- *Ping-based tools:* The tools to implement the Ping-related measurements and analysis are referred to as PingER [Matt00]. Ping-based tools can collect active measurements with ICMP messages (echo request/response). These tools have three components:

  - *Monitoring site module:* this component is installed and configured at the monitoring site. The Ping data collected needs to be made available to the archive hosts via HTTP. There are also PingER tools to enable a monitoring site to be able to provide short-term analysis and reports on the data it has in its local cache.
  - *Remote monitoring site module:* this component is installed at a passive remote host. It is associated with at least one monitoring site.
  - *Archive and analysis site module:* The archive and analysis sites may be located at a single site, or even a single host or they may be separated. The archive sites gather the traffic information, by using HTTP, from the monitor sites at regular intervals and archive it. They provide the archived data to the analysis site(s), that in turn provide reports that are available via the Web.
  - *Traceroute:* this tool prints out the intermediate hops between a source and destination pair and measures the round trip time between the source and each hop. Traceroute uses the IPv4 TTL field or the IPv6 hop limit field and two ICMP messages (i.e. 'time exceeded in transmit' and 'port unreachable'). It starts to send a UDP message to the destination with a TTL (or hop limit) of 1, which causes the first hop router to return an ICMP 'time exceeded in transmit' error. It continues to send a UDP message to the destination but increments the TTL by 1 each time. Finally, the destination receives the probe UDP message, and returns an ICMP 'port unreachable' as the UDP message is addressed to an unused port. In the default setting, it sends three probe UDP messages for each TTL setting. So the round trip time to each hop can be estimated by averaging the three measured times.
  - *SNMP:* one can use SNMP to collect local measurements from IP routers.
  - *Passive link measurements:* this approach requires special network devices such as protocol analyser or OCX-mon. An OCX-mon monitor is a rack-mountable PC running the FreeBSD or Linux operating system. In addition to the PC components (400 MHz PII, 128 Mbytes RAM, 6-Gbyte SCSI disk), two measurement cards are installed in the PC, and an optical splitter is used to connect the monitor to an OC3 (155 Mb/s) or OC12 (622 Mb/s) optical link. DS3, FDDI, and electrical interfaces are also available or being experimented with [McGr00].

Traffic engineering flow definition

The term 'flow' has been widely used but with various definitions such as in QoS provisioning, QoS routing, packet forwarding. A common definition of a 'traffic flow' in traffic engineering is important to data aggregation and to usage of traffic statistics collected by various router-vendor traffic collection schemes.

The definition of flow for traffic measurement is as follows.

- Directionality: Flows can be either unidirectional or bi-directional. With unidirectional flows, traffic from A to B and from B to A are considered separate traffic flows for aggregation and analysis purposes. Bi-directional data provides insights into the behaviour of individual protocols, including problems that may manifest themselves in core backbones, but which are harder to identify at endpoints. Obviously, bi-directional data is complicated and places further complexity on the traffic engineering algorithms. For simplification, one can assume that bi-directional flows no longer distinguish the difference between the two unidirectional flows of the same node pair. For example, the traffic volume or utilization on a bi-directional flow is just the larger traffic volume or higher utilization of the two unidirectional flows. Or, the bi-directional flow can be simply aligned to one of the unidirectional flows. As such, a bi-directional flow assumes the traffic between the node pair is always symmetrical.

- Flow endpoints: The fundamental criteria for flow specifications are the flow endpoints that describe the communicating entities. Flows can be referred to as traffic between:

  - *Applications:* identified by, e.g., < protocol-ID, source-port, source-IP-address, destination-port, destination-IP-address > .
  - *Hosts:* identified by, e.g., < source-IP-address, destination-IP-address > .
  - *Networks:* identified by, e.g., < source-IP-prefix, destination-IP-prefix > .
  - *Traffic sharing a common path on the network:* identified by, e.g., < ingress-router-interface, egress-router-interface > .

- Flow granularity: Each flow is associated with the granularity, which basically refers to the size of the flow.

Traffic monitoring sampling granularity

The efficiency and effectiveness of traffic monitoring is related to the traffic monitoring granularity, which can be specified in terms of sampling granularity and measurement granularity. When monitoring a network, a trade-off exists between precision of the monitoring and the overhead introduced.

There are generally two approaches to monitor traffic on a network. The monitoring can be part of the functionality of the monitored network element itself, or the monitoring is performed by a piece of dedicated equipment. In either case, the available processing power may not be powerful enough for carrying out a full monitoring task. Instead, the monitoring has to be conducted in a sampling manner. If this happens, one needs a rule to control the sampling such that the underlying traffic volume can be best estimated from the samples collected at an affordable/ economical rate.

While some measurement devices or high end routing/switching hardware may be capable of continuous monitoring of traffic flows, applications such as capacity planning do not require this level of detail. During the NSFNET era, ANS Communications

Inc. conducted tests on sampling at granularities of 1 in 50 packets, 1 in 100 packets, and 1 in 1000 packets. For the purposes of capacity planning, they found that sampling rates of 1 in N packets is acceptable for moderate values of N. But, sampling rates of as high as 1 in 1000 packets require confidence intervals that are too high to trust for capacity planning. What value of N is proper for network level reconfiguration, which by nature is capacity planning, in today's IP networks, needs additional research. While a few hundreds or less would be a good initial choice, the sampling granularity should be user configurable, an important feature required being to tune the trade-off among measurement precision, traffic characteristics and processing power. Sampling based on time intervals is discouraged since the results do not safely reflect the burstiness characteristics of traffic.

When network level reconfiguration is performance orientated (for example, optimising end-to-end latency), related measurement requires accurate packet transmission/arrival timestamps. Sampling granularity options that assist in filling these requirements include:

- A random choice of 1 of N where a random member out of the next N packets is chosen, thereby reducing biases introduced into the data by sampling.
- Selection of consecutive packets for inter-arrival time studies, for analysis of packet arrival times.
- Selection of every packet from a single flow (this 'flow' is defined from a higher layer's perspective).

Although additional research is needed regarding acceptable sampling rates and confidence intervals for specific application requirements, what is clear so far is the two properties of traffic monitoring for network level reconfiguration purpose:

- configurable sampling rates
- off-line data processing.

Measurement granularity and traffic matrix

In addition to sampling granularity, there is an issue of measurement granularity. The objective of the monitoring is a network traffic matrix by which the traffic demand of the near future is predicted. An IP flow for the purpose of network level reconfiguration consists of the traffic stream that arrives at an edge router destined to another edge router. The flow is characterised by the time trace of the bytes/second that arrives to the flow. The mechanisms for obtaining the time trace of the traffic stream are described in the following sections. The time trace of the traffic stream may comprise of average bytes in the traffic stream over a given period of time.

The primary measurement parameter of interest for network level reconfiguration is the traffic volume matrix between the edge routers in the IP/WDM network. Formally, we need to measure $V_T^P\ (i, j)$, which represents the volume of traffic of type P flowing from edge router i to edge router j in the averaging interval of size T. Therefore, these measurements provide a time trace of traffic volumes between each pair of edge routers for each traffic type. T is the measurement granularity parameter, which represents the time resolution of the traffic traces. Given V_T^P (i, j) , the IP network topology, and the routing algorithm used, the offered traffic volume to each IP link in each time period can be determined.

For example, the traffic stream may be described by a time trace consisting of 5-minute averages of the number of bytes per second seen on the stream. The time-granularity of the averages (parameter T) refers to the time-period over which the averages are taken. Fine-grained averages (over intervals of the order of seconds) are desirable, but this is traded off with the measurement capabilities and overhead (in terms of router processing time and storage space required for measurements) required to induce the averages. Coarse-grained averages over longer periods of time can be inferred from the fine-grained averages.

Ideally, a router's measurement export format should employ a system whereby each export packet is indexed to describe its contents, and routers are configured with knobs permitting users to specify which data fields they would like to be exported. The entry in the export packet would only write to those fields identified in the index. Using indexing should simplify flow collection and provide flexibility in terms of customer-specified tuple combinations and alternative aggregation schemes. In network level reconfiguration design, one can use a bit mask index to collect measurements. The bit mask is referred to as a network prefix that typically sets bits for IP addresses and netmask length of local networks attaching to a particular edge router.

7.3.5   Optical Signal Performance Monitoring

A sophisticated traffic engineering framework designs network and topology according to traffic demands as well as utilizes network resources subject to resource availability and signal QoS constraints. To some extent, optical layer characteristics influence network level decision making. For example, signal QoS can limit the number of wavelength channels supported over a fibre link and further the data rate supported on each wavelength channel. In addition, optical signal QoS possesses dynamic factors that cannot be found in conventional electrical signals. Without considering WDM layer characteristics, one can only expect that the WDM layer does not have any resource constraints. Once the virtual IP topology (based on traffic demand) is generated, the assumption is that the topology can always be supported using light circuits. Further sophistication arises when multiple lightpath route computations are intentionally processed simultaneously for optimality. In an integrated IP/WDM framework, these optical WDM characteristics need to be linked to the adaptive IP control protocols. A WDM network may have its own fault management system, but it should be integrated closely with IP control. So the integrated IP/WDM network still possesses the key features of IP, flexibility and adaptability.

One may argue that some optical characteristics should be taken care of by lower layers (i.e. layer 1 or layer 2). But the problem precisely is how this responsibility should be divided between lower and upper layers and how the network control should leverage or at least be aware of these optical related properties.

Performance monitoring in all-optical networks is an expensive process, which requires optical signal splitting using special devices or at the NE. However, after splitting, the original signal is likely to be degraded and therefore its travelling distance is limited without signal regeneration. Optical domain signal regeneration (for example, using optical transponders) is immature and very expensive, so in fact, at the current stage, performance monitoring for all-optical networks in a scalable fashion is still an open issue by itself. In O-E-O optical networks, signal QoS is less of a problem since at every hop, optical signals are 3R regenerated.

7.4 Teletraffic Modelling

There are two phases in modelling a communication or computer network: traffic modelling and system modelling. A traffic model is used to describe the incoming traffic to the system, such as traffic arrival rate, traffic distribution, and link utilization, whereas the system model is used to describe the networking system itself such as topology and queuing model. There are two types of system models: loss system model and queuing system model. A pure loss system model can be used to model the circuit-switched networks in which there is no waiting place. So if the system is full, when a customer comes, he/she will not be served but lost. The loss system relies on over-provisioning to address customer needs. A pure waiting system model is used to model the packet-switched networks, where the assumption is that there are an infinite number of waiting places. So if all the servers are busy, when a customer comes, he/she can occupy one of the waiting places in the system. There is no customer lost but the customer may need to wait for a certain time to be served. The wait model can set buffer size and queuing policy to address customer concerns. This section focuses on traffic modelling.

7.4.1 Classical Telephone and Data Traffic Model

Telephone traffic can be modelled using the Erlang model, which is a pure loss model. If the amount of traffic or traffic intensity is described as α,

$$\alpha = \lambda \times h$$

where λ represents the customer/call arrival rate and h represents the mean (call) holding time (service time). The unit of traffic intensity is called Erlang (erl). Traffic of one Erlang means that on average one channel is occupied. Blocking in the **Erlang model** refers to the event that a call is lost. There are two different types of blocking quantities: call blocking and time blocking. The call blocking is the probability that a call (a customer) finds that all channels have been occupied whereas the time blocking is the probability that all channels are occupied at an arbitrary time. Certainly, the call blocking, B_c, is a better measure of QoS from a customer point of view. Assuming a $M/G/n/n$ loss system, which has n channels on the link, calls arrival following a Possion process with rate λ, and call holding times independent and identically distributed according to any distribution with h, the relationship among call blocking, traffic intensity, and mean holding time is given by the Erlang blocking formula:

$$B_c = Erlang(n, \alpha) = \frac{\dfrac{\alpha^n}{n!}}{\displaystyle\sum_{i=0}^{n} \dfrac{\alpha^i}{i!}}$$

Data traffic can be described using queuing models, what are wait models. Data traffic is represented by packet arrival rate λ, average packet length L, and packet transmission time $1/\mu$. Assuming the system R represents the link's speed, i.e. the data units per time unit, packet transmission time is also equal to L/R. So the amount of traffic can be described by the traffic load ρ:

$$\rho = \frac{\lambda}{\mu} = \frac{\lambda \times L}{R}$$

From a customer point of view, QoS is an important feature. The QoS is presented by P_z, which is the probability that a packet has to wait longer than a preference value z. Assuming a $M/M/1$ queuing system, which has packet arrival following a Possion process with rate λ and packet lengths independent and identically distributed according to exponential distribution with L, the relationship among traffic load, system capacity, and QoS is given by the following formula:

$$P_z = Wait(R, \lambda; L, z) = \begin{cases} 1, \lambda L \geq R(i.e.\rho \geq 1) \\ \dfrac{\lambda L}{R}\exp\left(-\left(\dfrac{R}{L} - \lambda\right)z\right), \lambda L < R(i.e.\rho < 1). \end{cases}$$

7.4.2 Novel Data Traffic Models

In [Lela94], Ethernet LAN traffic was studied with high-accuracy recording of hundreds of millions of Ethernet packets including both the arrival time and the length. The study concluded that the Ethernet traffic seems to be extremely varying due to the presence of burstiness across a wide range of time scales from microseconds to milliseconds, seconds, minutes, hours, and days. In addition, the study also showed that the Ethernet traffic is statistically self-similar. This means that the traffic looks similar in all time scales and can use a single parameter, known as the Hurst parameter, to describe the fractal nature. These Ethernet traffic properties cannot be captured using the conventional traffic models such as the Poisson Model.

In [Paxs95], Internet WAN traffic was studied with both packet level and connection level measurements. The study concluded that at packet level, the empirical distribution of TELNET packet inter-arrival times is heavy tailed not exponential as traditionally modelled, and at connection level, for the inter-active TELNET sessions, the connection arrivals follow a Poisson process (with hourly fixed rates). However, the study also concluded that at the connection level, for connections within user-initiated sessions (FTP, HTTP) and machine generated, the connection arrivals are bursty and sometimes even correlated and no longer follow a Poisson process.

To capture the burstiness of Internet data traffic, subexponential distributions such as Log-normal, Weibull, Pareto distributions, and heavy-tailed distributions such as Pareto distribution with location and shape parameters can be used. For processes with long range dependence, self-similar processes such as Fractional Brownian motion (introduced later) can be used.

7.4.3 A Bandwidth Projection Model

Closed-loop traffic engineering can be performed according to feedback as well as bandwidth projection. Feedback-based closed-loop traffic engineering is described in Section 7.10. Bandwidth projection can be considered as a useful tool for traffic engineering. The prediction of bandwidths in the future can be used to trigger the network level reconfiguration. By predicting the bandwidth of the traffic flow, one can determine the capacity requirements of the IP/WDM link, and therefore make reconfiguration decisions.

An IP traffic stream is the unidirectional flow of IP packets (of the same traffic class) between two end-points. The end-points may be adjacent routers in which case the IP traffic stream is the traffic flowing on the link(s) between the two routers. Alternatively, end-points of a traffic stream may be non-adjacent routers. An IP traffic stream is unidirectional; this reflects the possible asymmetry of traffic between two end-points. Given a traffic stream, one would like to characterise and estimate the bandwidth of the traffic stream. Although the following method is generally applicable, it is expected that it will be applied to estimate the offered load to an IP link, and these estimates will then be used to perform reconfiguration decisions.

The prediction time-horizon is how far ahead into the future one predicts. The time-horizon for network level reconfiguration is determined by several factors. One would prefer network level reconfiguration to be able to respond to changes in traffic trends such as time-of-day load changes. On the other hand the reconfiguration time-horizon has to be at least as long as the time duration for the reconfiguration procedure. The reconfiguration time horizon consists of the following factors:

- time for making a prediction;
- time for computing the next topology;
- time for migrating the topology to the new topology.

The time to make the bandwidth predictions depends on the computational complexity of the prediction model. The time to compute the next topology depends on the complexity of the algorithms or heuristics utilized to perform the topology design. The time to migrate the topology to the new topology depends on the migration procedure employed. Assume the migration procedure consists of a sequence of establishing and taking down individual IP/WDM links. In this case, the migration time consists of the time to establish and tear-down WDM links, and the time for the routing protocol to stabilise after each topology change.

As determined from the above discussion, one assumes a certain reconfiguration time horizon, which is the time period that determines how frequently network level reconfigurations can be performed. This time period is referred to as the coarse-grained time period (as opposed to the fine-grained time period which is the time-period for traffic measurements). The coarse-grained time period will be a configurable parameter to the design. The impact of different values of the coarse-grained parameter is evaluated.

A prediction of the traffic stream bandwidth in the next time period depends on several factors including the following:

- *Time-of-day and day-of-week:* Chapter 3 presents evidence on the presence of time-of-day correlations in Internet traffic volume.
- *Correlations from the previous time samples:* the assumption is that traffic volume in the recent past will impact the traffic volume in the near future.
- *Traffic arrival process:* From the discussion in Chapter 3, one cannot assume Poisson arrival processes. In order to predict bandwidth requirements, the self-similarity characteristics of the traffic stream have to be taken into account.

The goal is to develop an empirical parameterised model for predicting the traffic bandwidth in the next time period. This empirical model will utilize historical traffic measurement information, and will assume a self-similar traffic arrival process. The model presented below was originally proposed by A. Neidhardt and J. Hodge at Bellcore, for predicting the capacity of an ATM VPC carrying IP traffic and extended in the NGI SuperNet NC&M project at Bellcore/Telcordia [Supe01]. Without loss of generality, the following discussion assumes that the size of the coarse-grained period is 1 hour.

Fractional Brownian motion process

Fractional Brownian motion (FBM) is a self-similar process that is described by three parameters, a mean arrival rate m, a variance coefficient a, and the Hurst parameter H. An IP/WDM network can model the arrival process as FBM to account for fine-grained traffic volume fluctuations within a coarse-grained period. [Norr95] defines the Fractional Brownian Motion (FBM) arrival process as follows:

$$A(t) = mt + \sqrt{am}Z(t) - \infty < t < \infty,$$

where $Z(t)$ is the normalised Fractional Brownian motion process with the following definition:

- $Z(t)$ has stationary increments
- $Z(0) = 0$ and $E[Z(t)] = 0$ for all t
- $E[Z(t)]^2 = |t|^{2H}$ for all t
- $Z(t)$ has continuous paths
- $Z(t)$ is Gaussian.

The variance of $A(t)$ is given as:

$$V[A(t)] = am|t|^{2H}$$

Consider a queue with the above FBM arrival process, and with constant service rate C. This system has four parameters: m is the mean arrival rate, a is a variance coefficient of the arrival process, H is the self-similarity parameter, C is the service rate. The overflow probability of the above queue i.e., $P(Q > B)$ where B is the buffer size, is given by the following approximation from [Norr94]:

$$P(Q > B) \geq \exp\left(-\frac{1}{2}(am)^{-1}(C-m)^{2H}H^{-2H}(1-H)^{-2(1-H)}B^{2(1-H)}\right)$$

Assuming that one requires that the overflow probability be bounded, i.e.,

$$P(Q > B) \leq \exp\left(-\frac{z^2}{2}\right)$$

An expression for the service rate of the queue C is obtained as follows:

$$C \quad \geq \quad m \;+\; m^{\frac{1}{2H}}\left(z^{\frac{1}{2H}}a^{\frac{1}{2H}}B^{-\frac{1}{1-H}}H(1-H)^{\frac{1}{1-H}}\right)$$

Traffic projection principles

The first principle is that traffic bandwidth in the next period depends strongly on the traffic seen on the traffic stream in the same time period in the previous week.

This principle reflects the strong time-of-day and weekly traffic volume patterns observed in traffic streams on links. Therefore, as a first-order estimate, the average traffic volume in the next period is exactly the same as the volume seen in the same period, of the same day of the previous week. This can be expressed as:

$$F_0 \cong F[h, d]$$

where $F[h, d]$ is the traffic volume seen in hour h of day d of the previous week. Assume that the growth rate of traffic from week to week is modelled by a function with one parameter γ. Also assume that the growth function is exponential as:

$$F_1 = F_0 \, e^{\gamma F_0}$$

where γ is a model parameter to be estimated from the traffic measurements. As an example of determining γ, we can perform the following simple fit procedure. Let W_0 and W_1 be the total traffic volumes seen in the previous two weeks on the traffic stream. Then γ can be determined from the equation:

$$W_1 = W_0 \, e^{\gamma W_0}$$

The second principle is that traffic bandwidth prediction in the next period will differ from the actual observed traffic in the same way that the prediction in the previous period did.

Let $A(h - 1)$ be the actual traffic volume measured in time period $(h - 1)$. Let $F(h - 1)$ be the predicted traffic volume in the period $(h - 1)$, then:

$$\left(\frac{A(h-1)}{F(h-1)}\rho + (1-\rho)\right)$$

is a scaling ratio to account for the difference between the predicted value and the actual value in the previous period. Therefore:

$$F_2 = F_1 \left(\frac{A(h-1)}{F(h-1)} \rho + (1-\rho) \right)$$

where ρ can be chosen by fitting against the past measured data. For example, one can choose a value of ρ such that the error due to scaling given by

$$\left[\frac{A(h)}{F(h)} - \rho \frac{A(h-1)}{F(h-1)} \right]$$

is minimised for past data. In other words, ρ can be chosen to minimise

$$E \left[\frac{A(h)}{F(h)} - \rho \frac{A(h-1)}{F(h-1)} \right]^2$$

where E is the expectation operator, which results in:

$$\rho = \frac{E\left[\dfrac{A(h)}{F(h)} \dfrac{A(h-1)}{F(h-1)} \right]}{E\left[\dfrac{A(h-1)}{F(h-1)} \right]^2}$$

Assuming an FBM arrival process with mean rate F_2, and given that the router buffer size is B and given that the packet loss probability should be bounded above by ϵ, a requirement on the capacity can be obtained as follows:

$$F_3 = F_2 + F_2^{\frac{1}{2H}} \alpha(a, H, B, z)$$

where

$$\alpha(a, H, B, z) = \left(z^{\frac{1}{2H}} a^{\frac{1}{2H}} B^{-\frac{1}{1-H}} H (1-H)^{\frac{1}{1-H}} \right)$$

Two methods are described below to estimate the parameters a and H from the measured traffic. The first method assumes that measurements of the traffic volume for each of N consecutive fine-grained intervals of size τ are obtained. Denote the traffic volume over interval i as $T(i)$. Then the mean traffic volume estimate $\hat{m}$ is obtained as:

$$\hat{m} = \frac{\sum\limits_{i=1}^{N} T(i)}{N}.$$

The variance estimate $\hat{V}_t$ is obtained as:

$$\hat{V}_t = \frac{\sum\limits_{i=1}^{N} (T(i) - \hat{m})^2}{N-1}$$

The measurements can be aggregated into k non-overlapping blocks of size kt each and obtain the variance $\hat{V}_{kt}$. Given these two variance estimates, i.e., $\hat{V}_t$ and $\hat{V}_{kt}$, the values for a and H can be obtained.

In the second method, the parameter H can be estimated from the variance-time plots as follows. Given a time trace X_k, $k = 1, 2, \cdots$, we construct an m-aggregated time trace $X_k^{(m)}$, $k = 1, 2, \cdots$, by averaging the original series X_k over non-overlapping blocks of size m; that is

$$X_k^{(m)} = \frac{1}{m} \left(X_{km-m+1} + \cdots + X_{km} \right).$$

Then for long-range dependent processes [Will95] we have:

$$V\left[X^{(m)} \right] \approx m^{-2(1-H)}$$

Therefore, if we plot

$$logV\left[X^{(m)} \right]$$

against log(m), the slope of the resulting graph is given by $-2(1 - \hat{H})$, where $\hat{H}$ is an estimate for H.

Model parameters

The following parameters are defined for the bandwidth prediction model:

- *Size of the coarse-grained period:* The traffic monitoring schemes result in traffic matrix data that contain traffic volume averages over a fine-grained period of time. The size of the coarse-grained period is used to obtain the traffic averages over the coarse-grained period by aggregating the find-grained traffic data.
- *Router buffer size:* The router buffer size is used in the model to predict the capacity of a traffic flow.
- *Packet loss probability bound:* This parameter is used to predict the bandwidth of a traffic flow.
- *Network topology, routing algorithm:* The network topology together with the routing algorithm help determine the traffic flow that is offered to a link, from the edge-router to edge-router traffic flow measurements.

The measured traffic parameter is the fine-grained edge-router to edge-router traffic matrix. From this measured parameter, all the other parameters are inferred. The following parameters are computed for the purposes of reconfiguration at the beginning of each coarse-grained period:

- The coarse-grained averages for traffic flows for each ingress-egress edge router pair: The coarse-grained averages are computed from the measured fine-grained traffic data.
- The complexity of computing the coarse-grained averages is $O\left(N^2\right)$ where N is the number of edge routers in the network.

- The coarse-grained and fine-grained averages for the traffic volume offered to each unidirectional link: from the edge-router to edge-router traffic matrix, the network topology, and the routing algorithm, the traffic volumes offered to each link can be computed for both the fine-grained and coarse-grained periods.
- The computational complexity is $O\left(E^2\right)$ where E is the number of links in the network.
- Bandwidth Prediction Parameters, F_1, F_2, and F_3: for each traffic flow, the bandwidth prediction parameters F_1, F_2, and F_3 can be computed from the above equations. This computation is performed at the beginning of each coarse-grained period.

The computational complexity is $O\left(N^2\right)$ where N is the number of edge routers in the network.

Since a reconfiguration event changes the topology of the IP network, several parameters need to be recomputed after a reconfiguration. In particular, the coarse- and fine-grained averages for traffic flows to each link will need to be recomputed after a reconfiguration. Fitted parameters a, H, ρ, and α are used in the bandwidth prediction model, and are fitted from the measured (and computed) traffic data. The fit for each of these parameters needs to be computed once for each reconfiguration event or periodically (for example once a week).

7.5 MPLS Traffic Engineering

IP provides a relatively simple yet scalable solution, in which packets are forwarded hop-by-hop based on the destination in the packet header and the local routing table. The goal of MPLS traffic engineering is to optimise the utilization of network resources by precise control over the placement of traffic flows within its routing domain. The control mechanisms such as those for signalling, information dissemination, and packet forwarding used in the MPLS traffic engineering are already presented in the previous chapter.

With regard to path selection, MPLS traffic engineering can be used for two purposes:

- **Load balancing**: used to balance the traffic flows across the network to avoid congestion, hot spots, and bottlenecks. It is especially designed to prevent situations where some components of the network are overutilized while other network components are underutilized.
- **Network provisioning**: used to provision the network in a global, optimal manner, for example, as a result of network planning.

7.5.1 Load Balancing

In an IP network, multiple equal cost paths may be formed between nodes. Without explicit routing or load balancing support, a path is chosen arbitrarily. Figure 7.6 shows the well-known 'fish' problem, in which all traffic are forwarded along one path. As a result, the path may be congested, but the other equal cost path sits idle. To

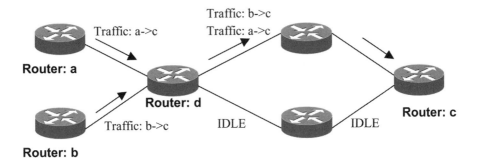

Figure 7.6 The 'fish' problem.

address this, OSPF introduced a technique, Equal Cost Multi Path (ECMP), by which loads are evenly divided over the multiple paths. Three methods have been proposed to split traffic onto multiple equal-cost paths.

- *Per packet round robin forwarding:* this method distributes packets in a round robin fashion among the multiple paths. Round robin forwarding destroys the traffic association or the packet serialisation, and results in a poor TCP performance. Hence, this method is applicable only if the delays on the multiple paths are almost equal.
- *Dividing destination prefixs among available next hops:* this is a coarse method to try to keep the traffic association by splitting traffic according to the prefix of the packet destination address. This can be applied to a high speed WAN, but short prefixs are problematic since the majority of the traffic is often destined to a single prefix.
- *Hashing to the source and destination pair:* this method employs a hashing function, for example, the CRC-16, which is applied to the source address and the destination address in the packet. The hash space is evenly divided among the available paths by setting thresholds or performing a modulo operation. As such, traffic between the same source and destination remains the same path. This method is applicable to high speed WANs.

MPLS traffic engineering is more sophisticated than ECMP in at least two aspects. First, MPLS offers the optimal path selection. Overall, ECMP only tries to evenly distribute load over multiple equal-cost paths, but it neither tries to allocate flows optimally to the multiple path nor has the knowledge of the availability and the dynamic loading condition of the multiple paths. MPLS traffic engineering through the OSPF opaque LSA flooding mechanism constructs and maintains a traffic engineering database that contains traffic engineering information regarding each link on the total bandwidth, the available bandwidth, the reserved, and the reservable bandwidth. According to the traffic engineering database, MPLS traffic engineering is able to make an optimal flow allocation decision in a dynamic network environment. Obviously, an equal load distribution over multiple paths is not always optimal. For instance, a part of an equal-cost path is significantly overloaded while other paths are either slightly loaded or even idle. An optimal load balancing algorithm should assign traffic flows to the paths in inverse proportion to the traffic already using the path. IETF OSPF-OMP (Optimised Multi Path) suggests the implementation of LSA_OMP_LINK_LOAD and LSA_OMP_PATH_LOAD opaque LSAs. The LSA_OMP_LINK_LOAD opaque LSA contains the following information.

- Link loading in each direction measured as a fraction of the link capacity.
- Packet drop rate due to queue overflow in each direction.
- Link capacity in kilobytes per second.

The LSA_OMP_PATH_LOAD opaque LSA contains the following information;

- The highest loading in the direction from the source toward the destination expressed as a fraction of the link capacity. Note the link with the highest load may not be the link with the lowest available capacity.
- Total packet drop in the direction from the source toward the destination due to queue overflow. This can be computed using

$$L_{path} = 1 - \prod_{links} (1 - L_{link}),$$

where L_{path} is the packet drop rate for the path and L_{link} is the packet drop rate for each link on the path.
- The smallest link capacity on the path in the direction from the source towards the destination.

To adjust equal path loading accurately, OSPF-OMP also defines the equivalent loading and the critical loaded segment. The equivalent load is derived using the actual fractional loading multiplexed by an estimated factor based on the loss of the extent to which TCP is expected to slow down to avoid congestion. For every set of paths, the part of the path with the highest equivalent load is defined as the critical loaded segment. In addition, each path in a next hop structure keeps three variables: traffic-sharing, move-increment, and move-count. The OSPF-OMP algorithm adjusts each path's loading in the following manner:

- The path move-increment is unchanged if the path contains the critical loaded segment.
- If the path does not contain the critical loaded segment but the critical loaded segment has changed, this path contains the previous critical loaded segment. The path is adjusted as follow:
  - The move-increment is set to the lowest move-increment from any of the paths containing the critical loaded segment.
  - Set the move-increment to half of its original value.
  - If the path does not contain the critical loaded segment and either the path does not contain the previous critical loaded segment, or the critical loaded segment has not changed, increases the move-increment.

Figure 7.7 shows an example of load balancing using OSPF-OMP, in which at router d, traffic heading to router c is split over the two available paths. By applying a hash function to the source and destination pair at router d, traffic from a to c is forwarded to one path while traffic from b to c is assigned to another path.

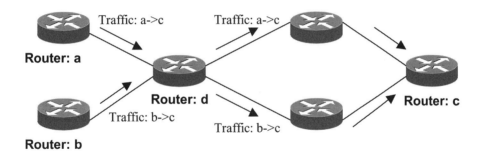

Figure 7.7 OSPF-OMP.

Comparing with ECMP, MPLS traffic engineering provides explicit path routing capability. As a result, MPLS traffic engineering is able to compute and establish LSPs, which can completely change the forwarding adjacency. In the case of under-utilized networks, routing decisions are dominated by minimising delays. In the case of non-underutilized networks, routing decisions have to consider the lower capacity links and the heavily loaded links. By load balancing, the network utilization can be maximised. However, as utilization grows further, load balancing by adjusting link costs is no longer adequate simply because the network has reached or closes to its maximum capacity. MPLS-OMP (Optimised Multi Path) uses the same load balancing algorithm as in OSPF-LMP. The key difference between MPLS-OMP and OSPF-OMP lies in the capability of MPLS for LSP setup/tear-down. By incrementally adding circuits to accommodate traffic growth, MPLS traffic engineering hopes to avoid hot spots or congestion in order to maximise network utilization or throughput. From an ingress point of view, once a LSP is established, the LSP reached egress becomes a virtual neighbour and its load status is set accordingly. If there are multiple paths between these two nodes (ingress and egress), load is split over these paths. Likewise, underutilized paths can be deleted from the forwarding adjacency.

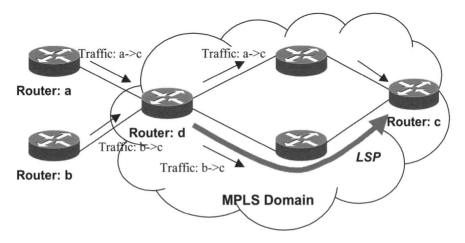

Figure 7.8 MPLS-OMP.

Figure 7.8 shows an example of load balancing using MPLS-OMP, in which router d is a MPLS ingress router and initiates a LSP from router d to router c. Once the LSP is setup, the forwarding adjacency (i.e. the next hop structure) is updated at router d so that router c becomes its virtual direct neighbour. By collecting the LSP path loading, the LSP is configured only to carry the traffic originated from node b. In such a way, traffic is distributed based on the resource availability and capacity. Note this example does not intend to show the necessity of adding a LSP, but it illustrates flow allocation under the scheme of MPLS-OMP.

7.5.2 Network Provisioning

MPLS traffic engineering can be used for network provisioning as a result of long-term or relatively short-term network planning. Through dynamic LSP setup/tear-down, MPLS traffic engineering can support a range of network applications, for example, static and dynamic VPN and virtual LAN. For traffic engineering purposes, virtual LSPs can be established or deleted based on measured and/or projected traffic distributions. Two tasks of MPLS traffic engineering (for network provisioning) are LSP design and flow allocation. LSP design determines the routing path and the life span of the LSP, whereas flow allocation maps network flows to the available resources including LSPs. Both tasks alone are classical mathematical optimisation problems. LSP design can be formulated as the optimal path problem. Based on the metric and optimisation objective, optimal path problems can be classified into the shortest path problem, the maximum capacity path problem, the fractional path problem, and the quickest path problem. Flow allocation can be formulated as the multi commodity problem, in which each commodity (i.e. network flow) has an associated demand and source-sink pairs. For an undirected graph G, a special node s, called the source, and a node t, called the destination, the multi commodity problem can be represented as a linear program of the following form:

$$\sum_j f_{i,j}^k = \begin{cases} F_k, & i = s_k \\ -F_k, & i = t_k \\ 0, & i \neq s_k, t_k \end{cases}$$

$$0 \leq \left| f_{i,j}^k \right| \leq u_{i,j}^k, 0 \leq \sum_k \left| f_{i,j}^k \right| \leq u_{i,j}, (i,j) \in A$$

In this formulation, k refers to the commodity, $f_{i,j}^k$ refers to the k commodity flows on arc (i,j), $u_{i,j}$ and $u_{i,j}^k$ are the positive numbers representing the total capacity of arc (i,j) and the capacity for the k commodity, and F_k represents the total flow across the network for the k commodity. The objective of the multi commodity problem is

$$\max \sum_k |f_k|.$$

Since this is a linear program (the objective function and the constraints are linear functions of the variables), it can be solved using the LP methods such as LP-relaxation and rounding. The detailed theoretical treatment of multi commodity problem can be found in the reference [Okam83].

The challenge here is also on how the two tasks interact. For instance, when to trigger a LSP setup? A global optimisation approach tries to integrate LSP design and flow allocation, but faces enormous complexity. A local optimisation approach divides the tasks into separate phases. Within each phase, optimisation method can be applied independently.

7.6 Lightpath Virtual Topology Reconfiguration

As we discussed in Chapter 5, in reconfigurable WDM networks, IP adjacency is likely to be built upon multi-hop WDM lightpaths. A cost-saving advantage of a WDM optical network is that it can support a relatively smaller operational footprint (at least in the backbone). This means that it is highly likely that various IP links will share the same physical fibre link and an IP virtual link may be routed over several WDM switch hops.

Figure 7.9 shows the layout of virtual topology design and routing in IP over reconfigurable WDM networks. There are three main components in the figure:

- traffic routing
- IP topology design
- lightpath routing.

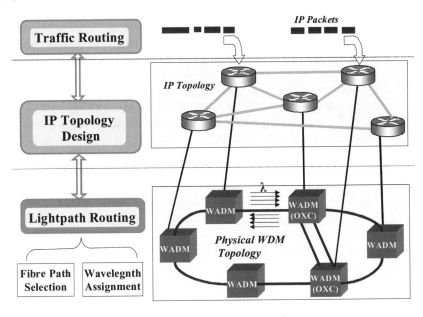

Figure 7.9 Virtual topology design and routing.

Traffic routing refers to the conventional packet routing, for example, OSPF, introduced in Chapters 2 and 3. IP topology design is the topic of this section. Lightpath routing provides the mapping from a virtual IP topology to the physical WDM topology. Lightpath routing consists of two correlated issues: fibre path selection and wavelength assignment, which are discussed in Chapters 4 and 6. In summary, light-path routing can be implemented in one of the following two flavours:

- *Static lightpath routing:* this approach pre-computes and stores the routing path(s). Alternate paths for each primary path can also be computed and stored. Wavelength assignment is conducted upon lightpath connection requests. Only simple heuristics are used for wavelength assignment. Wavelength assignment examples are random selected wavelength or first-fit selected wavelength channel.
- *Adaptive lightpath routing:* this approach uses the dynamic SPF (shortest path first) algorithm for path routing, which requires link state information dissemination. Due to the presence of local-maintained link state database, wavelength assignment can be more sophisticated. Examples of wavelength assignment heuristic algorithms include least-loaded, most-used, and best-fit connection data rate wavelength channel.

IP topology design and lightpath routing are control plane functions, whereas traffic routing is the only component used for data packet forwarding as well as packet routing.

Since both virtual topology design and lightpath routing are control plane functions, the two components can be either loosely connected or closely coupled. The former follows an overlay IP/WDM traffic engineering approach whereas the latter forms an integrated IP/WDM traffic engineering solution. In a standalone traffic engineering application, constraint-based lightpath routing can also be included as an evaluation tool for topology design algorithms. This way guarantees that the designed topology can be realised in the WDM layer under the current capacity.

In an overlay IP over WDM network, the server layer can belong to a transport service provider that serves multiple network service clients, such as VPN customers. In such a scenario, a customer at the IP layer subscribes transport services from the WDM network. In the service agreement, the customer specifies a fixed set of IP routers that directly access the WDM network. The WDM layer provides lightpath connections among these routers. However, unlike leased line connections in today's VPNs, the arrangement of these lightpath connections is not fixed. While it is possible that the total number of lightpath connections is fixed or limited, each lightpath connection can be reassigned to connect a different pair of routers in response to dynamic changing patterns of the traffic demands. This requires a virtual topology design algorithm at the IP layer. Here, the virtual topology is a graph consisting of nodes and links. The nodes are the routers, and the links are the WDM lightpath connections.

In the rest of this section, we formulate the problem first and survey some of the existing approaches. We will also present a set of heuristic algorithms that optimise network throughput and/or weighted hop distance. The ideas behind the heuristics can be used to develop new algorithms tailored to different objectives.

7.6.1 Regular vs. Irregular Virtual Topology

Regular topology refers to the topology with well-defined and patterned node connectivity, whereas irregular topology is usually constructed dynamically to optimise certain performance metrics. Regular topology is structured and it is likely there exists well-accepted definitions and systematic study. Routing and management of regular topology is simple, but adding or deleting arbitrary nodes from the structured pattern (while still maintaining the pattern) is difficult. Examples of regular topologies are:

- Ring
- Shuffle-Net
- Manhattan Street Network (MSN or 2D Torus)
- GEMNet
- HyperCube
- de Bruijn Graph.

These regular topologies usually have low routing complexity and are symmetric, but they differ in the number of transmitters required per node, scalability, and fault tolerance. For example, a n-ary hypercube with N nodes requires $(n–1)\log_n N$ transmitters per node with average hop distance $= \log_n N$ and network diameter $= \log_n N$. By comparison, a MSN requires only two transmitters per node but both average hop distance and network diameter are equal to $\sqrt{N}$. The Shuffle-Net also has a smaller average hop distance than that of MSN since for a given path length, it has more alternate paths from an end-node to any other end-node. However, it needs more transmitters per node, which is n. In the rest of this section, we focus on irregular topologies since irregular topologies can be optimised to support certain uniform or non-uniform traffic patterns.

7.6.2   Topology Design Problem Formulation

While designing the virtual topology, one can select different performance objectives, which is solely an administrative decision. There are basically two types of objectives. One is application orientated, which is normally related to QoS perception at the application, such as end-to-end latency. The other is network orientated, which is usually relevant to network resource utilization levels, such as overall throughput.

The input of topology design includes a traffic demand matrix. For an IP network with N routers, the traffic demand matrix is an $N \times N$ matrix T, whose element $T(i,j)$ is an aggregated traffic flow (in number of bits per second) from router i to router j. Values of the elements should be determined using certain prediction techniques based on current measurement. We consider traffic trend prediction a separate issue, and will focus on the topology design perspective of this problem. Thus, one can define the traffic demand matrix, $T(i,j)$, equal to the measurement of the traffic volume between router i and router j over a controllable time window. The topology design algorithm can also take input from other prediction software tools for IP traffic flow bandwidth demands.

Triggering conditions are closely related to optimisation algorithms and objectives. The topology design algorithm can take two types of triggering. The first type is parameters sensed from the network, such as link load status as a triggering criterion, for automated adaptation. The second type is administrative decisions from the exterior of the network, as in the case of using the algorithm for proactive provisioning topology design. According to [Rama96], the virtual topology design and routing problem can be formulated as a Mixed Integer Linear Program (MILP). The objective of MILP is to minimise the maximum congestion on any link of the virtual topology while constraining the average delay to certain levels. This formulation can be summarised as follows:

$$\min\left(\max_l c_l, \forall l\right),$$

where c_l is the traffic volume (or number of packets) on the link l. The MILP optimisation is subject to these constraints:

- *Virtual topology nodal degree constraints:* nodal degree vector, D, has n elements. Each element, d_i, $\forall\{i\} \subset N$, is the nodal degree of router i.
- *Wavelength continuity constraint:* during lightpath routing, the lack of wavelength conversion capability on the outgoing switch port represents the wavelength continuity constraint. A switch may only support a group of ports to be wavelength convertible and a wavelength interchangeable port may only convert into a wavelength without a specific wavelength band.
- *Limited delay:* for each node pair, the delay time is at most some multiples of shortest-path delay in physical topology.
- *Physical fibre topology and wavelength availability constraint:* lightpath routing is based on the physical fibre topology and available wavelength channels.

The above MILP formulation is targeted for global optimisation for any given delay constraints. It includes both virtual topology design and lightpath routing. Hereafter in the section, we only consider virtual topology design. In our discussion, we take persistent link congestion as a trigger to activate the topology design algorithm. Here both the number of congestion instances and duration of the congestion persistence can be defined by the user. Note that the reconfiguration algorithm can help only when network resources are not evenly utilized. In such a case, reconfiguration can rearrange the resources in a better manner. However, if a network were globally overloaded, reconfiguring the virtual network over the same WDM physical topology would not help. Therefore, in addition to the presence of congestion, triggering conditions also include divergence of the load distribution.

7.6.3 Heuristic Algorithms

Since MILP is computationally intractable, topology design can be solved using heuristic-based algorithms. An advantage of using heuristics over a rigorous optimisation algorithm in designing the virtual topology is flexibility. While designing a heuristic algorithm, one can steer emphasis onto different issues as well as saving computational cost.

Lightpath topology design algorithm survey

In [Rama96], three heuristics are proposed: heuristic topology design algorithm (HTDA), minimum-delay logical topology design algorithm (MLDA), and traffic independent logical topology design algorithm (TILDA). HTDA attempts to create lightpaths between node pairs in the order of descending traffic demands. Given the network nodal degree, a lightpath is generated in order to accommodate the most demanding traffic that has not been carried. This process continues until there is no more network resource. If all of the traffic demand has been allocated, the rest of network resources are randomly selected to form lightpaths until the resources are used up. The idea behind this heuristic is simply to route the heavy traffic on a single hop. MLDA sets up lightpaths between adjacent node pair and then HTDA is applied to allocate the rest of the resources subject to constraints. Hence, MLDA is really an extension to HTDA in the case when the logical network degree is higher than the physical network degree. TILDA ignores the traffic demand completely, but focuses on minimising the number of wavelengths to be used. TILDA constructs one-hop lightpaths first and then two-hop lightpaths. The heuristic continues with lightpath setup until the constraints are reached.

In [Bane97], two heuristics are proposed: LP-based one-hop traffic maximisation scheme (OHTMS) and link elimination via matching scheme (LEMS). OHTMS is similar to the HTDA, where it attempts to maximise the total one-hop traffic while maintaining connectivity of the virtual topology. LEMS first creates a fully connected virtual topology (a bipartite graph) in which each partition contains all the nodes in the physical topology. The weights of the edges are set according with their traffic demands. A minimum weight perfect matching is identified for the edges. The matched edges are deleted from the graph, and traffic is pushed over the updated topology for re-evaluation. The edges weights are updated, and to be deleted edges are searched and eliminated in the same manner. This process continues until the constraint limit has been reached.

Spanning tree topology design heuristic algorithms

In the literature, the 'one-hop traffic maximisation' orientated heuristics have been proven to provide better performance [Bane97]. Three different 'one-hop traffic maximisation' heuristics are introduced in [Liu02]. These methods differ in the way that the remaining resources are allocated after the initial virtual topology connectivity is provided.

The algorithms have the following definitions:

- The *lightpath topology* at IP level is denoted by $< N, L >$, where N is the set of routers, L is the set of IP links, $n = |N|$, and $l = |L|$. Within N, n_i represents the router identifier. An IP link is always bi-directional, so two unidirectional lightpaths in the WDM layer form a virtual link in the IP layer. Within L, l_{ij} is the link identifier and l_{ij-c} represents the link bandwidth. The following discussion does not distinguish l_{ij} and l_{ji}, so $l_{ij-c} = l_{ji-c} = \max(l_{ij-c}, l_{ji-c})$.
- *Nodal degree vector*, D, has n elements. Each element, d_i, $\forall\{i\} \subset N$, is the nodal degree of router i. Obviously, $l = \frac{1}{2}\sum_i d_i$ for any IP virtual topology.
- *Traffic demand matrix* is denoted by T. T is an n by n matrix, whose element, $T(i,j) \geq 0$, $\forall\{i, j\} \subset N$, is the volume of traffic in unit of time flowing from router i to router j.
- *Throughput X is an $n \times n$ matrix*, whose element $X(i,j)$ is the traffic volume (in bits) transported from router i to router j in a certain period of time.

The algorithms use the following notations:

- *Residual traffic demand matrix,* denoted as T^r, is an n by n matrix determined by $T^r = T - X$.
- A *flow vector,* F, is constructed from the traffic demand matrix T in two steps:
    - Symmetrising T to obtain T^s, whose element $T^s(i, j) = T^s(j, i) = \max\{T(i, j), T(j, i)\}$, $\forall\{i, j\} \subset N$.
    - Sorting the upper (or the lower, since the symmetry) triangle of T^s in descending order to a vector to obtain F.
- A *flow-hop vector,* F_h, is defined as: $\{F(i, j) \times H(i, j)\}$, $\forall\{i, j\} \subset N$, sorted in descending order, where $H(i,j)$ is the hop distance from node i to node j.
- *Connection matrix,* G, is used to represent the graph corresponding to the virtual IP topology. G is an $n \times n$ matrix. The value of its element $G(i,j)$ equals 1 if there is a connection between router i and router j, and 0 otherwise. Due to the symmetrical property of the IP virtual topology, G is a symmetric matrix.

Based on the above notations, the algorithms are, in particular, interested in the following metrics:

- *Normalised throughput,* $\eta = \dfrac{\sum X(i, j)}{\sum T(i, j)}$

- Average weighted hop-distance, $h = \dfrac{\sum(H(i, j) \times X(i, j))}{\sum T(i, j)}$

subject to Nodal Degree constraints,

$$\sum_i G(i, j) \le D(j), \forall j$$

and

$$\sum_i G(j, i) \le D(j), \forall j$$

The algorithm design and later performance evaluation aims at improving both **normalised throughput gain** (NTG – γ) defined as:

$$\gamma = \frac{\eta_{new}}{\eta_{old}} - 1$$

and Average Weighted Hop-distance Gain (AWHG - φ) defined as

$$\varphi = \frac{h_{new}}{h_{old}}$$

where h_{new} and η_{new} represent the average weighted hop-distance and **normalised throughput** for the new topology and h_{old} and η_{old} represent the two metrics for the old topology or a fixed topology.

The proposed heuristic algorithms design a lightpath topology subject to given physical nodal degree of each node. The output topology allows parallel paths between the same pair of nodes. A lightpath connection is bi-directional. The heuristics start with a graph of n isolated nodes that are corresponding to the n routers.

These heuristics have the same first phase, which is to create a spanning tree (a general tree not necessarily a binary tree) so that minimal connectivity is provided. $max[T(i,j)]$ is used as the spanning tree metric. The initial graph connectivity is achieved by iterating through the flow vector F, which is sorted in descending order of the loading demand. For each $F(i)$ corresponding to $T^s(p, q)$, a connection is assigned between router p and router q if there is no path between these two routers. Unlike the conventional 'one-hop traffic maximisation' heuristics, the proposed heuristics will not simply allocate the remaining connections using F. Instead, the heuristics evaluate the primitive topology using a simulated flow-based environment. The proposed heuristics differ in the way of allocating the remaining connections. Given the n node network, the initial connectivity topology always has n–1 connections. In an operational network, how to allocate the remaining resources is not a trivial issue as proved in the simulation. Three heuristic based algorithms are presented below.

Residual demand heuristic algorithm (RD)

This heuristic has the following steps:

- **Step 1.** Construct a spanning tree to provide initial connectivity between the nodes according to the flow vector F subject to the Nodal Degree constraints. Since F is sorted in the descending order according to the demand, the initial connectivity tree is constructed to accommodate the most demanding flows as a one-hop neighbour. There are no parallel connections between the same node pair. Each connection uses the maximum bandwidth as defined in the matrix L.
- **Step 2.** Apply the traffic demand to the incomplete topology. Two methods for traffic allocation are considered. One method is the conventional shortest path routing, i.e. SPF, and the other method employs SPF and a flow deviation technique, for example, ECMP (Equal Cost Multipath) or OMP (Optimised Multipath). The traffic allocation method in the algorithm emulates the routing policy in the operational network. For example, a conventional IP network only supports SPF such as the routing computation in OSPF; an optimal routing policy requires load balancing among alternate paths such as OSPF-OMP.
- **Step 3.** Since there are interfaces yet to be assigned, it is likely that not all demands can be supported by the incomplete topology. So, create a new flow vector using the residual demand matrix. The new flow vector is stored in the descending order of the residual demands.
- **Step 4.** Search through the new flow vector to find such a flow that a new connection can be assigned between its end points. Insert this new connection into the topology and set the connection to its maximum bandwidth according to F. Then loop back to Step 2. The algorithm stops if the new connection cannot be found.
- **Step 5.** If the number of free interfaces is not less than 2, apply the convergence heuristic presented below.

Since evaluation of the intermediate topology is based on the residual demand, the corresponding algorithm is named as RD. A variation of this stage is to look at the residual demand-hop count product. Accordingly, the corresponding algorithm is named as RDHP. The RDHP tends to favour flows with high values of the product of residual demand and hop count.

Residual demand hop-count product heuristic algorithm (RDHP)

There are four steps in this heuristic:

- **Step 1.** Construct a spanning tree to provide initial connectivity between the nodes according to the flow vector F. This step is the same as the Step 1 in the RD.
- **Step 2.** Apply the traffic demand to the incomplete topology. This step is the same as the Step 2 in the RD.
- **Step 3.** Compute hop distance between all node pairs based on the incomplete topology.
- **Step 4.** Create a new flow vector using the residual demand matrix of which each element is weighted by its hop count. This flow vector is stored in the descending order of the residual demand hop-count product. Search along the flow-hop vector to find such a flow that a new connection can be assigned between its end points. Insert this new connection into the topology and set the new connection to its maximum bandwidth according to F. Then loop back to Step 2. The algorithm stops when there is no new connection found. If the number of free interfaces is not less than 2, apply the convergence heuristic presented below.

Another variation of this stage is to simply look at the demand-hop count product. This is an obvious result of the overall optimisation objective, i.e. minimising the weighted hop count of the network. Hence, it is named as DHP. The DHP reduces computational cost incurred after each connection addition at the cost of reduced topology quality, as explained in the simulation study section.

Demand hop-count product heuristic algorithm (DHP)

This heuristic consists of the following steps:

- **Step 1.** Construct a spanning tree to provide initial connectivity between the nodes according to the flow vector F. This step is the same as the Step 1 in the RD.
- **Step 2.** Compute/re-compute hop distance between all node pairs based on the incomplete topology.
- **Step 3.** Create a new flow vector in the descending order using the demand matrix of which each element is weighted by its hop count.
- **Step 4.** Search along the flow-hop vector to find such a flow that a new connection can be assigned between its end points. Insert this new connection into the topology and set the new connection to its maximum bandwidth according to F. Then loop back to Step 2. The algorithm stops when there is no new connection found.
- **Step 5.** If the number of free interfaces is not less than 2, apply the convergence heuristic below.

The above algorithms may stop when there are still multiple interfaces available. An example of such situations is that all the unallocated interfaces belong to the same node. This could happen as a result of mismatch between the nodal degree and the traffic demand. Since the algorithms try to accommodate the heavy demands with direct connections, the less-traffic node may end up with several free interfaces but the rest of the network is fully occupied. A heuristic for convergence is provided, which has the following Steps.

Convergence heuristic

- **Step 1.** Identify the node with multiple (more than or equal to 2) open IP interfaces.
- **Step 2.** Apply traffic demand to the partially completed topology.
- **Step 3.** Sort all connections based on utilization levels in descending order. The connection utilization level is set to the higher one of the link utilization levels of the two links that construct the connection.
- **Step 4.** Select the least loaded connection that is not an instance to the node with open interfaces from the sorted list. Break this connection and create two new connections pair-wise between the two existing open interfaces. The two newly opened interfaces result from breaking the least loaded connection.
- **Step 5.** Go back to Step 2 if the number of open interfaces is still larger than or equal to 2.

The intuition behind this heuristic is that by using up two open interfaces, one can break an existing connection to liberate another two interfaces such that new connections can be set up between both the open and the liberated interface pairs. This modification to the designed topology results in traffic that used to travel the broken connection having to traverse, in the worst case, two more hops. One can select the least loaded connection for breakup since this limits performance degradation. Using up all the interfaces especially when the topology design is only based on the traffic demands such as DHP, tends to provide a better performance topology. In addition, more deployed resources tend to accommodate more demands in the future.

In an IP backbone network, each node usually has at least a few interfaces, which infers the total number of connections, l, can be accommodated in the network is much larger than the $n-1$ connections in the initial spanning tree graph. This implies the setup of the $l-n+1$ connections is in fact an important procedure during topology design.

7.6.4   Virtual Topology Migration

In an overlay IP over reconfigurable WDM networking model, IP and WDM can communicate through optical UNI or WDM NC&M (see Chapter 5). Optical UNI allows little topological information of the physical network (for example, WDM network configuration and connections) to be shared with the client network. WDM NC&M presents completely separate IP and WDM management systems. Hence, the assumption on overlay network reconfiguration is that the physical

WDM network can support the virtual topology (which is an output from the topology design algorithm in the previous section). If a virtual topology cannot be supported or a lightpath cannot be established due to WDM layer constraints, one solution during migration is backup to the original topology. A recent study shows that the resource bottleneck in IP/WDM networks is really the IP interfaces not the WDM interfaces [Wei00a]. IP interfaces are comparatively more expensive and scarce than WDM interfaces. In general, there are a number of WDM interfaces available. So if a virtual topology cannot be supported by a carrier (i.e. the server network provider), the carrier should consider either update its capacity or reject the request right away. The carrier may prefer this to sharing its topological information with its clients. The topology design heuristics and the migration heuristic are presented separately. They can be considered as separate tools and used when necessary. But a general assumption in the overlay IP over the reconfigurable WDM networking model is that the virtual topology can be supported in the WDM network.

Dynamically verifying physical WDM network constraints during topology reconfiguration requires a more integrated IP/WDM approach. For example, in the peer IP over reconfigurable WDM networking model, the WDM network topological information is shared with IP networks, which makes integrated IP/WDM reconfiguration strategies possible. A simple example is to verify each lightpath's feasibility in the topology design algorithm. As such, the topology design algorithm suggests feasible virtual topologies only. However, this approach relies too much on the accuracy and consistency of the topological information available at a local router's database. In a distributed and real-time environment, the local topological information may not be up-to-date and network convergence in response to, for example, link state changes, may take some time. In addition, flooding WDM network information to IP networks may not scale from the network control's point of view. Although the peer IP over reconfigurable WDM networking model was proposed, there is no (at least not yet) commercial implementation of such a networking model. Reconfiguration in IP over packet switched WDM networks is discussed later in this chapter.

Even if WDM layer resources are sufficient to support any migration sequence (a most demanding example can be adding all new connections before tearing down any unwanted one), there are other issues concerning the migration. Since WDM reconfiguration deals with large-capacity wavelengths (up to OC-192), changing allocation of resources in this coarse granularity has effects on a large number of end-user flows. In general, a migration procedure consists of a sequence of establishing and taking down individual WDM lightpaths. Traffic flows have to adapt to lightpath changes after each migration step. Depending on network structures, the implication can be extended to the routing perspective of the network, which in turn may affect more user flows. A typical case where this 'path-ology' appears is in an overlay IP/WDM network. Since WDM lightpath reconfiguration is viewed as a lightpath topology change at the IP layer, the IP routing protocol has to adapt to this change. It is known that topology changes can easily cause transient 'black holes' and forwarding loops with any existing IP routing protocol. Therefore, extra efforts are needed in order to eliminate side impact on users' traffic during routing protocol convergence after each reconfiguration.

Algorithms finding the optimised migration schedule involve complexity due to a variety of operational concerns. Heuristics can be employed to foster a smooth migration, although they cannot guarantee an optimal solution. A heuristic algorithm is provided below aimed at incremental lightpath topology migration in an overlay IP/WDM network.

Topology migration heuristic slgorithm

There are four primitive operations involved in the migration, and they are defined as below.

Operation A: Remove an edge

- Configure the two corresponding IP router interfaces down.
- Tear down the two unidirectional lightpaths in the WDM layer.
- Update the log file for edge removal (including the lightpath details such as client interfaces, name/address of WADMs, intermediate hops or WSXCs, wavelengths and the router details).

Operation B: Add an edge

- Compute and set up the virtual link in the WDM layer. Note a virtual IP link is implemented using two unidirectional lightpaths, which can follow the same fibre route but occupy different wavelength channels. If by any chance the edge insertion fails, the algorithm aborts by calling Operation R.
- Configure the IP router interfaces up with the updated interface IP addresses. Note by using unnumbered IP interface or pre-assign multiple secondary IP address and one primary IP address to one interface, it is no longer needed to update the interface IP address.
- Update the log file for edge insertion (including the lightpath and router details).

Operation E: Evaluate the network connectivity

- Fully connected is defined as starting from any node and able to travel to any other node. This can be implemented using the Depth-First Search algorithm.
- Return 0 if the network is disconnected and 1 if the network is fully connected.

Operation R: Restore the original topology
For example, migration has failed:

- Based on the log file, step-by-step tear down/set up lightpaths so that the original/old topology is restored.
- Send error messages or notifications to managers and/or clients.

The incremental reconfiguration migration heuristic algorithm consists of the following steps:

- **Step 1.** Construct two queues, *dequeue* and *enqueue,* by comparing the old and the new topology. The *dequeue* contains the edges to be removed, which are sorted in the ascending order of edge load. The *enqueue* lists the edges to be added. Maintain a log file to keep track of the edges removed/added.

- **Step 2.** Conduct an initial remove, which selects two edges from the *dequeue*. If an edge removal results in a disconnected graph (based on the Operation E), the edge is skipped. If all *dequeue* edges cause network disconnection, the two least loaded edges are selected. The selected edges are removed using the Operation A and deleted from the *dequeue*.
- **Step 3.** Try to add edge(s) from the *enqueue* if the interface and wavelength constraints can be satisfied. The queue is searched through and all eligible edges are added. An edge is added using Operation B, and deleted from the *enqueue*.
- **Step 4.** Search through the *dequeue* and locate the edge that is least loaded but does not result in network disconnection. Remove the edge and delete it from the *dequeue*.
- **Step 5.** Loop through Steps 3 and 4 until both *enqueue* and *dequeue* are empty.

Note Step 2 is performed before Step 3 to free connected interfaces. An operational network usually employs all the interfaces. So without tearing down existing connections, the network cannot establish new connections. This agrees with our previous statement that reconfiguration is to reorganise resource structuring according to demand but subject to the same physical network constraints.

7.7   Reconfiguration for Packet Switched WDM Networks

The packet switched WDM networks are still under research, but early photonic packet switching system prototypes have appeared [Chan00]. There is no doubt that these systems share similarities with other packet switched networks, for example, electrical IP routers. However, optical packets my have a different header format, which is similar to the MPLS label header. In addition, photonic packet switching systems prefer a large packet size comparing to the end user traffic supported by electrical IP routers. Certainly, this may change as the photonic packet switching continues to progress. For example, optical switching fabrics may be able to switch packets rapidly, for example, in nanoseconds, and optical memory may become available.

7.7.1   Packet Switched WDM Reconfiguration Overview

As proved in the IP networks, there is a need to support packet switching as well as circuit switching in packet switched networks. An IP/OLS network can be modelled in such a way that any wavelength on a fibre in the WDM layer can be dynamically set to either packet mode or circuit mode. In the packet mode, OLS works as MPLS label switching does in an electrical label switch router. But the same set of operations in OLS takes place in an optical domain. In the circuit mode, OLS works as an optical cross-connect network, which requires an explicit signalling stage to establish the communication channel.

Figure 7.10 shows a packet switched WDM network reconfiguration. As indicated in the figure, there is an integrated IP/OLS fibre topology, upon which there are lightpaths and MPLS LSPs. OLS reconfiguration refers to lightpath connection recon-

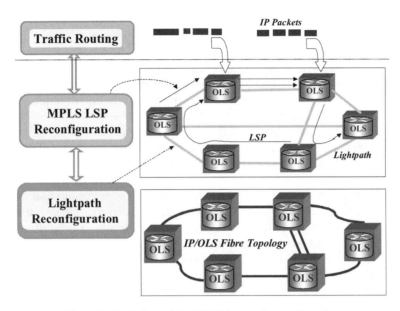

Figure 7.10 Packet switched WDM network reconfiguration.

figuration and MPLS LSP reconfiguration. At the current stage, OLS networks do not fully support IP destination-based forwarding, i.e. in the data plane, the OLSR does not read nor understand the IP datagram header.

In this section the reconfiguration algorithm that integrally engineers the IP layer as well as WDM layer is discussed. The algorithm is most suitable for integrated IP/WDM networks, where an IP-centric protocol is employed to control the physical network. All network elements are IP aware. IP addresses are assigned to switch/router interfaces. A link state IP routing protocol, such as OSPF with proper extensions, is used for network elements to discover the physical topology. Wavelengths within a fibre are controlled using a MPLS-based mechanism (i.e. local wavelength selection). Information regarding lightpath connectivity and operation mode of each wavelength of all fibres are also exchanged through OSPF extensions. Each network element maintains two network topologies. One is the physical topology that describes the physical network elements and the fibre connections between them. Another is the lightpath topology that is defined by the lightpath connections. When a network element decides to set up a new lightpath connection, the head end of the lightpath is responsible for routing the lightpath through the physical topology subject to network constraints. When a source node wants to send data to a destination node, there may or may not exist a direct lightpath between the two. Further, setting up a new lightpath may or may not be possible due to the wavelength channel availability and other constraints. In conventional electrical MPLS, LSPs are virtual channels so that they can be set up to support even the full mesh connectivity. Hence, label switched data in MPLS can be delivered in one LSP hop.

Given this less-than-full-mesh connectivity in OLS, data routing is needed at each network element and the corresponding routing space is the lightpath topology. Thus, there are two routing layers. This layered architecture is a natural result of embedding a packet paradigm (IP) in a circuit switching world (cross connect WDM). Consequently, traffic engineering can be exercised in each layer. While at the upper layer, i.e. in the lightpath topology, existing electrical MPLS traffic engineering solutions can be applied. The lower layer needs a novel algorithm to address configuration and reconfiguration of the lightpaths in the physical topology of the WDM network. In addition, it is also needed to co-ordinate interaction of traffic engineering operations between the upper and lower layers.

There are two approaches to establish a new path, which could be either lightpath or LSP. According to the first approach, whenever a node needs to set up an LSP to another node, the head end node first tries to set up a lightpath directly to the tail end node. If the physical layer cannot support this lightpath, the head end node tries to route the LSP through the current lightpath topology, i.e. set up an electrical LSP. If this fails too, lightpath reconfiguration is invoked. The second approach tends to take full advantage of already configured WDM resource before configuring additional ones. When a node needs to set up an LSP to another node, the head end node always tries to route the LSP through the current lightpath topology, i.e. set up an electrical LSP. If this fails, the head end node tries to set up a lightpath directly to the tail end node, i.e. set up an Optical LSP. If this is not successful either, lightpath reconfiguration is activated. Interesting enough, same lightpath setup and reconfiguration algorithms can be used in both approaches. The basic idea is shortest path routing over the physical topology subject to constraints such as wavelength availability, wavelength continuity and channel signal quality.

7.7.2   Reconfiguration Conditions

MPLS-based traffic engineering can be applied to the LSP reconfiguration in the upper layer. When an ingress edge router needs to set up an LSP to an egress edge router, the LSP will be computed in the residual topology obtained by applying all applicable constraints onto the lightpath topology. If the new LSP requires reserving certain bandwidth B, the residual topology can be obtained from the lightpath topology in which the links with available bandwidths less than B are removed. According to the residual topology, a shortest path from the ingress edge router to the egress edge router can be determined. In general, this path is different from the shortest path between the same router pair found in the lightpath topology. The MPLS signalling mechanism ensures the installation of the found path along the desired intermediate nodes. By combining constrained-based routing and explicit path setup, MPLS alone at this level can exploit the capacity between a given pair of nodes up to the limit defined by the min-cut between the two nodes. Once the min-cut point is reached yet more LSPs are required from the same ingress edge router to the same egress edge router, one of two actions can be taken:

- **Action 1:** Some existing LSPs are pre-empted to free up capacity on the bottleneck to accommodate the new LSPs should they have a higher setup priority.
- **Action 2:** New LSP setup requests are rejected.

To a certain extent, the first case can be considered as a specific constrained based routing problem, i.e. with pre-emption, whereas the second case simply suggests

MPLS load balancing has reached its limit. In such cases, the routing engine at the ingress edge router sees from the residual topology a disconnected graph with at least two components. The ingress edge router belongs to one component and the egress edge router belongs to the other. Without LSP pre-emption, if there is no WDM over the fibre link, the new LSPs will be definitely blocked.

In the presence of WDM technology, there is another tier of traffic engineering that backs up the upper layer LSP. Since the lightpath topology being composed of physical channel connections is reconfigurable, the chances are there might exist a lightpath topology in which those two disconnected components in the residual topology can be reconnected. This new lightpath topology at the same time should satisfy connection requirements for all existing LSPs. A reconfiguration heuristic algorithm designed to discover such a possible lightpath topology is presented later in the section.

7.7.3 A Case Study

Before presenting the heuristic algorithm, we use an example to illustrate the basic ideas behind the algorithm. Figure 7.11(a) shows the example WDM network with a 6-node ring topology. In the figure, the wide grey bands represent fibres, each of which supports two wavelengths. The dotted lines represent lightpath connections across the nodes. Different colours indicate different wavelengths. Figure 7.11(b) shows the corresponding lightpath topology.

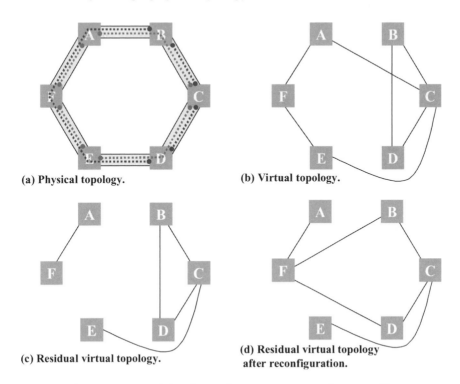

(a) Physical topology.

(b) Virtual topology.

(c) Residual virtual topology.

(d) Residual virtual topology after reconfiguration.

Figure 7.11 Using lightpath reconfiguration to accommodate more LSPs.

At one moment node C needs to set up a new LSP to node F, and the residual lightpath topology that node C sees when trying to set up the new LSP is shown in Figure 7.11(c). Clearly, the resources currently available to node C are insufficient to support the newly requested LSP. Now the question is whether it is possible to reconfigure some of the lightpaths to accommodate the new LSP. The answer is maybe. However, a definite answer is only a couple of test steps away. The first test is to see if a reconfiguration solution can connect the two disconnected components back together. The next test is whether all affected existing LSPs plus the newly requested one can be accommodated by the reconfiguration solution. In this example, a solution (also the only solution) that can pass the first test is to reconfigure the lightpath connection between node B and node D. An important relevant feature of this lightpath is that its end points belong to one component and the physical trajectory of the lightpath passes through the other component. Hence, a feasible reconfiguration solution is to 'break' the lightpath in the second component. One can choose to break the lightpath at node F so the original lightpath BD becomes two lightpaths BF and FD. Then the residual lightpath topology after the reconfiguration becomes what is shown in Figure 7.11(d).

The impact on all existing LSPs introduced by this reconfiguration is minimal. In particular, only LSPs that were using lightpath BD before the reconfiguration are exposed to limited impact. After the reconfiguration, each of these LSPs will travel one more hop provided no existing LSPs are rerouted to different paths. Furthermore, it is likely that the existing LSPs will stay on their original paths, which guarantees to pass the second test. Now setting up the newly requested LSP between node C and node F is straightforward. This example intentionally omitted the directionality of LSPs for the sake of a simple explanation. Taking directionality into account will not change any conclusions made in the example.

7.7.4   Heuristic Algorithm Description

The algorithm has the following definitions:

- The *physical topology* of a WDM optical network G is denoted by $< N, F >$, where N is the set of network nodes and F is the set of fibre connections. A fibre connection is a pair of fibre links pointing opposite directions between two network nodes.
- The *lightpath topology* P embedded in the physical topology is denoted by $< N, L >$, where L is the set of lightpaths. A lightpath is a directional optical channel starting from its head-end network node passing zero or more other network nodes and ending at its tail-end network node. So, the trajectory of a lightpath can be represented as a sequence of fibre links and/or a list of network nodes.
- The embedding of a lightpath topology into the physical topology is called the *mapping* of the lightpath topology. The mapping associates each lightpath in the lightpath topology with a list of network nodes that the lightpath traverses in the physical topology.

The algorithm uses the following notations:

- The *residual lightpath topology* (or simply *residual topology*) $R(A,Z)$ that a potential head-end, $A \subset N$, of a yet to be set up LSP with tail-end $Z \subset N$, sees is denoted by $< N, R >$, where $R \subset L$. A is not able to set up the LSP if $R(A,Z)$ is a disconnected graph, meaning that there is not always a path from any node to every other node.

- A disconnected R has two or more components. A *component C* of a graph G is a maximal connected subgraph of G, i.e. there is always a path from v_1 to v_2 $\forall (v_1, v_2)$ $\subset C$. A component can be a single node. By definition of residual lightpath topology, $R(A,Z)$ always satisfies that the potential head end, A, belongs to one component called *head component* denoted C_A, and the potential tail end, Z, belongs to a different component called *tail component* denoted C_Z.
- *Qualifier, Q*, is a LSP-specific lightpath qualifier. It specifies what the constraints are when routing the LSP through the current lightpath topology. When setting up a LSP to Z, A applies Q onto P to obtain $R(A,Z)$, i.e. $R(A,Z)$ is a subgraph of P obtained by $R(A, Z) = Q * P$. Specification and presentation of a qualifier is not presented here.

The heuristic algorithm has a basic version that is used when a single network node fails to locate resources to accommodate a new LSP, and an extended version that is able to solve the same problem for multiple network nodes when certain conditions are met.

The heuristic algorithm is based on the following lemma:

Lemma: Given a physical topology G, a lightpath topology P, a node pair A and Z of a LSP, the residual lightpath topology R(A,Z) with A and Z in disconnected graph components, a lightpath reconfiguration solution exists if and only if there is a lightpath that traverses the head component and the tail component according to the residual topology.

A lightpath qualifying for a lightpath reconfiguration solution is called a **feasible lightpath**. The algorithm does not have to search through all lightpaths in the residual topology in order to find a feasible lightpath. Rather, it can search within a smaller scope because a feasible lightpath between two points must traverse at least one fibre within the fibre minimum cut set per those two points. Therefore, it is sufficient to look at lightpaths that are instance to the minimum cut set of fibres between the head component and the tail component. Once a feasible lightpath is found, there may be more than a single way to reconfigure it. Which particular way is actually adopted is subject to additional considerations. A reconfiguration solution is said to be available if a feasible lightpath can be identified. When there are multiple feasible paths, the reconfiguration feasible path selection follows this preference order:

- The most preferable lightpath shares the same head end node.
- The second preferable lightpath belongs to either the head component or the tail component.
- The least preferred lightpath belongs to a third component.

The heuristic is involved when a node needs to set up a LSP to another node but finds insufficient resources in the current lightpath topology. Note that the heuristics can be generalised in cases, where additional resources are needed between two specified nodes. The input to the heuristics is the residual topology. In addition, the heuristic has to have full knowledge of the lightpath topology and its mapping to the physical topology as well as the physical topology itself. The heuristic first identifies the minimum cut set of fibres between the head component and the tail component, and then identifies all lightpaths embedded in this set of fibres. Next, the heuristic searches in the residual lightpath topology for any lightpaths that traverses the head component, a fibre in the minimum cut set and the tail component. A reconfiguration solution exists if such a lightpath is found. Pseudo code description of the lightpath reconfiguration heuristic algorithm is given below.

```
/* algorithm inputs */
G;   // physical topology
P;   // lightpath topology
M;   // lightpath to fibre mapping
A;   // head end node
Z;   // tail end node
R(A,Z);   // residual lightpath topology per A and Z
/* find the set of all nodes in the head component */
H: = getComponentSet(A,R(A,Z));
/* find the set of all nodes in the tail component */
T: = getComponentSet(Z,R(A,Z));
/* find the min cut set of fibres between H and T */
fibreMinCut = minCut(H,T,G);
/* find all candidate lightpaths */
candidateLightpath: = getLightpathId(fibreMinCut,M);
/* sort the candidate lightpaths according to the preference */
candidateLightpath: = sortCandidateLightpath();
/* retrieve the feasibleLightPath details for reconfiguration */
for each lightpath p in candidateLightpath {
 /* find the set of nodes that p traverses */
 L: = nodeList(p);
 if intersection(L,H,T) ! = 0 {
  feasibleLightpath: = p;
  return feasibleLightpath;
  }
 }
```

The above reconfiguration heuristic algorithm can be extended to cope with m potential LSPs with $n < m$ reconfigurations. The need for m instances of reconfiguration can be met by sequentially applying the basic version heuristic algorithm to each instance. However, a better, more optimised solution may exist for multiple LSP requests. The ideas are explained, using the example shown in Figure 7.12. Assume the head end *A1* of a potential LSP with tail end *Z1* sees a disconnected residual topology *R(A1,Z1)*, and the head end *A2* of a potential LSP with tail end *Z2* sees a disconnected residual topology *R(A2,Z2)*. The extended lightpath reconfiguration heuristic algorithm can formulate an equivalent reconfiguration problem, where the objective solution satisfies the needs of both *A1* and *A2* (subject to certain conditions).

The equivalent reconfiguration problem can be formulated in the following steps. Every step proceeds only if the previous steps have been successful. Failing in any step means the equivalent reconfiguration problem cannot be defined, and therefore,

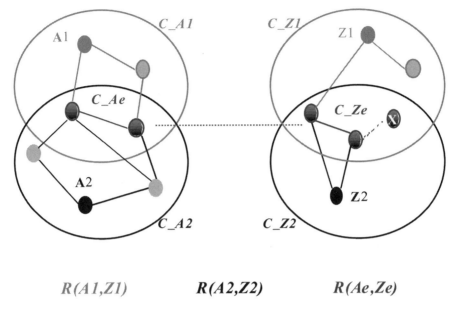

$$R(A1,Z1) \qquad R(A2,Z2) \qquad R(Ae,Ze)$$

Figure 7.12   An example where an extended heuristic algorithm can be applied.

a single reconfiguration solution does not exist for the set of individual reconfiguration needs. In such cases, each reconfiguration problem has to be solved separately. These steps are listed below.

- **Step 1.** The combination of qualifier of the first reconfiguration problem, $Q1$, and that of the second reconfiguration problem, $Q2$, are well defined, i.e. the equivalent qualifier exists, $\exists Qe = Q1 \vee Q2$. (This ensures that $Q1$ and $Q2$ are not specifying something contradictory to each other.)
- **Step 2.** Apply Qe to the intersection of the residual lightpath topology of the first reconfiguration problem, $R(A1,Z1)$, and that of the second reconfiguration problem, $R(A2,Z2)$, and the resulting equivalent residual lightpath topology is not empty, i.e.

$R(Ae, Ze) = Qe * (R(A1, Z1) \cap R(A2, Z2)) \neq \phi.$

- **Step 3.** Within $R(Ae,Ze)$, the equivalent head component, C_Ae, is not empty, i.e.

$C\_Ae = Qe * (C\_A1 \cap C\_A2) \neq \phi,$

and the equivalent tail component, C_Ze, is not empty, i.e.

$C\_Ze = Qe * (C\_Z1 \cap C\_Z2) \neq \phi.$

- **Step 4.** Eliminate unreachable sub-components.

In the third step, the equivalent head component and/or the equivalent tail component may become disconnected. Step 4 is especially designed for this situation, according to which a sub-component must be removed from the equivalent head/ tail component if there are no disjoint paths from the sub-component to all implied head/tail nodes. One such example is node X in Figure 7.12. Though belonging to the intersection of C_Z1 and C_Z2, node X becomes a disconnected component after Qe is applied to the intersection. It has to be eliminated from C_Ze because in the union of R(A1,Z1) and R(A2,Z2) there are no disjoint paths from this sub-component (i.e. node X) to Z1 and Z2. When this point is reached, a single reconfiguration solution can be sought by applying the basic version of the lightpath reconfiguration heuristic algorithm to the equivalent problem. In this example, there are only two reconfiguration problems/requests. When the number of reconfiguration problems is larger than 2, it can be proved that defining an equivalent reconfiguration problem is NP-complete. Nonetheless, this approach works well when the number of reconfiguration problems is small (e.g. 3 or 4). To be more effective, the order of the formulation steps can be changed. A necessary condition for an equivalent problem to exist is the existence of the equivalent head and tail components, which in turn relies on the existence of non-empty intersections of head components and tail components. When the number of reconfiguration problems is 3 or larger, these intersections may not exist for all elements of the problem set. However, they still can be applied for a subset of the problems/requests. Hence, the extended algorithm should be the first test to make when there are multiple simultaneous reconfigurations. As a consequence, qualifier combination should be performed after the set of mergeable problems is known. Pseudo code of the enhanced version presented below has taken these into account./* algorithm inputs */

```
G;   // physical topology
P;   // lightpath topology
M;   // lightpath to fibre mapping
{Q};   // qualifier set
{A};   // head end node set
{Z};   // tail end node set
{R({A},{Z})};   // residual lightpath topology set per corresponding pairs in
{A} and {Z}
```

/* compute intersection of head/tail components of {R({A},{Z})}, note xInstanceList contains IDs of problem with which x instances associate */

```
assign two empty lists, headInstanceList and tailInstance-
List, to every node in P
  for each R(i) in {R({A},{Z})} {
   for each node in P {
    if node is in the head component of R(i)
     headInstanceList(node) + = i;
    if node is in the tail component of R(i)
     tailInstanceList(node) + = i;
   }
  }
```

/* identify the set of mergeable problems NP-complete, use a heuristic. So solution may not be unique

```
  sort headInstanceList by length in descending order; */
  sortedHead: = sort(headInstanceList);
  m = 1; // number of possible mergings
```

```
for each headList in sortedHead {
 if length(headList) > = 2 {
  for each tailList in tailInstanceList {
  mergeable(m): = searchLongestMatch(headList,tailList);
  m + +;
  }
 }
}
e = 1; // number of equivalent problems
for each list in mergeable {
 Qe: = combine(Q(mergeable));
 /* compute equivalent head component and tail component */
 mergedA: = Qe * intersection(A(list));
 mergedZ: = Qe * intersection(Z(list));
 /* mergedA and/or mergedZ can be disconnected */
 for each subComponent in mergedA {
  if disjointPath(subComponent,{relatedHead}) exist
   Ae(e) + = subComponent;
 }
 for each subComponent in mergedZ {
  if disjointPath(subComponent,{relatedTail}) exist
   Ze(e) + = subComponent;
 }
 if (Ae ! = empty) AND (Ze ! = empty) {
 Re(Ae(e),Ze(e)): = Qe * intersection(R(A(list),Z(list)));
 e + +;
 }
}
```

In a single reconfiguration case, the potential head end and tail end nodes are known but the head and tail components need to be calculated. Then, reconfiguration is sought between these head and tail components. In a multiple reconfiguration case, both (equivalent) head component, Ae, and (equivalent) tail component, Ze, are known. Therefore, the algorithm does not need to calculate the head and tail components.

7.7.5 Heuristic Discussion

The reconfiguration example and proposed heuristic algorithms discussed in previous sections focus on the 'break a lightpath' scenario when upper layer MPLS failed to route a new LSP through the current lightpath topology. Reconfiguration in fact can follow a different direction, i.e. 'join two lightpaths'. Assume a LSP spanning multiple lightpaths is the only LSP using those lightpaths. These lightpaths can be reconfigured to eliminate intermediate hops by concatenating these lightpaths. Whether to concatenate two lightpaths at a joint node can be simply a local decision by that node. For a joint node, if all LSPs on an incoming lightpath, L_in, are continuing on a same outgoing lightpath, L_out, and all LSPs on L_out are extensions of LSPs coming in from *L_in*, i.e. no LSPs are dropped from *L_in* and/or added to *L_out* at this node, this *L_in* — *L_out* lightpath pair may be concatenated. In practice, many other considerations can be combined in determining whether to join these two lightpaths. For example, recent history of LSP add/drop to/from these lightpaths at this node, loading level of these lightpaths, etc.

Reconfiguration in either break or join scenario requires certain migration control so that services supported by the concerned LSPs can gracefully undertake the reconfiguration. We discuss the reconfiguration migration in the following section.

7.7.6   Lightpath Reconfiguration Migration

In IP over reconfigurable WDM networks, reconfiguration migration refers to IP topology migration from an old virtual topology to a new virtual topology. As indicated in the heuristic algorithms for virtual topology design, the assumption is that the topology design algorithms start from a dark network and outputs an optimal or near-optimal lightpath topology. In contrast, in packet switched WDM networks, reconfiguration migration refers to individual lightpath configuration/reconfiguration.

The purpose of exercising migration control in lightpath reconfiguration migration is to minimise side effects of reconfiguration on applications. With the presence of MPLS, however, migration control is relatively easy to achieve.

Let us first look at a 'break a lightpath' case. There may be zero, one, or more than one established LSP on each lightpath. Assume all established LSPs remain on their current paths, i.e. no rerouting is involved in case of emerging better paths. Then the migration problem becomes how to install proper entries in the forwarding label table at the breaking point. Suppose lightpath *orange* between node *A* and node *C* is being broken at node *B* (see Figure 7.13). This lightpath is currently carrying two LSPs. Before lightpath breaking, these two LSPs edge node *A* and node *C* as adjacency nodes although the lightpath traverses node *B*. The forwarding label tables at node *A* and node *C* have been set up accordingly. In order for these LSPs to work properly after lightpath *orange* is broken, node *B* must install proper entries in its own forwarding label table. If label significance is local to per lightpath port, a simple yet efficient way to select labels for these LSPs at node *B* is to reuse the same labels as assigned by node *A*. In so doing, node *C* will not need to make any changes to its label table as far as the reconfiguration is concerned.

When joining two lightpaths at a node, the fundamental operation is to let the downstream node learn the LSP label mapping at the joint node. Figure 7.13 can be used in a reverse style. Assume that originally, a lightpath *orange* exists from node *A* to node *B*, and another lightpath *orange* presents from node *B* to node *C*. These are two individual lightpaths but happen to have the same colour. There are two transit LSPs coming in from the *AB orange* lightpath continuing onto the *BC orange* lightpath. Now node *B* decides to join these two lightpaths. When the lightpaths are joined, without migration help, node *C* does not understand the labels carried by packets coming in through the *orange* lightpath from node *A*. This is because label significance is a mutual consensus reached at LSP setup time between adjacent nodes. Therefore, node *B* must let node *C* know its mapping from an *A*'s label to

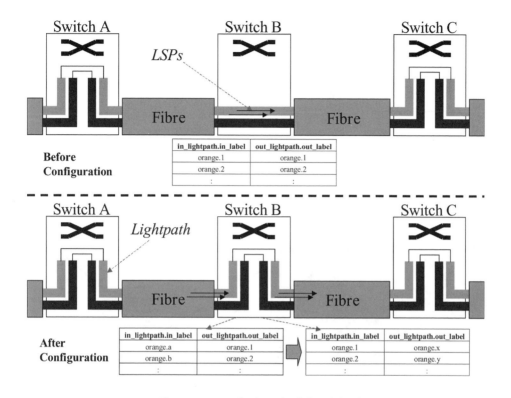

Figure 7.13 LSP healing after lightpath break.

its own label for those LSPs. In this example, the two lightpaths to be jointed happen to have the same colour. But this is not required when planning to join two lightpaths provided the node has full wavelength interchange capability.

7.8 Simulation Study of IP/WDM Reconfiguration

To verify the performance of the reconfiguration algorithms, the simulation study is needed. We focus our discussion on virtual topology design in IP over reconfigurable WDM networks. In this section, we present some performance results to illustrate and evaluate the performance of the topology design heuristic algorithms presented in Section 7.6.3. The related simulation is extracted from [Liu02].

The simulation is based on a flow-based model and the simulator is written in the C++ programming language. There are three different simulation scenarios supported by the simulator. The first scenario specifies the number of nodes and links and generates the topology randomly. This is used to evaluate the routing algorithm and the bandwidth allocation scheme under different traffic demands and topologies. The second scenario reads the initial topology from a file. The third simulation scenario designs new topologies based on current traffic demands according to three heuristic algorithms, RD, RDHP, and DHP. The output of the simulator includes the total network throughput, the total demand, the total loading, the final residual demand matrix, the network topology, and the final link utilization and loading. The input of the simulator consists of the initial topology, the node interface constraints, the skew factor, and the traffic-loading factor. The last two factors are used for traffic pattern generation. To reduce the variation of randomly generated traffic, each experiment is conducted 100 times using different seeds. The result presented is based on the average of the 100 trials. A fair bandwidth allocation scheme has been used in case of resource contention in the simulation.

The simulation uses an operational US carrier IP backbone network as the fixed topology. This network has 11 nodes and 19 connections as shown in Figure 7.14(a). A reconfigurable topology will differ from the topology only in the way that these 19 connections are rearranged. The reconfigured topology is computed using the heuristics presented in Section 7.6.3.

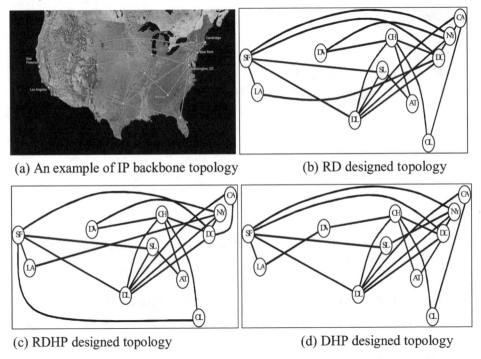

| (a) An example of IP backbone topology | (b) RD designed topology |
| (c) RDHP designed topology | (d) DHP designed topology |

Figure 7.14 Simulation topology (i.e. the fixed topology) and topology design examples using heuristic algorithms for the traffic pattern shown in Table 7.2.

7.8.1 Traffic Generation

For the flow-based simulation, the traffic is represented by a node-to-node flow matrix. Traffic pattern refers to the value distribution of a flow matrix. Every simulation is repeated for a large number of randomly generated traffic patterns. In order to justify different nodal degrees at different nodes in the fixed topology, the simulation used the average (across different traffic patterns) of the flow from node i to node j proportional to the nodal degrees of both node i and node j. Hence, the higher nodal degree of a node the more traffic it would source and sink. The particular traffic pattern that leads the heuristic algorithms to designing topologies is shown in Table 7.2. Numbers in the table are based on a standardised unit. All links are assumed to have the same capacity of 10 units in all topologies. Parallel links are allowed in the simulation as long as the nodal interface allows.

7.8.2 Simulation Results

Figure 7.14(b), (c) and (d) show example topologies designed by the heuristic algorithms RD, RDHP and DHP, respectively. The heuristic algorithm designed topologies and the fixed network have the same number of nodes subject to the same nodal degree and wavelength constraints. This infers that each of the topologies require the same amount of physical network resources, i.e. nodes and links. However, as we will see later in this section, they show different performances.

Figure 7.15 shows comparison of network throughput, normalised to the total offered load in different reconfigurable networks designed by different heuristic algorithms, as well as that in the fixed network. In the figure, topologies designed by the heuristic algorithms outperform the fixed network under a number of traffic patterns. In several patterns such as 9 and 11, RD and RDHP increase the normalised throughput by more than 20% over the fixed topology. Among different heuristic algorithms, RD usually provides higher network throughput than that of RDHP or DHP. Between RDHP and DHP, RDHP is likely to perform better for the given traffic patterns.

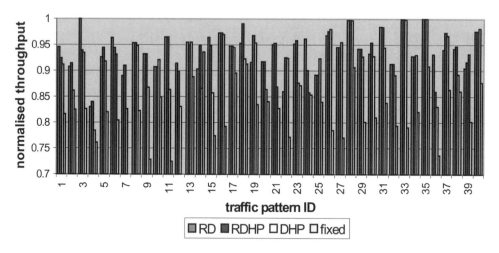

Figure 7.15 Normalised network throughput comparison for different networks under a number of traffic patterns.

Figure 7.16 shows the comparison of weighted hop distance in different reconfigurable networks designed by different heuristic algorithms, as well as that in the fixed network. Weighted hop distance is a criterion that describes how effectively a network uses its resources. Normally, a lower weighted hop distance is desirable

Table 7.2 A typical traffic pattern.

	SF	LA	DV	DL	SL	CH	AT	OL	DC	NY	CA
SF	0	0.549486	0.463462	2.16849	0.575856	2.82773	0.229423	0.432905	0.645896	2.11753	0.569666
LA	1.19533	0	0.009277	1.06417	0.084089	0.320559	0.041071	0.000897	0.062143	0.77727	0.163634
DV	0.386059	0.108151	0	0.466109	0.11617	1.22406	0.013141	0.124992	0.264557	0.782226	0.085995
DL	10.2344	0.58883	1.06043	0	2.84156	10.507	0.062129	1.39882	10.3082	6.49246	1.35553
SL	2.55097	0.266756	0.068721	0.836772	0	0.495859	0.237587	0.161092	0.183254	1.0779	0.377708
CH	5.88006	0.788903	1.10569	8.15519	1.40501	0	0.98693	0.0755333	7.09627	3.88836	1.17565
AT	0.218398	0.047663	0.1377	0.23907	0.418152	0.514161	0	0.114982	0.147137	0.466323	0.057136
OL	1.0572	0.108176	0.194177	1.36196	0.40501	1.24042	0.196472	0	0.825564	0.195465	0.026429
DC	2.66489	0.845782	0.497108	6.87902	0.092442	5.44517	0.846278	0.008201	0	1.11194	0.787645
NY	2.28568	0.388855	0.004004	0.794775	1.79912	2.14423	0.668502	0.063795	2.21274	0	0.569549
CA	1.22406	0.142283	0.009662	1.59708	0.372077	0.805307	0.059700	0.193139	0.525147	0.796924	0

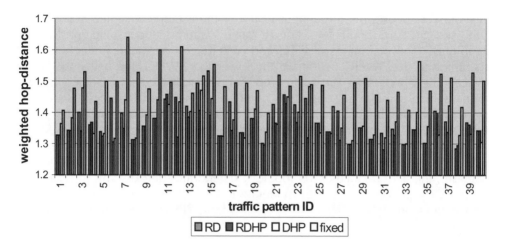

Figure 7.16   Weighted hop-distance comparison for different networks under a number of traffic patterns.

since traffic will be delivered in few hops on average. Preference level of different heuristic algorithms under this criterion conforms to what is shown in Figure 7.16. In several traffic patterns such as 7 and 10, RD and RDHP reduce the weighted hop distance by more than 15% over the fixed topology. Among different heuristic algorithms, RDHP usually provides a lower weighted hop distance than that of RD and DHP. Between RD and DHP, DHP is likely to perform better for the given traffic patterns.

As a means of traffic engineering, reconfiguring network topology is most effective when traffic is highly skew. Skewness represents the mismatch between a traffic demand and the virtual topology configuration. To study how the reconfiguration benefit increases as the traffic skewness increases, topologies designed by the heuristic algorithms were compared against the fixed network. Figure 7.17 shows examples of traffic patterns with different skewness (note the difference in scales). Although each traffic instance may vary, the traffic skewness is closely related to the nodal degree constraints.

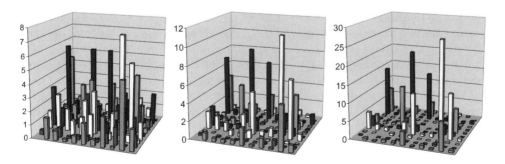

Figure 7.17   Traffic patterns with different skewness.

Figure 7.18 shows the throughput gain and the average weighted hop distance gain (see Section 7.6.3) for various traffic patterns under different heuristic algorithms. The definitions of NTG and AWHG can be found in Section 7.6.3. According to the AWHG definition, the lower the AWHG, the better the performance. The abscissa in the figure represents a traffic-loading factor, which is normalised to a minimal congestion point using the fixed topology. Hence, the figure shows the heuristics performance in the presence of congestion in the fixed topology as the traffic load steadily increases. In Figure 7.18(a), RDHP-designed topology constantly outperforms that of DHP. When the traffic load or the demand is low, RD topology gives the best performance; as the traffic load increases, the performance of RD topology deteriorates quickly. This indicates that RD topology performs well with moderate traffic loading and an expected traffic pattern. In Figure 7.18(b), RDHP outperforms both RD and DHP. There is a performance crossover between RD and DHP. DHP performs better with a higher load. Figure 7.18(c) and Figure 7.18(d) show the throughput gain in the case of skewed and very skewed traffic. Again, RDHP gives the best performance. DHP performs better when the loading factor is high, which implies that fewer congested links favours RD. Figure 7.18(e) shows the AWHG under the uniform traffic. RD topology gives the worst performance, and DHP topology provides the best performance when the network is slightly congested. As the congestion increases, RDHP topology outperforms those of the other two algorithms. Figure 7.18(f), (g) and (h) show the AWHG under lightly skewed, skewed, and very skewed traffic. Under skew and very skewed traffic patterns, when the network is slightly congested, reconfiguration gain is increased rapidly. When the network is congested to a certain extent, the percentage of the AWHG due to reconfiguration is bounded and may even decrease under different algorithms. Figure 7.19 shows the NTG and AWHG under different traffic patterns including uniform, lightly skewed, skewed, and very skewed, for ECMP. Nearly in all experiments, the proposed algorithm-designed topologies present higher throughput and lower weighted hop distance than that of the fixed topology.

One can also conclude from Figure 7.18 and 7.19 that the reconfiguration algorithms perform better when the traffic patterns skewness increases. This agrees with the prediction that reconfiguration is especially meaningful when the traffic pattern changes dramatically. This does not mean once there are heavy demands, reconfiguration can be used to increase the network capacity. Instead, it infers that reconfiguration can help to police the skewed traffic therefore to improve the total network utilization. For example, in the case of a special event such as a football game, there will be heavy traffic near the stadium. Reconfiguring the roadblocks from the normal configuration increases the total road usage. In terms of network throughput, RD is a good heuristic when there are a small number of congested links. RDHP gives the best performance when the traffic pattern is skewed. DHP is a good alternative when there are a large number of congested links despite the traffic pattern skewness. In terms of hop count, both DHP and RDHP designed topology perform better than that of RD simply because hop count is not taken into account in the optimisation process of RD. When the network is lightly congested, RDHP topology presents the best performance.

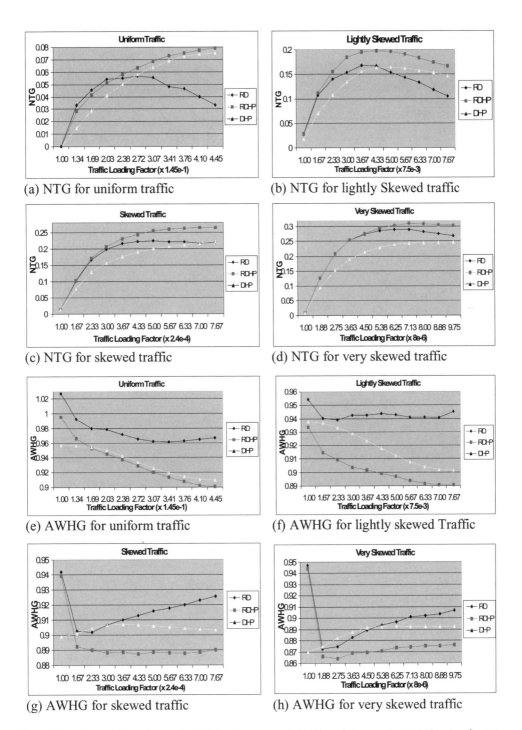

Figure 7.18 Network throughput gain (NTG) and average weighted hop distance gain (AWHG) using shortest path routing; every point plotted using the average over 100 trails.

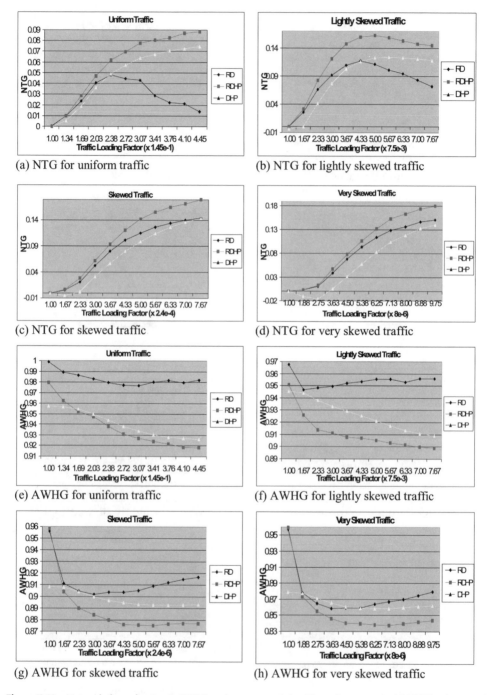

Figure 7.19 Network throughput gain (NTG) and average weighted hop-distance gain (AWHG) using equal cost multi-path routing; every point plotted using the average over 100 trails.

In the presence of optimal routing such as ECMP, all algorithms and the fixed topology provide an increased network throughput and reduced average hop counts (comparing with the performance of their counterpart under SPF). However, the NTG and AWHG using ECMP may not necessarily be larger than the NTG and AWHG using SPF since ECMP over one topology has higher throughput and lower average hop-distance than SPF over the same topology. However, since the simulation is only flow-based, it does not reveal the complexity, the network impact, and the negative effects introduced by ECMP.

It is also worth pointing out that the DHP algorithm itself is much simpler than RDHP and RD since it does not keep track of residual demands. A real operational network never lets its loading level reach 100% or not even 60%. It usually reserves a certain percentage of the network for emergency situations to ensure network availability and performance. This is an operational issue, which will be taken care of by the reconfiguration migration module in the IP/WDM traffic engineering framework or the network operation policy. Nevertheless, the simulation presented here proves that indeed there is a need for reconfiguration and the proposed algorithms present a better topology especially when the traffic pattern is skewed.

7.9   IP/WDM Traffic Engineering Software Design

The traffic engineering framework can be structured in two ways, overlay and integrated. Though the major functional components are the same, the software architecture varies slightly. The overlay approach features a client-server relationship, where the IP layer requests transport services from the WDM layer. Consequently, traffic engineering is performed in each layer individually. Thus, at each layer, there are corresponding traffic engineering and network control components. The two traffic engineering functional blocks at the IP layer and WDM layer are connected through a special interface to exchange pertinent information. The integrated approach forms a peer-to-peer relationship, whereby each network node consists of an IP router and a switching fabric. The detail of IP/WDM integration is still under research and development. In one example, the IP router (or IP control functionality) only provides the control plane, and the data traffic flows through the all-optical switching fabric directly. In another example, the IP router is expanded with the WDM switching fabric (usually O-E-O switching fabric) so the IP router has multi-wavelength interfaces. Finally, there are research groups working on pure photonic packet routers, where they attempt to implement IP control functions (such as header processing) in the pure optical domain. In integrated traffic engineering, every WDM switch/router is an IP addressable device, and each network node prefers to be equipped with one traffic engineering entity.

7.9.1   Software Architecture for Overlay Traffic Engineering

Figure 7.20 shows the software components and interfaces for the overlay traffic engineering approach. The IP layer exhibits a virtual topology that is an abstraction of physical network connectivity. The WDM layer manages the physical topology whose connectivity is based on wavelengths as well as fibres. The traffic engineering

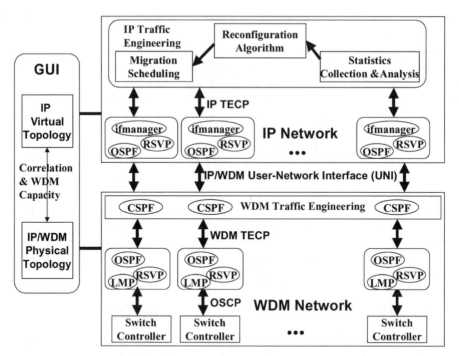

Figure 7.20 Software architecture for overlay traffic engineering in IP/WDM networks.

control part in the WDM layer is embedded in WDM performance and connection management functions. WDM-specific management functions are introduced in Chapter 4. The overlay traffic engineering software architecture is composed of IP network components and WDM network components (see Figure 7.20).

Each of the IP and WDM layers includes corresponding network control and traffic engineering related functions. IP network control consists of routing protocols such as OSPF, signalling protocols such as RSVP, and an interface manager (*ifmanager*); IP traffic engineering consists of statistics collection and analysis, a reconfiguration algorithm, and migration scheduling. IP-centric WDM network control includes routing protocols such as OSPF with optical extensions, signalling protocols such as RSVP with optical extension, LMP, and OSCP; WDM traffic engineering includes WDM traffic engineering algorithms such as CSPF. The traffic engineering to network control protocol (TECP) specifies the interface between network control and traffic engineering. In overlay traffic engineering, TECP has two groups of messages, IP TECP and WDM TECP. The TECP messages format is presented in the following sections. OSCP provides the interface between the WDM network control and the switch controller (see Section 6.9.3).

The IP interface manager has two tasks. First, it takes commands from the IP traffic engineering module to enable and/or disable proper router interfaces when under-lying lightpaths are being reconfigured. Second, it is responsible for the linkage between router interface and WDM add/drop port and provides the IP and the WDM address translation. IP/WDM address translation can also be provided using dedicated server(s), such as ARP and RARP. In the implementation, the IP router interface manager can issue SNMP commands to IP routers to query or modify the state of their interfaces. Or, it can be implemented using scripts to use the 'ifconfig' to reconfigure router interface state and configuration.

IP traffic engineering has three main functional components: reconfiguration algorithm, statistics collection and analysis, and migration scheduling. Statistics collection is responsible for monitoring the network to collect traffic statistics and setting performance thresholds; statistics analysis is capable of traffic pattern derivation and future traffic prediction. In the implementation, one can use the SNMP protocol to monitor IP routers for traffic measurements, and to collect data from IP routers. There is also third party software available for statistics collection, e.g. *libpcap*.

The reconfiguration algorithm takes inputs from the statistics collection and analysis component and the current topology from the routing protocol. It outputs a new topology to the migration scheduler. The reconfiguration algorithm is designed as a set of heuristics in supporting various reconfiguration objectives such as minimising network latency. The optimisation objectives can be specified from the GUI. Migration scheduling calculates a sequence of migration steps to minimise the reconfiguration impact for user traffic. Different strategies are available for migration scheduling. First, the scheduling can be performed with minimum knowledge of the WDM layer. With this strategy, constraints to the scheduling are predefined, so that the migration sequences are more likely to be accommodated by the WDM layer. A typical example of such constraints is the total occupied WDM resources (i.e. number of lightpaths) bounded in any migration step. The resultant migration schedule consists of a sequence of steps. Each step is a single setup or tear-down of a lightpath. Second, the scheduling can be done with full knowledge of the WDM layer. With this strategy, the resultant migration is guaranteed to be accommodated by the WDM layer. The price for a non-blocking migration sequence is that the migration scheduler has to poll the WDM layer for its state information. Alternatively, a simple migration sequence can be explicitly specified from the GUI.

In conventional WDM optical networks, traffic engineering related functions are provided by the connection and performance managers. In IP-centric WDM networks, OSPF and LMP provide the default routing, information dissemination, and topology and neighbourhood discovery. As in IP networks, the default routing in OSPF provides only the best-effort fibre path, which neither takes into account wavelength availability nor wavelength continuity constraints. Therefore, efficient, non-blocking lightpath computation requires WDM traffic engineering algorithms. A simple example of the WDM traffic engineering algorithms is CSPF. CSPF is more complex when more constraints are considered. CSPF can be implemented as a centralised explicit lightpath routing manager. To improve availability, CSPF can be distributed to each network node. A distributed implementation of CSPF requires synchronisation among WDM traffic engineering entities. Based on traffic engineering requests, lightpaths can be set up or removed using RSVP.

Finally, to facilitate users, IP/WDM GUI is designed to manage the layered network. The GUI provides configuration, connection, fault, and performance management interfaces. In overlay IP/WDM networks, the GUI also provides topology correlation; for example, a virtual lightpath on the IP topology is linked to the corresponding wavelengths and fibre links in the physical WDM topology.

7.9.2   Software Architecture for Integrated Traffic Engineering

Figure 7.21 shows the software architecture for integrated traffic engineering. In this approach, each IP/WDM node can have complete support for network control and traffic engineering. Each node possesses full topological knowledge, i.e. a W-MIB (wavelength MIB), which consists of not only the fibre topology but also embedded lightpaths and wavelength assignment status.

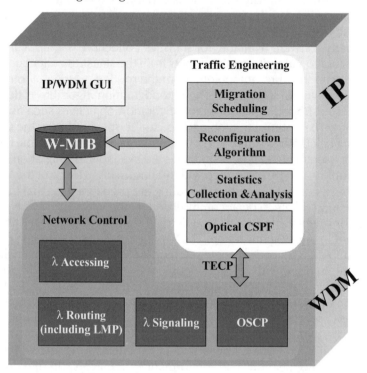

Figure 7.21   Software architecture for integrated traffic engineering in IP/WDM networks.

The main software components are identical to those found in the overlay traffic engineering architecture, but are structured in an integrated fashion. Network control is composed of four components: wavelength routing, wavelength signalling, wavelength accessing, and OSCP.

Wavelength routing (including LMP in the case of separated control channels from data channels) enables all network nodes to establish a consistent view of the network, and co-ordinates wavelength routing decisions to facilitate required light-path connections. The specific functions include physical/virtual network topology discovery, link (fibre, wavelength, and lightpath) state information dissemination and maintenance, and default best-effort routing. Wavelength routing functions can be supported through enhanced link-state protocols, such as OSPF with extensions. WDM-related link state information could be either implemented in a standalone TE database or integrated into the standard router link state database. Traffic engineering required information can be collected through OSPF opaque LSA flooding.

Wavelength signalling fulfils wavelength routing decisions by executing at least some of the following functions: wavelength assignment, lightpath priority arbitration, wavelength pre-emption, lightpath protection/restoration, and cross-connect setup/tear-down. A basic function to be supported by signalling is cross-connect setup and tear-down. Wavelength assignment requires the local signalling protocol daemon to have certain knowledge of wavelength usage and continuity constraints. An advanced function of the signalling protocol is to support the concept of prioritised lightpaths. In that case, wavelength signalling is responsible for resource arbitration should it occur that two lightpaths contend for the same wavelength of the same fibre. Furthermore, it is also possible to pre-empt a wavelength being used by a lightpath of lower priority in order to support a new lightpath of higher priority. In addition, signalling can be used to support adaptive QoS during lightpath setup. For example, when a segment of the routing lambda path does not support the required bit rate, QoS issues can be negotiated using a wavelength signal protocol.

Wavelength access control is used to map IP packets to wavelength and manage the IP router to WDM NE linkage. An important concern is how to design the packet to wavelength mapping function. In general, multiple lightpaths may exist or will be constructed for a given destination. So, a decision must be made, for each IP packet, on which lightpath to take. The mapping function shall map individual IP packets characterised by IP attributes (information found in IP headers, such as destination address, source address, and type of service, etc.) to wavelength channels or lightpaths featured by WDM physical layer properties (for example, connection data rate, OSNR of the optical channel, fibre hops, and channel loads). Wavelength access control is also responsible for managing the linkage between IP functions and WDM, for example, during reconfiguration to avoid impacts on users traffic.

As far as traffic engineering is concerned, every node can be equipped with the TE functional components: statistics collection and analysis, reconfiguration algorithm, migration scheduling, and optical CSPF algorithm. The functionality of these components is similar to their counterparts in the overlay approach. However, they can be configured in a more distributed manner. For example, a node can focus on the state of the lightpaths (initiated by itself) during migration scheduling.

7.9.3 IP Traffic Engineering to Network Control Protocol (IP TECP)

In the next three sections (Sections 7.9.3, 7.9.4 and 7.9.5), we detail the interface between network control and traffic engineering (Section 7.9.3 IP TECP and 7.9.5 WDM TECP), and between IP traffic engineering and WDM traffic engineering (Section 7.9.4 IP/WDM UNI) in the case of overlay traffic engineering. These interfaces are presented in the context of overlay IP/WDM networks.

Figure 7.22 shows the traffic engineering modules and interactions for the IP layer. The traffic engineering modules communicate with IP routers that participate in the traffic engineered IP virtual network to get router inventory information and traffic statistics from each router. The router inventory information includes detailed descriptions about each IP edge router of all its IP interfaces that can connect to a WDM network edge device. Example attributes of a router inventory database are as follows:

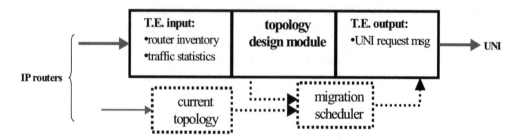

Figure 7.22   Traffic engineering modules and interactions for the IP layer.

- router ID;
- interface ID;
- interface IP address, if available;
- address for the WDM hook up point, if available;
- wavelength preferred;
- signal formats supported;
- bandwidth supported.

The above list represents the basic attributes with respect to an IP interface. More service-related and policy-orientated parameters can be defined. When an IP router that has already connected with another IP router over the WDM network participates in a traffic-engineered IP virtual network, the traffic engineering module needs information about the existing connections. In an established, closed control loop, however, full knowledge about the current IP virtual topology does not rely on any entity exterior to the traffic engineering module. This is because the current topology is the last output of the topology design algorithm. Communications between the traffic engineering module and IP routers can follow SNMP.

IP traffic engineering also needs to interface with WDM edge nodes or the WDM traffic engineering entity. The IP traffic engineering module will issue UNI messages on behalf of IP routers to different WDM edges nodes in a sequence determined by the migration scheduler. These UNI messages are intended either for a lightpath deletion or for a lightpath creation. A set of messages is defined for this purpose in Section 7.9.4. Since conventional IP networks require two-way connectivity between adjacent routers and the client layer is IP, lightpath manipulations requested by the UNI messages will always apply to lightpath pairs. The two unidirectional lightpaths of a lightpath pair provide connections in opposite directions between the same pair of IP routers. The same assumption is supported by the lightpath signalling mechanism of the WDM layer.

IP TECP is used by IP traffic engineering for collecting information from each IP router connected to the WDM edge switch. Three types of information are collected: router inventory information, traffic statistic information, and current virtual connection information. The information collection is supported by IP TECP messages that are defined in this section. All TECP messages use the common header shown in Figure 7.23.

01 2 3 4 5 6 7   01 2 3 4 5 6 7   01 2 3 4 5 6 7   01 2 3 4 5 6 7

Version	PDU length
Message Type	Transaction ID
(cont.)	
Sender ID	
Recipient ID	

Figure 7.23   TECP common header format.

Message field definitions in the TECP header are as follows:

- **Version:** the version number. This field uses two octets.
- **PDU length:** the total length of the PDU in octets. This includes the header itself. This field uses two octets.
- **Message type:** this field specifies the message type using two octets. IP TECP has defined the following message types:

  – *InventoryReq*: This message type indicates that the message sent by the IP traffic engineering module to a router is for a router inventory inquiry.
  – *InventoryResp*: This message type indicates that the message, sent by a router to the IP traffic engineering module in response to an *InventoryReq* message, is for reporting the inventory information of the sending router.
  – *TrafficReq*: This message type indicates that the message, sent by the IP traffic engineering module to a router, is for a router traffic-demand inquiry.
  – *TrafficResp*: This message type indicates that the message, sent by a router to the traffic engineering module in response to a *TrafficReq* message, is for reporting traffic-demand information of the sending router.
  – *ConnectionReq*: This message type indicates that the message, sent by the traffic engineering module to a router, is for the current virtual connection inquiry.
  – *ConnectionResp*: This message type indicates that the message, sent by a router to the traffic engineering module in response to a *ConnectionMessage*, is for reporting the current virtual connection information of the sending router.

- **Transaction ID:** this field specifies a unique identifier representing the transaction to associate responses to their corresponding requests. This field uses six octets.
- **Sender ID:** this field specifies the identification of the message sender. This field uses four octets.
- **Recipient ID:** this field specifies the identification of the message recipient. This field uses four octets.

Inventory request and response message

Inventory request has the message type code of *InventoryReq*. The message contains only the header (i.e. there is no message payload). Inventory response has the message type code of *InventoryResp*. In addition to the common message header, the message payload has the following payload format (see Figure 7.24). Message field definitions are as follows:

- **Interface ID:** this field represents the identification of the reported router interface that connects to the WDM network at WDM accessing point address. This field uses four octets.
- **Interface IP address:** this is the IP address of the interface. This field uses four octets.
- **WDM accessing point address:** this is the address of the WDM accessing point that the sender's interface with Interface ID and interface IP address connects. This field uses four octets.
- **Signal formats supported:** This field represents all signal formats supported by the interface with a bit vector. This field uses four octets.
- **Bandwidth supported.** This field specifies the maximum bandwidth supported by the interface in Mbps. This field uses four octets.
- **Upper lambda ID:** This field, together with the next field, specifies a range of preferred wavelengths. This field indicates the upper limit of the range using a lambda ID. If there is no preference, this field is set to '0' and the next field is ignored. This field uses two octets.
- **Lower lambda ID:** This field, together with the previous field, specifies a range of preferred wavelengths. This field indicates the lower limit of the range using a lambda ID. This field uses two octets.

```
0 1 2 3 4 5 6 7   0 1 2 3 4 5 6 7    0 1 2 3 4 5 6 7    0 1 2 3 4 5 6 7
```

Interface ID	
Interface IP address	
WDM accessing point address	
Signal formats supported	
Bandwidth supported	
Lower Lambda	Upper Lambda

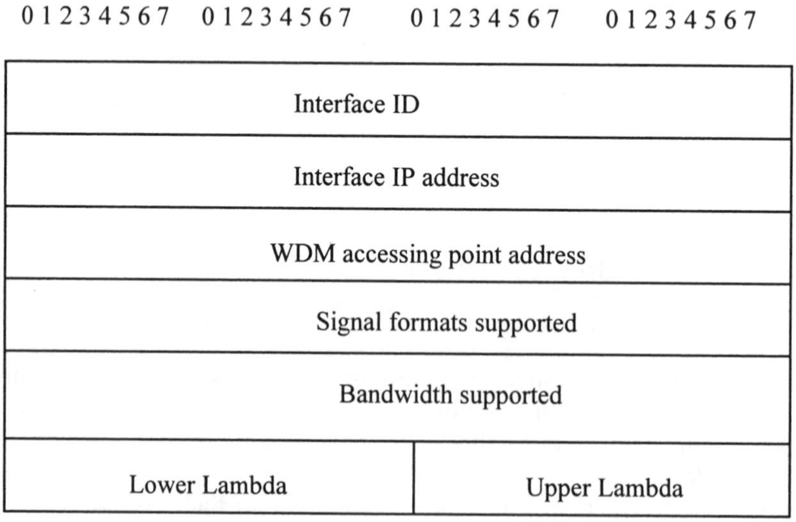

Figure 7.24 Message format for inventory response messages.

Traffic statistics request message

The message type code is *TrafficReq*. The payload message body format is shown Figure 7.25. The message fields are defined as follows:

- **Statistic type:** this field tells the recipient what to monitor by specifying the statistic type. It can be, for example, total traffic density or prioritised traffic density when *diffserv* is supported. This field uses two octets.
- **Sampling rate:** this field tells the recipient how the statistics should be collected through specifying a suggested sampling rate. The recipient can choose to use any sampling rate that is no coarser than the suggested one. This field uses two octets.
- **Average window:** routers are expected to preprocess the statistics they locally collected. This field tells each recipient (router) a proper average window size in time. This field uses four octets.
- **Report interval:** this field tells the recipient how often statistics reports are due at the traffic engineering module by specifying the time interval. This field uses four octets.

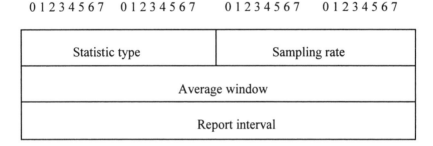

Figure 7.25 Message format for traffic statistics request messages.

Traffic statistics response message

The message type code is *TrafficResp*. The message body is composed of two parts: the first half is the received *TrafficReq* message, and the second half is the statistics information. The format of the second half varies with the type of the statistics. In traffic engineering, a traffic demand matrix is required. Its corresponding message body format is shown in Figure 7.26.

Message field definitions are as follows:

- **Egress Router ID i:** this field specifies the egress router for the traffic, where the ingress point of the traffic is the sender of this message. This field uses four octets.
- **Statistics:** this field attaches the statistics data. This field uses four octets.

Virtual connection status request and response message

Virtual connection status request message has the type code of *ConnectionReq*. The message contains only the common TECP message header. Virtual connection status

01234567   01234567   01234567   01234567

Statistic type	Sampling rate
Average window	
Report interval	
Egress router ID 1	
Statistics	
● ● ●	

Figure 7.26 Message format for traffic statistics response messages.

response message has the type code of *ConnectionResp*. The response message payload format is shown in Figure 7.27.

Since virtual connections are in fact related to the WDM edges nodes and interfaces and the UNI messages, the message field definition of Figure 7.27 is given in the following section of IP/WDM UNI (also see Figure 7.28).

7.9.4   IP/WDM User to Network Interface (UNI)

IP/WDM UNI is the interface between the IP traffic engineering module and the WDM traffic engineering module. In WDM layer, traffic engineering can be a centralised entity or a group of distributed modules located at WDM network edge nodes. In the latter case, the IP traffic engineering module is actually interfacing with multiple entities.

The UNI message format is designed to support the minimal set of UNI signalling functions. The message format can be easily extendable to include more fields and functions, such as parameters for service level agreement, identifications for different client groups, and security and accounting attributes (which a real network operator needs). All UNI messages include a version number indicating the version of UNI protocol followed by message type-specific data. The length of UNI messages is fixed. Figure 7.28 depicts the UNI message format.

UNI message field definitions are as follows:

- **Version:** this is the version number of the UNI protocol being used.
- **UNI Msg type:** five types of UNI messages have been defined: Lightpath Create Request (LCReq), Lightpath Create Response (LCResp), Lightpath Delete Request (LDReq), Lightpath Delete Response (LDResp) and Trap (Trap), respectively.
- **Initiator ID:** the ID of the message initiator. For a LCReq/LDReq message, it can be the A-end router or an authorised third party such as the traffic engineering

0 1 2 3 4 5 6 7   0 1 2 3 4 5 6 7   0 1 2 3 4 5 6 7   0 1 2 3 4 5 6 7

Lightpath owner ID	
A-end router ID	
A-end interface ID	
A-end IP address	
A-end WDM accessing point address	
Z-end router ID	
Z-end interface ID	
Z-end IP address	
Z-end WDM accessing point address	
Wavelength preferred	Signal formats
Bandwidth supported	

Figure 7.27   Message format for current virtual connection status response messages.

module. For a LCResp/LDResp or a Trap message, it can be the A-end WDM edge device or an authorised third party such as the traffic engineering module.

- **Lightpath owner ID:** any existing lightpath can be considered as the implementation of the LCReq message. For a given lightpath, this field is originally assigned by the initiator of the LCReq message (that created the lightpath). Therefore, for a first party initiated LCReq message, this field equates the Initiator ID of the message; for a LCReq message initiated by an authorised third party, this field may be either equal to the Initiator ID, or set to A-end Router ID. The value of this field is used by all other types of messages pertinent to the specified lightpath.
- **Request sequence number:** a request sequence number is assigned to each new request type message to distinguish among outstanding requests. This number is copied to the corresponding response type message to associate it with the original request.
- **A-end router ID:** the head-end router ID of the lightpath.
- **A-end interface ID:** the head-end router's physical port of the lightpath.
- **A-end IP address:** the head-end router's logical port of the lightpath.

0 1 2 3 4 5 6 7   0 1 2 3 4 5 6 7   0 1 2 3 4 5 6 7   0 1 2 3 4 5 6 7

Version	UNI message Type
Initiator ID	
Lightpath owner ID	
Request sequence number	
A-end router ID	
A-end interface ID	
A-end IP address	
A-end WDM accessing point address	
Z-end router ID	
Z-end interface ID	
Z-end IP address	
Z-end WDM accessing point address	
Wavelength preferred	Signal formats
Bandwidth supported	

Figure 7.28 IP/WDM UNI message format.

- **A-end WDM accessing point address:** the add port of the WDM edge device at the head end of the lightpath.
- **Z-end router ID:** the tail-end router ID of the lightpath.
- **Z-end interface ID:** the tail-end router interface ID of the lightpath.
- **Z-end IP address:** the tail-end router's logical port of the lightpath.
- **Z-end WDM accessing point address:** the drop port of the WDM edge device at the tail end of the lightpath.
- **Wavelength preferred:** this field is optional and used only by LCReq messages to specify the wavelength(s) preferred.

- **Signal formats:** this field is optional and used only by LCReq messages to specify the signal format(s) supported.
- **Bandwidth supported:** this field is optional and used only by LCReq messages to specify the bandwidth or bandwidth range supported.

The five basic types of UNI messages are Lightpath Create Request, Lightpath Create Response, Lightpath Delete Request, Lightpath Delete Response and Trap. We present each message type below.

Lightpath create request

A lightpath create request message is sent from an IP entity to a WDM entity through the control channel (either in-band or out-of-band). An IP entity in this context can be either an IP router that has at least one interface directly connected to a WDM edge device, or an authorised third party in the IP layer like the traffic engineering module. Similarly, a WDM entity can be either an edge WDM device such as WADM, or a controlling device in the WDM layer like a lightpath route computation engine. A lightpath create request message must define the following fields of the UNI message (i.e. the defined parameter set):

- version
- UNI msg type
- initiator ID
- lightpath owner ID
- request sequence number
- A-end router ID
- A-end IP address
- Z-end router ID
- Z-end IP address.

And optionally, the lightpath create request message defines the following fields of the UNI message (i.e. the optional parameter set):

- A-end interface ID
- A-end WDM accessing point address
- Z-end interface ID
- Z-end WDM accessing point address
- wavelength preferred
- signal formats
- bandwidth supported.

Lightpath create response

A WDM edge device generates a corresponding lightpath create response message as a result of processing every lightpath create request message. The lightpath create response message is sent to the lightpath route computation engine and the lightpath create request message initiator. A lightpath create response message has a defined parameter set (i.e. values must be set by the message initiator) that has the following attributes:

- version
- UNI msg type
- initiator ID
- A-end WDM accessing point address
- Z-end WDM accessing point address
- Z-end router ID
- Z-end interface ID
- Z-end IP address.

Also, the response message includes a copied parameter set (i.e. values are copied from corresponding fields of the request message) that has these attributes:

- lightpath owner ID
- request sequence number
- A-end router ID
- A-end interface ID
- A-end IP address.

In addition, the lightpath create response message uses an optional parameter set (i.e. values may be set by the message initiator) that has these attributes:

- wavelength preferred
- signal formats
- bandwidth supported.

The above parameter set is mandatory if these parameters were specified in the corresponding LCReq message.

Note that all Z-end parameters in the parameter set will be set to '0' if the attempt of creating the requested lightpath fails. Then, the values of all parameters in the copied set and the optional set become irrelevant.

Lightpath delete request

A lightpath delete request message is sent from an IP entity to a WDM entity through the control channel (in-band or out-of-band). An IP entity in this context can be either an IP router that has at least one interface directly connected to a WDM edge device, or an authorised third party in the IP layer like the traffic engineering module. Similarly, a WDM entity can be either an edge WDM device such as WADM, or a controlling device in the WDM layer like a lightpath route computation engine. A lightpath delete request message has a defined parameter set that includes these attributes:

- version
- UNI msg type
- initiator ID
- request sequence number
- A-end router ID

- A-end interface ID
- A-end IP address
- Z-end router ID
- Z-end interface ID
- Z-end IP address.

The lightpath delete request message also has an optional parameter set that includes these attributes:

- lightpath owner ID;
- A-end WDM accessing point address;
- Z-end WDM accessing point address.

Note that a *LDReq* message initiator may not know the owner of the lightpath it intends to delete. Then, it is up to the policy definition that whether such a request will be processed or not. One possible configuration is that the traffic engineering module can delete any lightpath, but an IP router can only delete the lightpaths it owns.

Lightpath delete response

A WDM edge device generates a corresponding lightpath delete response message as a result of processing every lightpath delete request message. The lightpath delete response message is sent to the lightpath route computation engine and the lightpath delete request message initiator. A lightpath delete response message has a defined parameter set, which includes:

- version
- UNI msg type
- initiator ID.

The lightpath delete response message also need to include a copied parameter set, which has these attributes:

- request sequence number
- A-end router ID
- A-end interface ID
- A-end IP address
- Z-end router ID
- Z-end interface ID
- Z-end IP address.

Its optional parameter set includes these attributes:

- lightpath owner ID;
- A-end WDM accessing point address;
- Z-end WDM accessing point address.

This parameter set is mandatory if these parameters were specified in the corresponding *LDReq* message.

Note that all A-end and Z-end parameters in the copied parameter set will be set to '0' if the attempt of deletion the requested lightpath fails. The values of all parameters in the optional set then become irrelevant.

Trap

A trap message enables a WDM edge device to notify all appropriate entities of significant events occurred on a lightpath. A trap message is always sent to the A-end IP entity, and the IP traffic engineering module and/or the lightpath routing engine, wherever is appropriate. A trap message has a defined parameter set as follows:

- version
- UNI msg type
- initiator ID
- lightpath owner ID
- request sequence number
- A-end interface ID
- A-end IP address
- A-end WDM accessing point address
- Z-end interface ID
- Z-end IP address
- Z-end WDM accessing point address.

Its optional parameter set includes:

- A-end router ID
- Z-end router ID
- wavelength preferred
- signal formats
- bandwidth supported.

Every network entity (that initiates a trap message) maintains an individual counter to mark the field of request sequence number of each trap message. As such a recipient can arbitrate conflicting trap messages from the same initiator based on the value of this field.

7.9.5   WDM Traffic Engineering to Network Control Protocol (WDM TECP)

In the WDM layer, what traffic engineering does is constraint-based routing in a broad sense. The common and mandatory constraints in WDM networks are wavelength availability and capability of wavelength interchanges. The lightpath route computation can be implemented in a centralised manner or a distributed fashion. A centralised method implies that the route computation is carried out at one place, so the control commands from routing are also issued from the same place. In distributed routing, route computations and control are performed at every network node. Centralised and distributed methods have their own merits and drawbacks. A centralised method permits simple control but may become a potential bottleneck for the operation. On the other hand, the distributed method can be scalable in the sense of availability, but has to exchange a large amount of state for synchronisation. WDM networks require exclusive access and hard-state reservation of wavelengths. Another characteristic of the WDM network is the complexity of the network switching element and the QoS of the optical physical signal. All these are exacerbated by the lack of vendor interoperability.

IP-centric WDM network control conforms to the IETF GMPLS framework. The constraint-based distributed lightpath route computation is a standalone entity (although it can be integrated into the conventional OSPF routing). Route computation information is made available to each WDM node through OSPF with optical extensions. Opaque LSAs can be used to carry WDM-specific state information. MPLS signalling protocol can be used for lightpath setup or tear-down. Communication between the lightpath route computation module (the WDM layer traffic engineering module) and the signalling module is supported by the message formats defined in this section.

WDM TECP uses the common TECP header defined in Figure 7.23. As we discussed in the architecture section, TECP provides the interface between IP/WDM network control and IP/WDM traffic engineering. In the overlay model, TECP is presented in two message groups, IP TECP and WDM TECP, since there are separate IP traffic engineering and WDM traffic engineering corresponding to the IP layer and the WDM layer. In the integrated model, there is one TECP because the integrated traffic engineering takes care of the integrated IP/WDM network.

WDM TECP message types

WDM TECP defines two classes of messages: trail management messages and event notification message, each of which in turn contains a number of different types of messages.

- Trail management messages can be of a type listed below:

  create and update trail request
  – explicit route trail request
  – trail response.

- Event notification messages can be of a type listed below:

  – wavelength event
  – port event
  – NE event
  – fibre event.

Trail management messages are used by WDM TE to request lightpath-related operations. WDM TECP has defined these operations: create, delete, query, protect, and reroute trails. WDM systems also accept explicit routed trails, where the entire lightpath is specified in the request message. When the size of the message is larger than the MTU size, multiple segment messages should be sent with the same transaction ID. All request and related response messages also have the same transaction

ID. When appending a piece of lightpath information (for example, a cross-connect) causes the message size to exceed the MTU size, this piece and the remaining information should be transmitted in another message.

Trail request messages can be grouped into three message types discussed in detail below:

- create and update trail request;
- explicit route trail request;
- trail response messages.

Create and update trail request message (*TReq*)

The message type code for create and update trail request messages is *TReq*. Messages of this type can be used to request WDM edge devices to establish new trails or update existing trails. Delete trail is considered a special operation of update trail. The message format for create and update trail request messages is shown in Figure 7.29.

Message field definitions are as follows:

- **Trail ID:** this field specifies the trail ID when it is required. This can be used in cases of trail updating and trail route detail querying. The trail ID is per WDM edge device unique. It is used together with the A-end NE address to uniquely identify a trail in the entire WDM domain. This field uses two octets.
- **Operation:** this field is encoded to indicate one of the following defined operations:

  - *TrailCreate*: this operation requests a trail setup. To set up a trail, the requester must specify the A-end NE address and add port ID, and Z-end NE address and drop port ID fields, and optionally specifies the signal type and protection

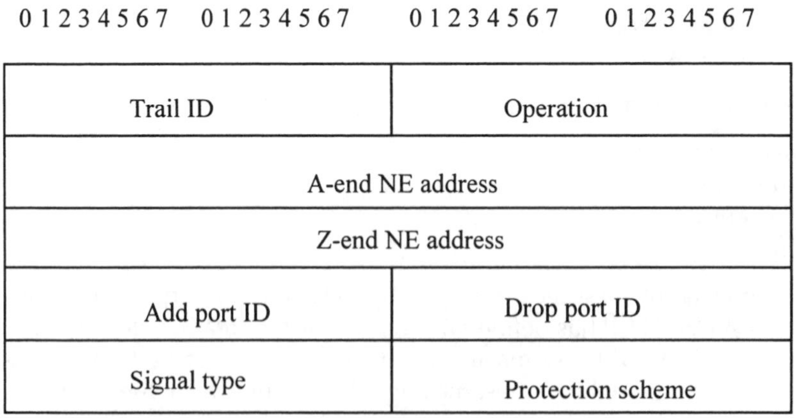

```
0 1 2 3 4 5 6 7    0 1 2 3 4 5 6 7    0 1 2 3 4 5 6 7    0 1 2 3 4 5 6 7
```

Trail ID	Operation
A-end NE address	
Z-end NE address	
Add port ID	Drop port ID
Signal type	Protection scheme

Figure 7.29 Message format for create and update trail request messages.

scheme fields. The trail ID field is unspecified. Its value will be determined once the setup is successful.

- *TrailDelete*: this operation deletes an existing trail. The trail to be deleted is identified by the trail ID. Other fields of the message are optional.
- *TrailProtect*: this operation protects an existing trail that is not currently being protected, or change the protection level of an existing protected trail. A protected trail means, for example, there is one or more backup paths. Thus, in case of primary path failure or signal quality deteriorating, signals can be transmitted using the alternate path. The trail to be protected is identified by the trail ID. The field of protection scheme indicates the protection level to be fulfilled/updated. Other fields of the message are optional.
- *TrailReRoute*: this operation reroutes an existing trail. The trail to be rerouted is identified by the trail ID. Other fields of the message are optional. If the reroute is unsuccessful, the existing trail will not be altered.
- *TrailRouteDetails*: this operation is used to query details of an existing trail as seen by the recipient of the message. The trail to be queried is identified by the Trail ID. Other fields of the message are optional.

- **A-end NE address:** the address of the source NE. This field uses four octets.
- **Z-end NE address:** the address of the sink NE. This field uses four octets.
- **Add port ID:** the port number, i.e. WADM signal add port, of the source NE. This field uses two octets.
- **Drop port ID:** the port number, i.e. WADM signal drop port, of the sink NE. This field uses two octets.
- **Signal type:** this field is used only when requesting a trail setup. A preferred signal type, for example, OC-48, can be specified. This field uses one octet.
- **Protection scheme:** this field is used when requesting a trail or asking protection for an existing trail. This field uses one octet. The following protection scheme types have been defined in WDM TECP:

  - Type 0: do not care
  - Type 1: 1:1 dedicated protection
  - Type 2: 1 + 1 dedicated protection
  - Type 3: shared protection
  - Type 4: multiple protection paths
  - Type 5: no protection.

Explicit route trail request message (*ETReq*)

The message type code for explicit route trail request messages is *ETReq*. TE has complete topological and link state information on the WDM network. Therefore, it has the choice and the ability to set up an explicitly routed trail. The format for this message type is shown in Figure 7.30. This message format is composed of a *TReq* portion and an explicit route list.

Message field definitions are as follows:

- **Operation:** this field is encoded to indicate one of two possible operations:

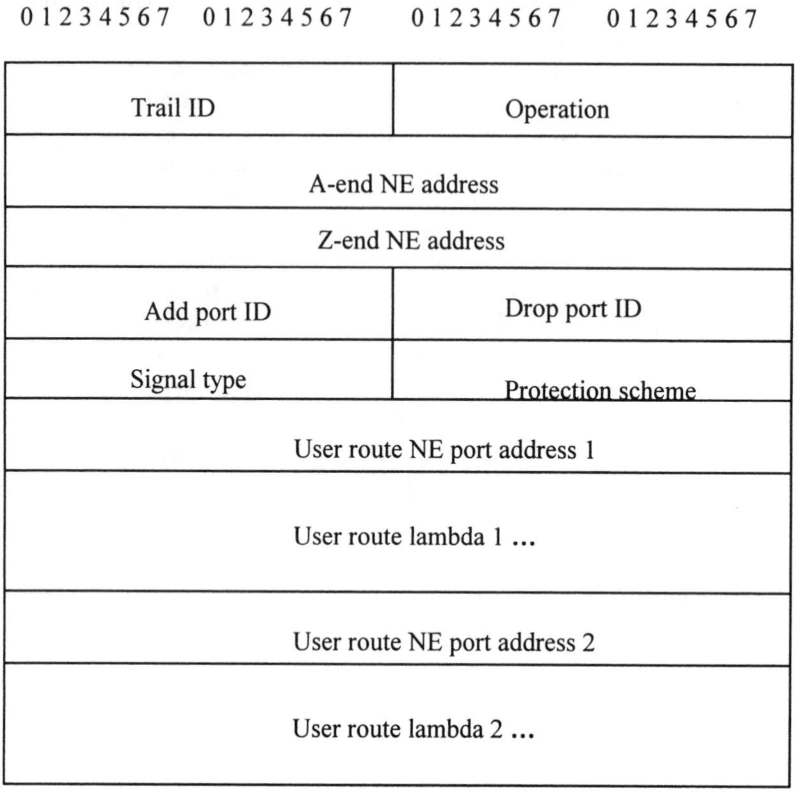

0 1 2 3 4 5 6 7   0 1 2 3 4 5 6 7   0 1 2 3 4 5 6 7   0 1 2 3 4 5 6 7

Trail ID	Operation
A-end NE address	
Z-end NE address	
Add port ID	Drop port ID
Signal type	Protection scheme
User route NE port address 1	
User route lambda 1 ...	
User route NE port address 2	
User route lambda 2 ...	

Figure 7.30 Message format for explicit route trail request messages.

- *TrailUserRoute*: this operation requests an explicit trail setup, which requires a list of wavelength cross-connect segment from the source to the sink. For each such segment, the user route NE address field must be specified. Filling in the user route port ID and user route lambda is optional. In the presence of wavelength continuity constraints, the lambda ID must be the same for all connection segments from the source to the sink. In this case, one can assume that the wavelength selection is conducted or to be conducted at the A-end NE. When the lambda ID is not selected and there are multiple available wavelength channels, it is up to the local switch to select a wavelength. A rule of thumb is if some of the fields are not specified, they will be determined by each intermediate NE based on resource availability.
- *TrailUserReRoute*: this operation reroutes the existing trail identified by the trail ID. Other fields of the message are optional. If the reroute is unsuccessful, the existing trail will not be altered.

- **User route NE port address i:** This is the NE incoming port address of the *i*th hop of the explicit trail. As in regular NE port addresses, this field uses four octets.

- **User route lambda i:** When setting up an explicit trail, a wavelength or a set of wavelengths can be specified for each hop. This field uses a bit-map representation for selected wavelengths. If a wavelength is selected for the explicit trail, the bits representing that wavelength are set to 1; bits corresponding to unselected wavelengths are set to 0. This field uses sixteen octets so it can work with WDM density as high as 128 wavelengths per fibre.

Trail response message (*TResp*)

The message type code for trail response messages is *TResp*. A *TResp* message takes the format as shown in Figure 7.31. A trail response message uses the same transaction ID and message ID as those found in the request message. All other fields in the message must be filled out in accordance to the actual setup of the trail. In particular, if a request message's operation is set to *TrailCreate*, *TrailProtect*, *TrailReRoute*, *TrailUserRoute*, or *TrailUserReRoute*, the trail ID field must be filled in with the proper value to represent the newly created trail; if a request message whose operation is set to *TrailDelete*, or *TrailRouteDetails*, the value of the trail ID field must be

```
0 1 2 3 4 5 6 7    0 1 2 3 4 5 6 7      0 1 2 3 4 5 6 7      0 1 2 3 4 5 6 7
```

Trail ID	Action
A-end NE address	
Z-end NE address	
Add port ID	Drop port ID
Signal type	Protection scheme
NE port address 1	
Lambda 1 ...	
NE port address 2	
Lambda 2 ...	

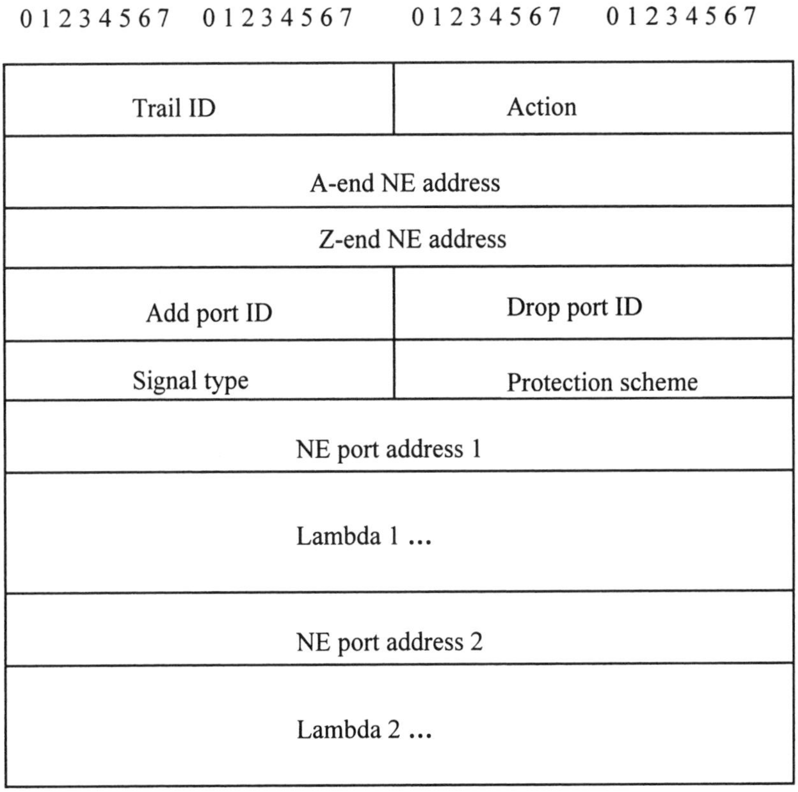

Figure 7.31 Message format for trail response messages.

copied from the request message. The value of the operation field must be set to that of the message being responded to.

Message field definitions are as follows:

- **NE Port Addr i:** this is the NE incoming port address of the *i*th hop of the trail identified by the trail ID. This field uses four octets.
- **Lambda i:** this field uses a bit map representation of the selected wavelength. This field uses sixteen octets so it can work with WDM density as high as 128 wavelengths per fibre. But for a valid trail response message one and only one bit of this field is set to 1.

Event notification message (EN)

Event notification messages allow the traffic engineering module to be informed of certain extraordinary conditions in the WDM network. Event notification messages are not acknowledged; and event messages are not sent during initialisation. All event notification messages are of the same type whose format is shown in Figure 7.32. The message type code for event notification messages is *EN*.

```
0 1 2 3 4 5 6 7    0 1 2 3 4 5 6 7      0 1 2 3 4 5 6 7    0 1 2 3 4 5 6 7
```

Event type	Port ID	
Lambda ID	NE ID	
(cont.)	Status	Severity
Event	Event Description	
(cont.)		

Figure 7.32 Message format for event notification messages.

Message field definitions are as follows:

- **Event type:** this field (2 octets) can be encoded to indicate one of the following types:

  - *LambdaEvent*. This event type indicates that the event is related to a particular wavelength or a wavelength channel, for example, signal QoS events.
  - *FibreEvent*. This event type is used to refer to events associated with fibre links, for example, link cut.
  - *PortEvent*. This event type indicates that the event is related to a particular switch port, for example, port or the corresponding circuit is burned.

- *NE Event*: This event type indicates that the event notification is about a particular NE, for example, NE down.

- **Lambda ID:** This field must be specified if and only if the event type is LambdaEvent. This field uses 2 octets.
- **Port ID:** This field must be specified if and only if the event type is *PortEvent*. This field uses 2 octets.
- **NE ID:** This field must be specified if and only if the event type is *NEEvent*. This field uses 4 octets.
- **Status:** This field indicates the status of the notification. The field uses 1 octet. The following statuses have been defined:

  –Type 0: OK
  –Type 1: Cleared
  –Type 2: Alarm.

- **Severity:** This field specifies how serious the event is, and uses 1 octet. The following severity types have been defined.

  –Type 0: Informing Notice
  –Type 1: Warning
  –Type 2: Minor
  –Type 3: Major
  –Type 4: Critical.

- **Event:** This field specifies what has happened to the component identified by the ID field (lambda ID, port ID, or NE ID). This field uses two octets. The following events have been defined:

  –Type 0: running
  –Type 1: ready for service
  –Type 2: down
  –Type 3: down because of neighbour state changes.

- **Event description:** This field further documents the occurred event. This field uses 6 octets.

IP/WDM address resolution

IP routers access the WDM network through WDM edge devices, i.e. a WADM. The physical connection between an IP router interface and a pair of WDM add/drop ports will not be altered during the WDM layer reconfiguration. The IP adjacency is defined by the way lightpaths established between access points. Two IP routers are neighbours of each other if and only if a lightpath has been set up between the router-to-WDM access points, i.e. the add/drop ports, of these two routers. An explicit routed lightpath determined by the lightpath route computation module in a WDM edge device in accordance to the client layer's request spans from an outgoing transport fibre port of the ingress WADM to an incoming fibre transport port of the egress WADM. Each WDM edge control node is responsible for setting up an internal mapping table that associates each add/drop port to the IP address(es) of the attached router interface so as to cross-connect each lightpath termination point to the right add/drop port.

A GMPLS-controlled WDM network has IP address assigned to each WDM NE, or every interface of each NE, in order to compute routes for lightpaths. Although routing interior to one layer is invisible to another layer, each WDM edge device must know which routers connect to which WADM. To achieve this, we have a few alternatives. First, we can run a BGP among the WDM edge devices, or define a new opaque LSA message to take advantage of the OSPF instance running for the WDM layer. This approach is basically an IP approach by extending IP control protocols. Second, we can construct a centralised server or management module for IP/WDM address translation. This approach is similar to the ARP and RARP, for example, in IP over ATM. In addition, we can make use of manual or static configuration. These approaches require constructing and maintaining an internal IP/WDM address-mapping table. An example attribute in such an internal mapping table is shown in Table 7.3

Table 7.3 IP/WDM access control mapping table.

Entry	WDM NE ID	WDM add/drop port ID	Router IP Addresses	Router interface type
Sample value	128.96.50.122	/shelf=1/slot=2/cp=3/ port=4	128.96.71.134 128.96.72.134 128.96.73.134	SONET, or OC-48

7.9.6 IP/WDM Traffic Engineering Tools

Figure 7.33 shows an example of the TE tools GUI. In the top-left frame of the figures, there is a physical IP/WDM topology map that shows the IP and WDM nodes, the access links (in black), and the WDM links (in blue). Each WDM link consists of a number of wavelength channels, for example, eight wavelengths per fibre in the figures. The number of wavelengths free and in-use are dynamic information and represent the constraints in the SPF routing algorithm. The top-right corner of the figures shows the virtual IP topology map, where each line between two IP routers represents an IP link. Clicking on the link on the virtual topology map triggers the end-to-end physical lambda path display in the IP/WDM physical topology map. The traffic demand matrix is displayed in the bottom-left frame of the figures. IP link load distribution can be viewed in the bottom-right frame. The TE tools are started by parameter settings for the simulation, which includes the degree of skewness from the current traffic pattern, traffic loading factor, reconfiguration algorithm selection, and IP routing selection (SPF or ECMP). Once the traffic demand is generated, a topology is computed according to the specified reconfiguration algorithm. Then, the virtual IP topology map is updated and the IP link load utilization under the new topology is displayed. On the physical IP/WDM topology, the wavelength constraints

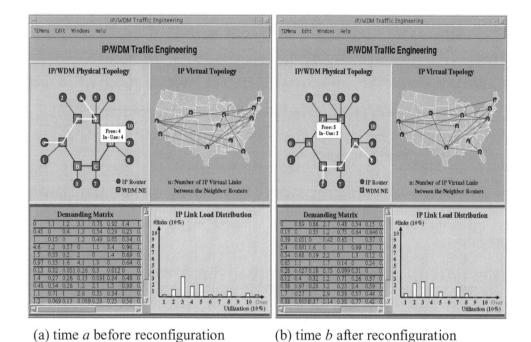

(a) time *a* before reconfiguration   (b) time *b* after reconfiguration

Figure 7.33   Example of traffic engineering tools GUI.

are also updated. Once a WDM link on the physical IP/WDM topology map is clicked, the link wavelength availability window is displayed.

7.10   Feedback-Based Closed-Loop Traffic Engineering

The IP/WDM traffic engineering capability presented in this chapter can be used for network planning and/or network control and management. Closed-loop traffic engineering refers to the automatic process of traffic engineering over operational IP/WDM networks to adaptively control the network so that network resources can be highly utilized. Adaptive control has been an identifiable topic in control engineering for a number of years. It has mainly been concerned with feedback systems shown in Figure 7.34. The description 'adaptive' in control engineering signifies that the controller in Figure 7.34 performs two simultaneous functions:

- First, it controls the controlled process.
- Second, it adapts itself to that process and its disturbances so as to give better control.

Adaptive control in control engineering is a generalisation of classical manual configured feedback control in systems where the controller in Figure 7.34 employs, for example, a linear control law. In classical linear control, coefficients of the linear control law are all time-variant and pre-specified statically. In adaptive control, some

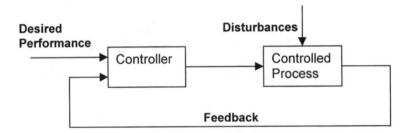

Figure 7.34 A feedback control system in control engineering.

or all of the coefficients are automatically adjusted in response to online measurements of process or disturbance variables. The effect is to produce a linear feedback controller that automatically tunes itself to match the controlled process and its disturbances.

The closed-loop traffic engineering in IP/WDM networks can be based on traffic statistics collection (i.e. feedback) and/or bandwidth projection (i.e. estimate). Figure 7.35 shows the feedback-based closed-loop traffic engineering in which the network is the controlled process and the traffic engineering server is the controller. The controller includes bandwidth projection tools, topology design algorithms, and migration scheduling policies. Once the controller outputs a new topology, the controlled process (i.e. the network) implements this topology. Before another topol-

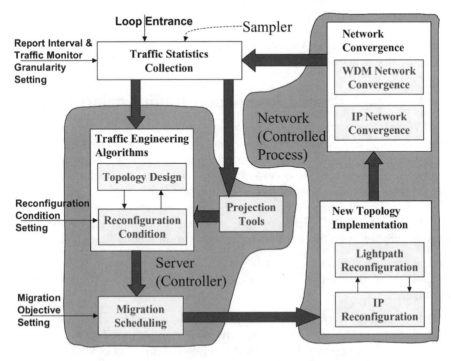

Figure 7.35 Feedback-based closed-loop network traffic engineering.

ogy reconfiguration is planned, the network must physically implement the new topology and then disseminate the link state information network wide. Therefore, network topology implementation and network convergence are important as they play an important role in network stability. We discuss the reconfiguration process involved in network topology implementation and network convergence for IP over WDM networks below.

7.10.1 Network Topology Implementation Process

In an overlay IP/WDM network, there are two tasks associated with IP topology reconfiguration, WDM reconfiguration and IP reconfiguration, respectively. WDM reconfiguration instructs the OXC and OADM to set up the desired lightpath topology and has the following components:

- **Lightpath routing** t_{lr}: if the detail hops of a lightpath are not given in the reconfiguration trigger, the end-to-end path has to be computed dynamically. An example approach for wavelength routing and assignment is to use the Dijkstra SPF algorithm subject to constraints. Constraints that have to be considered include wavelength availability and wavelength continuity.
- **Lightpath topology setup** t_{setup}: this includes a distributed signalling procedure and switch setup. Depending on the implementation, signalling may be responsible for local lambda selection as exploited in MPLS. Switch setup may require a reset operation before adding a new connection to the fabric.
- **Routing convergence** t_{wdm-c}: this represents the time for the WDM routing information base to resynchronise after update. If a link state protocol is used in wavelength routing, this is the time for the link state database to converge. If WDM network uses a single and centralised connection manager to compute lightpaths, this represents the connection database update time once changes occurred.

WDM reconfiguration time T_{wdm} can be defined as:

$$\left(t_{lr} + t_{setup} + t_{wdm-c} \right) - \beta \times t_{wdm-c}, 0 \leq \beta \leq 1$$

where β represents the overlapping factor between lightpath computation and setup time and WDM convergence time.

IP reconfiguration alters the IP interface status and address if necessary, and then waits for the routing protocol to converge. Hereafter, we use OSPF as the IP routing protocol as the link state protocol not only supports multiple metrics but also promises a faster convergence time. IP reconfiguration, T_{ip}, includes the following components:

- **Interface reconfiguration** t_{if}: the time to change the IP interfaces as specified in the new topology.
- **Routing protocol convergence** t_{ip-c}: the OSPF convergence time. This includes the time for detection, propagation, and SPF recalculation. The number of calculations that must be performed given n link-state packets is proportional to $n \log n$ in a modern SPF algorithm. OSPF convergence time is related to the size and the type

of the network, for example, the number of routers within each area, the number of neighbours for any one router, the number of areas supported by any one router, and the designated router selection.

T_{ip} can be written as:

$$\left(t_{if} + t_{ip-c}\right) - \gamma \times t_{ip-c}, 0 \leq \gamma \leq 1$$

where γ represents the overlapping factor between interface configuration and OSPF convergence.

7.10.2   Network Convergence

When IP network topology changes, IP traffic must reroute quickly based on the new lightpath topology. IP convergence time describes the time it takes for an IP router to start using a new route after a topology changes. Reconfiguration convergence refers to the time that an IP/WDM reconfiguration has completed and IP and WDM network has converged. That is, after a reconfiguration time interval, the new IP/WDM network is ready for another reconfiguration. Reconfiguration convergence time T_r can be written as:

$$T_{ip} + (1 - \alpha)T_{wdm}, 0 \leq \alpha \leq 1$$

where α is the overlapping factor between IP reconfiguration and WDM reconfiguration. To minimise T_r, IP and WDM reconfiguration should be conducted in parallel. However, migration schedule may require serialisation between certain IP and WDM reconfiguration processes to reduce instability and/or avoid traffic loss. Application impact due to reconfiguration in overlay IP/WDM networks is within the interval of T_r − t_{wdm-c} since applications do not require WDM network convergence.

7.10.3   A Testbed Study on IP/WDM Traffic Engineering

In [Liu02], an IP over a reconfigurable WDM testbed network is constructed using Bellcore/Telcordia MONET WADMs and Linux-based IP routers. The WADMs support wavelength-selective cross-connections. They support eight wavelength channels at the physical transport interfaces and corresponding single wavelength channel client interfaces for add drop. The client interfaces are used to connect to the electronic IP routers. Each WADM has a corresponding NE controller that connects to its internal IP control router. WDM signalling messages that trigger connection setup and release are translated into NE add, drop or cross-connect messages conveyed across this interface. The internal IP control routers are connected by a point-to-point Ethernet that mirrors the point-to-point topology of the WDM network (see Figure 7.36).

IP/WDM testbed description

The input signals from the electronic IP routers are translated into MONET-compliant wavelengths external transponders. The four electronic IP routers connected to the

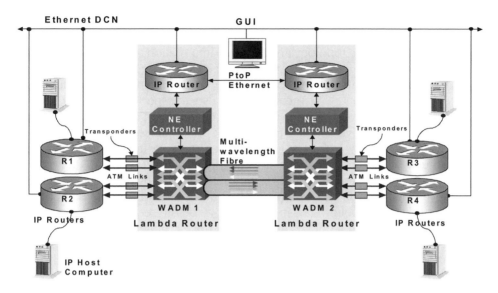

Figure 7.36   An IP over reconfigurable WDM network testbed.

WADMs are actually PCs running Linux. These IP routers are each equipped with two single-mode ATM network interface cards. The routers are connected using IP/ATM OC-3c point-to-point PVC connections. All these managed elements are connected to another Ethernet subnet, which serves as the data communication network (DCN). The DCN is used by the GUI and traffic engineering (TE) components to communicate with the managed elements.

In the IP network, each of the Linux-based PC router runs OSPF to support IP packet routing. To facilitate IP topology reconfiguration, an interface configuration manager is developed which interacts directly with the Linux kernel and the OSPF routing protocol to modify the state and the address of an IP interface. The WADMs are each controlled by corresponding NE controllers that export the NE-NC&M interface that can be used to provide element-level management functionality. To enable reuse of IP control protocols for WDM layer control, the OSPF and RSVP protocols are extended and modified to address WDM specific concerns and to support lightpath routing and setup. To handle the wavelength constraints of a WDM network, a CSPF algorithm is implemented to compute the explicit lightpath.

Testbed experimentation

The traffic engineering server has two modules: **topology design algorithm** and **migration scheduling**. IP traffic monitoring is based on the *libpcap* package that provides packet level capture and monitoring. Raw interface-related and source-destination data are first processed, analysed, and aggregated before being fed into a topology design algorithm. The topology design algorithm in turn is able to compute an optimal or near-optimal IP topology according to the collected traffic demand data. To minimise the impact due to reconfiguration, the migration module schedules topology migration into a series of steps or phases. Although both the IP and the WDM layers utilize IP addressing, the implementation follows an overlay networking model where topology and resource information are not shared between the layers. Rather a user-network interface (UNI) is defined to bridge the IP layer and the WDM layer. Lightpath request, deletion, and query interfaces are supported in the UNI implementation. UNI messages are also transported over TCP sockets. Figure 7.37 shows the workflows in the traffic engineering experimentation.

To evaluate the effectiveness and the network performance of IP topology recon-figuration, the IP layer was loaded with selected traffic patterns and performance compared under different topologies. To study the application impact, the network convergence, and the network performance due to reconfiguration, the following scenario (see Figure 7.38) was designed. The traffic demand is made of four flows. There does not exist a better way, in the topology before reconfiguration, to route/reroute any flows to exploit unused link capacity (for example, links between R3 and R4 in Figure 7.38(a)). Therefore, without reconfiguration, there is no room to further increase the total throughput. The same set of flows, however, can each alone occupy a full link in the reconfigured topology. This scenario clearly shows the additional advantage in reconfiguration-based traffic engineering over fixed topology multi-path routing. The experimentation shows that traffic engineering can indeed trigger IP topology reconfiguration to increase overall network throughput due to a reduced average *weighted-hop-distance,* defined as:

$$WHD = \sum_{\forall(i,j)\in N} flow_{ij} \times hop_{ij} / \sum_{\forall(i,j)\in N} flow_{ij}$$

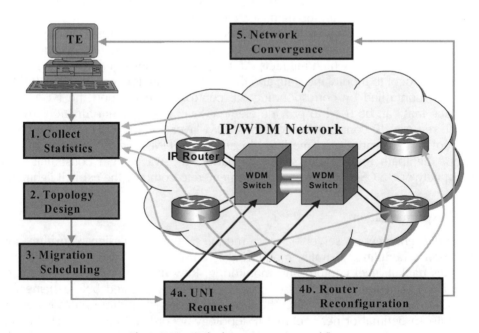

Figure 7.37   Testbed experimentation workflows.

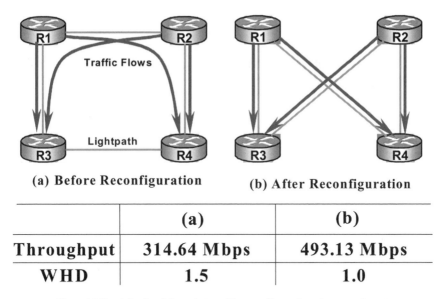

	(a)	**(b)**
Throughput	**314.64 Mbps**	**493.13 Mbps**
WHD	**1.5**	**1.0**

Figure 7.38   A feedback-based closed-loop traffic engineering experiment.

An example trigger condition in this experiment can be:

$$1 - \frac{WHD_{new}}{WHD_{old}} \geq \frac{1}{3}$$

which means that a reconfiguration is performed if the reconfigured topology can reduce average weighted hop distance by one third.

Table 7.4 shows the reconfiguration convergence time on the testbed. OSPF convergence time is related to its timer tuple setting:

(hello interval, router dead interval, poll interval, retransmission interval, transit delay interval).

By default, the timer tuple has these values: *(10 s, 40 s, 120 s, 5 s, 1 s)* as suggested in the OSPF specification [Moy98]. By suitably tuning OSPF timers, the routing convergence time can be reduced. Based on these tuple settings: *(5 s, 15 s, 10 s, 5 s, 1 s)*, the reconfiguration convergence time observed is between 19 and 20 s. Setting the timer tuple into a much smaller interval, for example, hello interval in milliseconds, will result in more control traffic in the network. Without QoS mechanisms to queue the control packets in front of data packets, the network may be hard to converge.

Table 7.4   Reconfiguration time over the IP/WDM testbed.

Per switch setup time (s)	t_{setup} (s)	t_{lr} (s)	$t_{wdm\text{-}c}$(s)	t_{if} (s)	$t_{ip\text{-}c}$ (s)
0.002	1.008	0.02	2.01	2	14

7.11   Summary

In this chapter, we presented a detailed discussion on IP/WDM traffic engineering. We started to answer what is IP/WDM traffic engineering. In the discussion, IP/WDM traffic engineering is to optimise the network resource usage to achieve the highest utilization. The scope of IP/WDM traffic engineering in this book covers IP/MPLS traffic engineering and WDM traffic engineering.

IP/MPLS traffic engineering copes with optimal routing and flow allocation. The traffic engineering in IP networks is designed to complement the default distributed control mechanisms implemented in the IP networks. The IP distributed control mechanisms aim for simplicity and scalability. Therefore, it only employs the best effort SPF routing and collects and stores the best effort routing-related information. However, to provide differentiated QoS requested by customers, traffic engineering functionality is needed. IP/MPLS traffic engineering collects and stores traffic engineering information in addition to the conventional network control information such as the link state information. The traffic engineering-related network information is specified according to traffic engineering needs/functions but collected and maintained through the distributed control mechanisms. An example of IP/MPLS traffic engineering is to compute and set up an explicit path to avoid network congestion/hot spots. The discussion on MPLS traffic engineering can be used for two purposes: load balancing and network provisioning. For load balancing, we summarised the IETF effort on OSPF-OMP. We formulated the network-provisioning problem using multi-commodity theories. In is worth pointing out that IP/MPLS traffic engineering is regarded as a *value-added service* to IP.

The chapter has focused on WDM traffic engineering in IP/WDM networks. WDM traffic engineering is to take advantage of the fact that a physical WDM network can support a number of virtual lightpath topologies. Each virtual topology in IP over WDM networks is an IP topology. However, a topology reconfiguration on an operational IP/WDM network is an expensive operation. Reconfiguration is 'painful' since it impacts user and application traffic and may break or interrupt ongoing services completely.

We approach the IP traffic engineering by formulating a framework according to which, first, IP/MPLS traffic engineering is applied to balance the loading (i.e. flows) among network components, and, second, WDM traffic engineering is used to reconstruct the lightpaths to adapt to the traffic pattern subject to the physical WDM network constraints.

The presented IP over a WDM reconfiguration framework consists of these components:

- traffic monitor
- traffic analysis
- bandwidth projection
- signal performance monitor
- traffic-engineering reconfiguration trigger
- lightpath topology design
- topology reconfiguration
- lightpath reconfiguration.

We discussed each component in the framework. In particular, we covered traffic monitoring and statistics collection, signal performance monitoring, bandwidth projection, topology design, and reconfiguration migration.

We emphasised virtual topology design in the discussion, where we formulated the topology design problem using MILP, surveyed the literature, and presented a group of heuristics using spanning tree and flow-based evaluator. To verify the performance and show the insight in simulation, we discussed traffic generation, two types of routing schemes (SPF and ECMP), and performance results by comparing the reconfigured topologies with the fixed topology. The simulation results can be summarised as follows:

- Reconfigured topologies outperform the fixed topology in terms of both improved throughput and reduced average hop distance. Hence, the simulation results prove that reconfiguration is beneficial.
- With respect to SPF vs. ECMP:
    - all topologies perform better in terms of throughput under ECMP;
    - when the network is heavily loaded, ECMP may increase the average hop distance because optimal routing may loop traffic when network is heavily loaded;
    - reconfigured topologies outperform the fixed topology despite routing with/ without flow deviation methods.

- The improvement in the simulation is in following ranges:
    - throughput gain is 10~20% for low skew traffic patterns and 20~30% for high skew traffic patterns;
    - average hop distance reduction is 10% for low skew traffic patterns and 15% for high skew traffic patterns.

- Among the reconfiguration algorithms:
    - in terms of throughput, RD performs well when there are a small number of congested links; RDHP performs the best when the traffic pattern is skewed; RDHP's performance is independent of skewness when the network is heavily congested;
    - in terms of hop distance, DHP and RDHP performs better than RD, because RD did not take hop-distance into account in its optimisation process;
    - when the network is lightly congested, RDHP performs the best.

We also presented a discussion on reconfiguration for packet switched WDM networks. Between IP over reconfigurable WDM and IP over switched WDM, reconfiguration in IP over switched WDM is more flexible and, to some extent, more dynamic because it expects that lightpath reconfiguration takes place frequently. Reconfiguration in IP over reconfigurable WDM is basically network level provisioning (for certain virtual lightpath topology), and therefore it is easier to implement and is reliable. Reconfiguration for switched WDM requires parallel decision-making dynamically, which in turn may demand synchronisation among decision makers. Hence, it is complex and less optimal.

IP/WDM traffic engineering can also be described in two models: overlay model and integrated model. The overlay traffic engineering model is suitable for overlay-controlled IP over WDM networks; the integrated traffic engineering model can only be effectively implemented in a closely coupled IP to WDM networks. In terms of implementation, there are two approaches for traffic engineering: centralised and distributed. Mixing the two models and the implementation approaches gives four alternatives: centralised overlay traffic engineering, centralised integrated traffic engineering, distributed overlay traffic engineering, and distributed integrated traffic engineering. In general, the overlay and the centralised approaches are simple and can leverage the existing techniques developed in IP, WDM, ATM, or frame relay networks; the integrated and the distributed approaches are flexible and can be more scalable. We discussed IP/WDM traffic engineering software design in the chapter. We presented the corresponding architectures for overlay and integrated traffic engineering models. Since there are four alternatives for traffic engineering, we presented a TECP protocol to generalise the interface between different traffic engineering approaches and various network control mechanisms. The interface details the parameters needed in the communication.

Finally, we reviewed a recent study on an IP over a WDM testbed constructed in the NGI SuperNet NC&M project. The testbed is used to demonstrate a feedback-based closed-loop traffic engineering. Within the loop, the sampler, i.e. the traffic monitor, takes traffic measurements from the network; the controller, i.e. the traffic engineering server that has topology design algorithms and topology migration scheduler, decides when and how a reconfiguration should be performed; the controlled process, i.e. the network, implements the new lightpath topology, reconfigures the IP interfaces, and starts to disseminate the related information to converge the network. Through the performance measurements collected in the testbed, one can conclude the following:

- Although WDM networks can support multiple virtual topologies, IP networks (at least IPv4) assume a static topology and do not have topology reconfiguration schemes in place.
- Existing link state routing protocols such as OSPF are slow in convergence. Some of the reasons are listed below:
  - no standard way to convey a WDM link removal to the IP layer, so the IP/WDM network has to rely on multiple Hello protocol message lost to detect the link removal;
  - ordering of the routing update computation vs. the link state update propagation is not defined;
  - delayed computation of multiple routing updates delays network convergence;
  - routing update could be computationally intensive for large networks.

- Reconfiguration convergence time is an important measure since it indicates the controlled process (i.e. network) stability. The testbed with four IP routers and two WDM switches record around 20 s for a topology reconfiguration instance.
  - Further reduction may be possible with more aggressive tuning, for example, by setting the OSPF timers. However, this does have side effects, i.e. more control

traffic. It also requires a more powerful router to process and sophisticated QoS mechanisms to differentiate the control traffic from data traffic.
- More fundamental modifications are needed for really fast, i.e. milliseconds, convergence.

• Setting the reconfiguration trigger condition is an interesting topic, especially in closed-loop traffic engineering. The trigger condition should consider at least two factors:

- *cost efficiency:* whether the reconfiguration should be performed through analysing the gain, the cost, and the side effects (introduced by reconfiguration);
- *reconfiguration convergence:* whether a reconfiguration can be performed or whether it will lead to further network instability.

In addition to traffic engineering, WDM reconfiguration can also be triggered by fault, protection/restoration, or network maintenance. The reconfiguration can also be employed to avoid malicious attacks. In the information warfare, network elements or components might be compromised, which can be detected through third party software and/or tools; through reconfiguration, the compromised components can be isolated while the rest of the network can still provide services.

8

Other IP/WDM Specific Issues

- IP/WDM group communication
- IP/WDM network and service management
- TCP over optical networks

8.1   IP/WDM Group Communication

Group communication is required in some applications such as video conferencing, where packets are delivered to a specified subgroup of network hosts. A well-known, in-use multicast backbone is the MBone, whereby a logical network is imposed over the Internet through tunnelling implemented by the multicast-enhanced routers to forward multicast datagrams. In optical networks, an optical signal can split into multiple light paths/branches in the physical layer and this splitting can be regarded as physical layer multicasting. However, wavelength splitting degrades optical signal quality, so to achieve certain QoS parameter settings, the splitting is constrained by several factors. IP multicasting support over IP/WDM networks is related to the type of WDM switch, for example, switched WDM or reconfigurable WDM, and the IP and WDM inter-networking architectures, e.g. overlayed or peer-to-peer.

8.1.1   IP Multicasting

In conventional IP networks, multicast is desirable to support group communication because it not only reduces host or server processing load, but also saves bandwidth by reducing network loading.

Figure 8.1 presents a comparison between unicasting and multicasting a packet. Using unicasting, the server has to send four identical packets to four other hosts. Hence, four packets are inserted into the network. Using multicasting, the server inserts only one packet into the network and routers forward this packet only to multicasting members. The multicast group is identified by a multicast address. Unlike a unicast IP address that uniquely identifies a single IP host, a multicast IP address identifies an arbitrary group of IP hosts that have joined the group. In IPv4, class D addresses, spanning from 224.0.0.0 through 239.255.255.255, are reserved

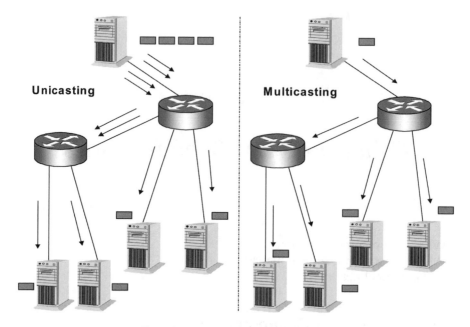

Figure 8.1   IP unicasting vs. multicasting.

for multicast addresses. A multicast address is denoted by its first 4 bits (i.e. with the
pattern 1110) in the first octet of an IP address.

When a router receives a multicast packet, the router can perform a reverse path
forwarding (RPF) check on the packet according to a multicast routing table. The
packet is discarded if the check fails, and forwarded to other interfaces only if the
check succeeds. The RPF examines the source address of the arriving multicast
packet to determine whether the packet has arrived via an interface that is on the
reverse path to the source.

Multicast traffic is delivered over a multicasting distribution tree connecting multicast group members. A multicasting tree can be formed in two ways: source SPF tree
and shared tree. In the first method, each multicast source computes its own source
SPF tree to other multicast members just as in IP unicasting. In the second method,
there is a shared distribution tree, which is pre-configured and has been established
to connect all members. Every multicast source sends its multicast traffic to the root of
the shared distribution tree, which will forward it to all the members in the multicast
group.

The current multicast routing protocols can be classified into three categories:

- dense mode protocols such as DVMRP (Distance-Vector Multicast Routing Protocol) and PIM-DM (Protocol Independent Multicast Dense Mode):
- sparse mode protocols such as PIM-SM (Protocol Independent Multicast Sparse
  Mode) and CBT (Core-Based Tree);
- link state protocol such as MOSPF (Multicast OSPF).

A dense mode protocol uses a SPF tree to 'broadcast' the existing multicast traffic in the format of < source, multicast group > to every node. This works as a push operation. A sparse mode protocol usually uses the shared multicast tree to distribute the existing multicast traffic to the multicast receivers. This works as a pull operation, in which a receiver needs to request a copy of the multicast traffic. Finally, a link state protocol floods special multicast, link state information that identifies the whereabouts of group members in the network. The routers in the network use this information to compute their own SPF tree to reach all the receivers in the group.

8.1.2 IP Multicasting in Presence of GMPLS

As introduced early in the book, GMPLS can precisely control traffic flow through LSP setup. The LSP can be regarded as layer 2 switching so it changes the IP forwarding adjacency. In the presence of GMPLS LSPs, IP multicasting will treat LSPs as virtual links and establish neighbourhood adjacency just as in IP unicasting. In addition, for certain multicasting applications such as those that have long-duration and high-bandwidth requirements, IP multicasting can leverage the traffic engineering capability provided by GMPLS. For example, a source can compute and establish LSPs to reach some of or all of the multicast members. Then, the multicast traffic is forwarded using LSPs.

Figure 8.2 shows an example of IP multicasting with LSPs. After congestion is detected in the network, LSPs are established to avoid hot spots. The IP multicasting tree is computed based on the updated topology with LSPs.

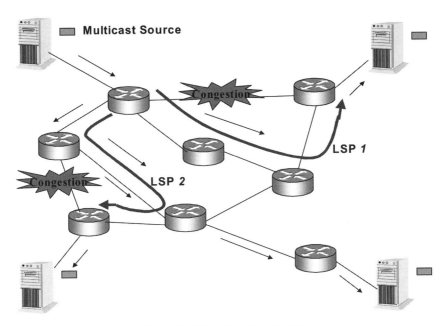

Figure 8.2 IP multicasting with LSPs.

8.1.3 IP over WDM Multicasting

IP over reconfigurable WDM and IP over switched WDM are the two important IPs over WDM networking architectures. The reconfigurable WDM is always layered as IP over lightpaths in the data plane. The switched WDM, in the data plane, forms an overlay network in the case of OBS or OLS, and a peer network in the case of OPR. In a data-plane overlay IP over WDM network, a WDM network has a virtual lightpath topology in addition to the physical fibre topology. IP multicasting is computed over the existing virtual lightpath topology. In addition, LSPs can be established to share the channel capacity provided in the lightpath. These LSPs carry fine granularity traffic and provide time-sharing of the wavelength channel. In the physical layer, point-to-multipoint lightpaths can be set up.

IP multicasting can be considered as an application that can take advantage of the IP/WDM multi-layer infrastructure (see Figure 8.3).

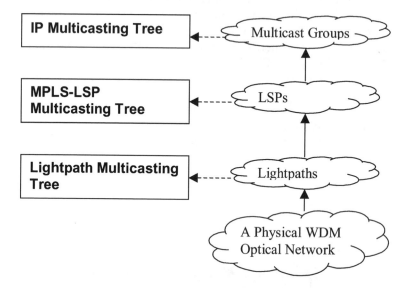

Figure 8.3 Multicasting for a data-plane overlay IP/WDM network.

In a data-plane peer-to-peer IP over WDM network, OPRs can be regarded as IP routers with multi-wavelength interfaces. Therefore, IP multicasting over OPRs can be performed the same as in IP multicasting over conventional IP routers. Again, GMPLS can be employed to support cut-through circuits in the packet switched network.

There are several issues arising from WDM optical networks:

- WDM optical switches may support signal splitting. As a result, WDM networks support point-to-multipoint circuits in addition to point-to-point circuits. However, signal splitting degrades the signal quality, so it basically limits the number of times a signal can be split. Signal splitting is an expensive operation so not every port/interface on the switch fabric can support splitting and not all switches in an operational network can afford signal split ports. Sparse splitting means only a subset of switches in a network supports wavelength splitting; limited splitting means an input signal can only be split into a limited number of outputs and only a limited number of inputs can be split simultaneously. All these have to be translated into constraints for the multicasting algorithm.

- Due to the nature of the overlay network (i.e. in the data plane), the virtual WDM topology can be configured and/or reconfigured to support IP multicasting. Hence, an optical multicast routing solution requires the multicast distribution tree computation and lightpath routing are considered in an integrated fashion.
- Being IP multicast-capable can be interpreted as running multicast routing protocols, whereas WDM multicast-capable is translated into signal splitting. In IP-controlled optical networks, the IP controller certainly can be IP multicast-capable, but the corresponding switch may not support all-optical wavelength splitting. This can be overcome using O-E-O conversions, but the benefit of all-optical networking is greatly reduced. Another solution is to align the IP controller multicasting capability with the corresponding WDM switch multicast capability.
- Wavelength routing in WDM networks has to cope with both wavelength availability and wavelength continuity. Wavelength conversion capability affects multicast tree formation.
- IP multicast routing is designed under the assumption of unlimited buffer space, so once a contention occurs for an interface, packets can be buffered at the router. In all optical networks especially in OPR, there is no optical memory (i.e. random access memory) available, so buffer space limitation has significant impact on blocking and packet dropping probability and, in turn, on multicast routing. This can be addressed using GMPLS LSPs, but how to set up LSPs effectively in a dynamic environment is a challenge in network control.

8.2 IP/WDM Network and Service Management

The TeleManagement Forum (TMF) is a non-profit organisation providing network management standards to improve the management and operation of communications services. TMF was formed in 1988, and was initially named as the Network Management Forum (NMF). TMF technical work is organised into modelling/specification teams and associated catalyst teams that test the specifications (www.tmforum.org). The TMF Multi-Technology Network Management (MTNM) team defined a CORBA-based network view for SONET/SDH, ATM, and DWDM by providing the definition and specification of a CORBA-based NMS-EMS interface (see Figure 8.4). This specification includes requirements, case studies, a protocol-independent information model (UML-based) and OMG IDL.

In terms of network management functions, MTNM covers:

- configuration management (such as inventory management, point-to-point and multipoint connections, network level protection, open SONET rings, and equipment inventory management);
- fault management (such as alarm reporting);
- performance management (such as current and historical data for SONET/SDH and DWDM).

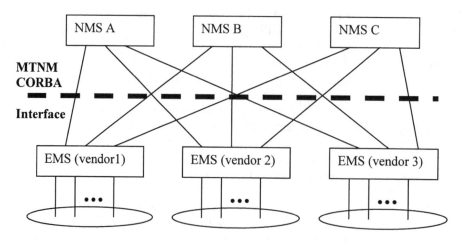

Figure 8.4   MTNM NMS-EMS architecture.

MTNM defines these basic UML objects:

- connection termination point
- physical termination point
- subnetwork connection
- topological link
- managed element
- subnetwork
- EMS.

It has adopted a coarse-grained approach with respect to modelling, i.e. there is many-to-one mapping between information model and interface specification. In MTNM, there is no general support for scoping and filtering, i.e. data retrieval operations are defined on an as-needed basis. From a vendor point of view, so far, most interest in MTNM NMS-EMS interfaces comes from SONET/SDH and DWDM vendors, and little interest if any from ATM vendors.

8.2.1   CORBA Reference Model and Telecom Facility

CORBA stands for common object request broker architecture, which is a standard-based framework for distributed computing. The goal of the CORBA standard is to provide interoperability, portability, reusability, and mapping interfaces to multiple programming languages, e.g. C + + , Smalltalk, Java, and COBOL. Use of CORBA can simplify the application development effort and modularise application components for reuse. In particular, CORBA automates many common network programming tasks such as:

- object registration
- location and activation
- request demultiplexing

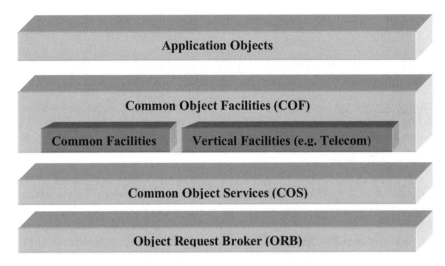

Figure 8.5 CORBA reference model.

- framing and error handling
- parameter marshalling and demarshalling
- operation dispatching.

Figure 8.5 shows the CORBA reference model. ORB (Object Request Broker) represents the enabling mechanism providing the capabilities that allow objects to

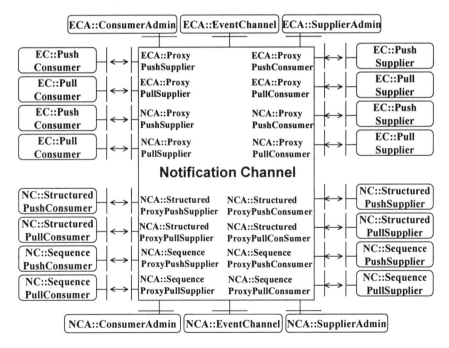

Figure 8.6 Telecom notification service.

interact in a distributed and possibly heterogeneous environment. The ORB works as a broker for objects, so it separates the object implementation from its invoking interface.

COS (Common Object Services) is a collection of basic capabilities (i.e. services) that support functions for using and implementing objects. Examples of COS are:

- a naming service
- a trading service
- an event service
- a messaging service
- a security service.

COF (Common Object Facilities) includes common facilities and vertical facilities. The common facilities provide services that many applications may share, but which are not as fundamental as the COS. The vertical facilities, also known as domain facilities, are defined for specific industry segments such as telecommunication. Application objects at the topmost of the framework represent applications, which utilize the services and facilities provided in the distributed environment.

For our interest, the OMG Telecom Task Force defined two services in the Telecommunication domain facility, the notification service and the Telecom log service, respectively. The notification service supports all of the COS event service's capabilities via inheritance, allowing interoperability between event service clients and notification service clients. In addition, it enhances the event service by adding support for event filtering, subscription information sharing between notification channels and their clients, and quality of service properties configuration. The notification service can be configured in relation to delivery guarantee, event ageing characteristics, and event prioritisation on a per connection, message, or notification channel basis. The log service supports logging and forward events to applications or other logs. The log service can be implemented using either event or notification channel, although its basic version does not need a channel.

Figure 8.6 shows the notification service architecture. The IDL module names of the interfaces by the service are abbreviated in the figure and listed below:

- EC: CosEventComm
- ECA: CosEventChannelAdmin
- NC: CosNotifyComm
- NCA: CosNotifyChannelAdmin.

An instance of the notification service event channel or notification channel is that it inherits all of the interfaces supported by the event service event channel. Therefore, an instance of the former can be widened to an instance of the latter, e.g. to the EventChannel, ConsumerAdmin, and SupplierAdmin interfaces. The EventChannel interface is used for the instances of ConsumerAdmin and SupplierAdmin interfaces. The ConsumerAdmin interface is a factory used to create various types of supplier proxies; the SupplierAdmin interface is a factory used to create various types of consumer proxies.

As indicated in the figure, the ProxySupplier interfaces are offered to consumers on behalf of suppliers while the ProxyConsumer interfaces are offered to suppliers on behalf of consumers. Events transported over the interfaces can be in the form of structured events, sequences of structured events, or 'anys' (un-typed event). A structured event is created because the typed event is hard to apply and implement. It is basically a standard data structure into which any event type can be mapped.

Figure 8.7 shows the log service architecture. A log object is both an event consumer and an event supplier. So, events can be written to a log as log records, forwarded from a log to any applications or other logs, or generated from this log when the state changes. Once the event/notification channel is used for a log service, it can no longer support filtering for consumers and suppliers. However, the log service provides a log filter interface that can select events before writing to the log store. Applications that are not aware of the event/notification channel can use the log interface directly writing events to the log store. The log store is comprised of log records, each of which represents an event. A log record has the following structure.

```
struct LogRecord {
  RecordId id;   // load record unique identifier
  TimeT time; // time when event is logged
  NVList attr_list; // user defined name/value pair
  Any info; // the event data itself
}
```

Typed events can be store in the typed log record, which has the following structure.

```
struct TypedLogRecord {
  DsLogAdmin::RecordId id;
  DsLogAdmin::TimeT time;
```

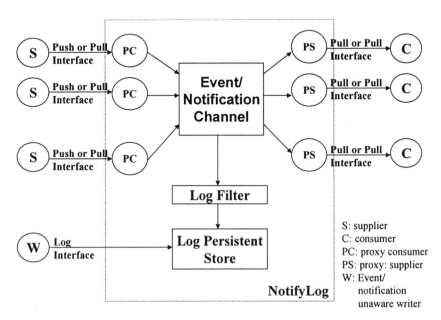

Figure 8.7   Telecom log service.

```
DsLogAdmin::NVList attr_list;
CORBA::RepositoryId interface_id; // repository ID of the interface
```
that sent the typed event

`CORBA::Identifier operation_name;` // name of the operation that emitted the typed event `ArgumentList arg_list;` // argument list that contains the event data

```
}
```

Both log record and typed log record can be queried, retrieved, and deleted based on the record identifier, time, attributed, or contents through query constraint language, e.g. SQL. The log also supports monitoring for its operational state, current log size, and quality of service. The log interface supports the monitoring and setting of these parameters:

- administrative state
- availability status
- maximum log size
- log full action
- log duration
- log scheduling
- log capacity alarm threshold
- log record compaction
- quality of service (the flush operation).

8.2.2   Connection and Service Management Information Modeling (CaSMIM)

CaSMIM is designed by the MTNM for connection management across multiple network types, discovery of network topology, and service assurance. Support of CaSMIM can allow an operator to automate service provisioning and monitoring end-to-end across multi-technology, multi-vendor, multi-layered networks. The model is designed to be service-oriented but technology independent, so technology-specific information is passed transparently and is not modelled. It is service-oriented modelling as only connection services and impact on customer services are modelled.

The interface layout between MTNM and CaSMIM is presented in Figure 8.8. CaSMIM provides the interface between Service Management System (SMS) and Cross Domain Manager (XDM) and between XDM and the technology-specific Domain Manager (DM). MTNM offers the interface between the technology DM and its corresponding EMS. So far, MTNM has defined interfaces for SONET/SDH, ATM, and DWDM. The IP Network Management (IPNM) team provides the interface between IP DM and IP EMS. The ADSL equivalent DM-EMS interface is not defined yet. The CaSMIM model describes the characteristics of connections, termination points, and links. It supports general features that are not specific to a technology such as DWDM line rate and DWDM port availability, as well as technology-specific features that are specific to a technology. CaSMIM uses OMG services of notifications, logging, and naming.

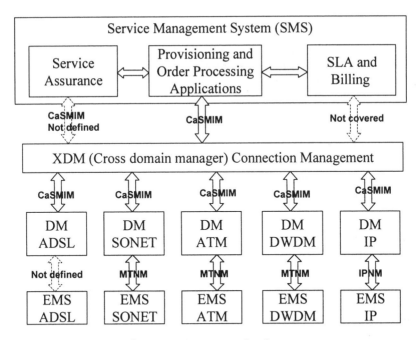

Figure 8.8 CaSMIM interface layout.

8.2.3 Optical Network Service Management

An important step towards optical network commercialisation is service management. Network operators and carriers have been increasingly focusing on a variety of service management functions provided over the optical network. With the agile IP/WDM network, the service management functions can be provided in a timely and flexible fashion. In particular, these functions enable the customer to create dynamic SLAs with variable QoS and CoS (Class of Service) parameters, monitor ongoing service in real time themselves, modify existing service on-the-fly, and order customerised new services (e.g. enforcing routing and connection constraints).

Some key service management functions are listed below.

- **Customer relationship management (CRM):** CRM presents a single, consolidated system image to the customer and allows closer interaction between customer and the optical network service management system, e.g. through customer profiling, order history, and online service status monitoring and analysis. CRM can be customised to reflect individual interest. Recently, web-based CRM systems allow customers to manage the optical network services themselves. Through a web portal, a customer is able to negotiate and establish dynamic SLAs with optical network providers, provide client access policies, and access security and authentication management.
- **Billing:** a flexible billing system will not only provide incentives to customers and maximise network infrastructure investment, but also increase optical network resource utilization. Billing and accounting management used to be network

management functions, but today's service management requires dynamic network management and real-time service response. Hence, once the order is received, the customer credit is verified and the honoured request is forwarded to network management. The connection is established with the required bandwidth. The connection start and termination times are recorded. The customer will be billed accordingly.

- **Provisioning:** based on optical network operation history and service planning or projection, optical network services are provisioned in seasonable fashion, which in order, triggers network provisioning.

- **SLAs:** this represents the contractual agreement between customer and service provider regarding services offered by the optical network. Once a service provider accepts SLAs, SLAs need to be translated into policies. Service management has access to the optical UNI that will pass policies to optical network control and establish the service according to the specification.

- **Service monitoring:** as more and more customers have technical knowledge of networks and computing facilities on site, clients prefer to have service monitoring facilities at their premise. In such a way, they assure the service quality is provided from time to time according to the SLA. For example, they might be interested in the total bandwidth provided in the optical connection and the optical signal quality, e.g. in terms of wavelength power and OSNR, of a wavelength channel.

- **Service activation:** an optical network service has its own life span. In order to offer dynamic optical network services, the services must be able to activate automatically. In optical networks, clients can activate a service through a proxy that will signal the network manager for connection setup with the specified parameters.

8.3   TCP over Optical Networks

An overview of TCP is provided in Chapter 3. TCP offers end-to-end data transport management, which guarantees data delivery and integrity across the Internet. The 16-bit checksum field of the TCP packet is computed for the entire TCP segment including TCP header and payload. Figure 8.9 shows an example of TCP adaptive transmission over a conventional IP network, in which the time out parameter of the sender is dynamically computed based on the acknowledgement from the receiver. The sliding window size of the sender is also adjusted accordingly.

However, IP/WDM optical networks possess some unique characteristics that challenge the conventional TCP algorithm (see Figure 8.10):

- *Parallel paths and explicit routing:* WDM-enabled optical networks possess a smaller operational footprint compared with the single mode fibre networks. This in fact indicates the existence of parallel paths (not just fibre lines but also wavelength channels), or equal-cost multi-paths between a pair of nodes. For better utilization, optimal routing is introduced for load balancing among multi-paths. GMPLS introduces an explicit routing mechanism through LSP setups. The unidirectional LSPs and optimal routing over equal cost multi-paths, to some extent, destroy the traffic symmetry, which causes the packet delivery and its

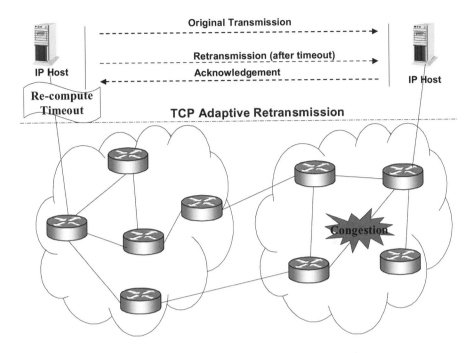

Figure 8.9 TCP over buffered conventional IP networks.

acknowledgement to take different paths. This will delay the TCP acknowledgement. The dynamic routing scheme may also cause TCP segments to arrive out-of-order.

- *No optical buffer:* until now, there is no optical memory (i.e. random access memory) available, so once a contention occurs for an interface at an optical packet router (OPR), only one packet can get through while others have to be dropped. The no-buffer packet network in optical packet switching challenges the conventional TCP congestion control algorithm.
- *Optical signal QoS:* any losses due to errors in transmission on a less than perfect channel, rather than congestion, cause TCP to reduce its overall sending rate to help alleviate the 'perceived congestion'. For example, all-optical network transmission requires no O-E-O conversion at the core node, so data or signal on the wavelength channel may be corrupted without anyone or any device noticing. In the absence of verified cross-connect, a wavelength channel (possible cross-multiple domains and networking technologies) is not fully tested before the application traffic flows through. Separating control channel from data channel in optical networks makes the health of the operating data channel vulnerable.

Figure 8.10 shows the TCP over optical Internet, which consists of GbE/WDM network, OXC, and OPR network. In single-mode fibre networks, fibre topology is the IP topology, so TCP over an optical network is similar to TCP over conventional IP.

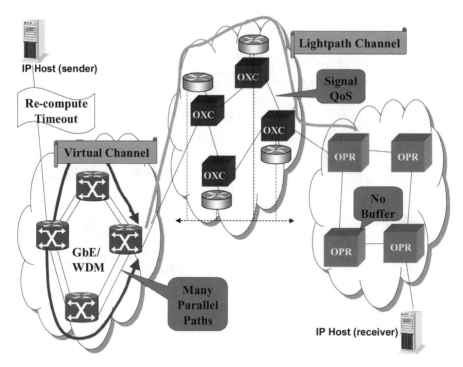

Figure 8.10   TCP over IP/WDM networks.

IP over WDM networks can be based on either reconfigurable or switched WDM network equipment supporting several inter-networking models. However, from the data plane point of view, IP/WDM forms either an overlay or a peer network. In the overlay data-plane network such as IP over OXC, an IP topology using the virtual lightpaths is established before the TCP traffic flows through. In such a case, WDM is no different from any other layer 2 technology such as ATM. WDM may introduce some specific link layer problems, but it can be taken care of by layer 2 and/or 3 mechanisms or algorithms. In the peer network such as IP over OPR, there is a single, integrated topology, which embeds wavelength channels into the fibre links.

In all-optical packet switched networks, there is no optical memory available. Hence, congestion control algorithms offered by TCP need to be modified. However, network planning for bufferless packet switched network suggests the best method to avoid congestion is over provisioning. This means if congestion occurs, other parts of the network are likely to be underutilized. GMPLS can come to the rescue by setting up explicit LSPs.

TCP over optical networks requires further study.

9

Concluding Remarks

- Book summary
- IP/WDM network applications
- Future research

9.1 Book Summary

In this book, we have presented a detailed study and technological overview of WDM optical networks taking advantage of adaptive and scalable IP network control mechanisms. In particular, we have focused on the network transport issues: network control and traffic engineering. The main motivation behind IP over WDM is that *optical networks with WDM technology can provide abundant bandwidth while nearly all application data traffic in the networks is IP and IP control protocols are widely deployed and understood.* The key advantages of IP over WDM networks can be reiterated as follows;

- IP/WDM inherits the flexibility and the adaptability offered in the IP control protocols.
- IP/WDM can achieve or aims to achieve 'real-time' (on demand) provisioning in optical networks.
- IP/WDM hopes to address WDM or optical NE vendor interoperability and service interoperability (i.e. QoS) with the help of IP protocols.
- IP/WDM can achieve dynamic restoration by leveraging the distributed control mechanisms implemented in the network.

In Chapter 1, we motivated the audience by listing the driving forces behind IP over WDM research. We articulated the concepts of WDM-enabled optical networks and IP over WDM. We reviewed the worldwide NGI effort and the current IP/WDM standardisation. Chapter 2 presented an overview of the IP over WDM background information that includes optical communications, WDM network testbeds and product comparison, network protocols, Internet architecture, IPv4 addressing, gigabit Ethernet, MPLS, and distributed systems. Chapter 3 gave an in-depth study of traffic analysis and engineering and IP routing protocols. Chapter

4 described the optical modulation, the recent optical component, switching technology, and software development in WDM optical networks. An overview of optical networking components and switching technology was presented. We introduced a framework for WDM network control and management. An information model for WDM optical networks was presented and examples of WDM NC&M functionality were discussed. Chapter 5 presented IP over WDM architectural models, IP/WDM inter-networking models, and IP/WDM software architectural models. Chapter 6 introduced IP/WDM network control which focused on these issues: IP/WDM network addressing, topology discovery, IP/WDM routing, signalling, WDM access control, IP/WDM restoration, GMPLS, inter-domain network control, and WDM network control and management protocol. Chapter 7 introduced IP/WDM traffic engineering in which a framework is presented and the related functional components within the framework are detailed. MPLS traffic engineering is discussed. The chapter emphasised the IP/WDM traffic engineering, in particular the lightpath topology design, the lightpath and topology reconfiguration, and topology migration. IP/WDM network-specific issues are discussed in Chapter 8. The chapter included topics on group communication, IP/WDM network and service management, and TCP over optical networks.

9.2 IP/WDM Network Applications

IP/WDM networks can be employed to support several groups of applications:

9.2.1 MAN and WAN Network Transport

WDM as an exploring fibre bandwidth technology can be used wherever is needed, i.e. WAN, MAN, or LAN (see Figure 9.1). But a business plan or model must be developed to justify the potential revenue from the cost. WDM-enabled optical networks provide a relatively smaller operational footprint. Hence, it saves operational and maintenance costs. However, WDM network equipment is expensive especially using long-reach optics, so there is a trade-off between the operational cost saved and the extra cost to update to the WDM equipment. As such, the majority of the WDM applications will be in the WAN and MAN area, where the fibre spans a relatively long distance.

WDM can also be used in the enterprise networking, where high bandwidth applications are hosted.

9.2.2 Layer 2 or Layer 3 VPN, VLAN, Leased Fibre Line or Wavelength Channel

A WDM network can provide VPN services to customers, who could be either end users or service providers themselves. Depending on the WDM network configuration, VPNs can be set up in layer 2 or layer 3 according to the requested connection duration and bandwidth, for example, wavelength channels or subwavelengths (i.e. TDM access). WDM networks can also provide leased line or dark fibre service to

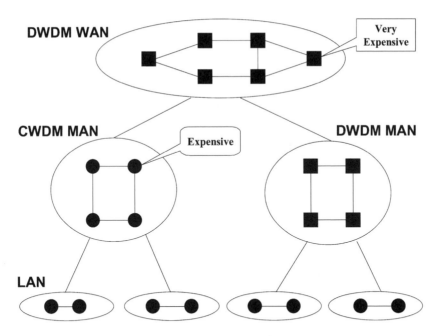

Figure 9.1 WDM WAN and MAN.

customers. In fact, this approach is getting popular since users or service providers may want to have their own switching equipment and policies. As such, they have more knowledge and control of the network, and are able to offer more flexible application-orientated services.

9.2.3   Optical Interconnect

WDM networks can be used as an interconnect technology to provide connectivity for distributed entities. These entities can be designed for the following purposes:

- **Parallel computation:** since WDM-enabled fibre provides abundant bandwidth with fast transmission time, scientific computation can be partitioned and conducted in parallel, utilizing the existing distributed system resources such as CPUs and memory. The massive parallel systems trade off the communication time and co-ordination overheads with parallel system (i.e. CPU) computation power. Given a complex and time-consuming task, parallel processing improves the response time and increases the system and network utilization.
- **Distributed applications:** WDM networking devices can be used to support a range of distributed applications from financial and trading systems, distributed databases, to military applications. An example is the distributed collaborative application, in which users co-operate from remote locations to plan, design, and implement a common task or mission such as groupware and military mission planning.
- **Storage area network (SAN):** as the popularity of the Internet grows, end users and companies would like to have their own web sites (without maintaining the web

pages and server themselves). This requires public data storage for renting. Data storage networks can be formed by applying WDM technology interconnecting data storage centers.

9.2.4 Bandwidth Brokers and Traders

A key advantage of IP/WDM optical networks is the flexibility provided through IP control protocols. An IP-centric WDM network is able to (or will) provide on demand provisioning. The flexibility and fast provisioning time offered in an IP/WDM network is giving birth to a new range of bandwidth broker and trader applications that are going to further improve the efficiency and increase the bandwidth utilization of optical networks.

9.3 Future Research

As the communication and computer networks are quickly converging, next-generation networks, presumably based on IP, are going to replace the conventional circuit and packet switched networks. However, the current IPv4, designed decades ago, lacks sufficient capacity to address the next generation Internet scalability. Example limitations include:

- the limited and inefficient use of address space;
- the lack of QoS support, poor security;
- the lack of support for mobile applications and wireless devices.

IPv6 has started to address these issues, and has proposed facilities for autoconfiguration and renumbering. The current IETF effort on next-generation IP has its focus on supporting mobile hosts and plug-and-play Internet appliances.

However, to better support the desired features lacking in IPv4 and traffic engineering, the next-generation IP must maintain and extend one of the key advantages of the Internet – the flexibility with dynamic connectivity and discovery even in the access and backbone network. IPv4 relies on static IP topology, and using GMPLS, it allows cut-through paths using virtual IP addresses over the static topology. However, GMPLS is not as scalable as IP control protocols, and IP is not designed for dynamic topology reconfiguration. The existence of a large number of unidirectional LSPs and wavelength channels may have deep impact on traffic symmetry, routing behaviour, and network stability. In addition, separating data channels from the control channel in IP-controlled optical networks makes the health of the operating data channel vulnerable. For example, all-optical network transmission requires no O-E-O conversion at the core node, so optical signal quality may be degraded or data on an operational wavelength channel may even be corrupted. In the absence of verified cross-connect, a wavelength channel (possibly cross-multiple domains and networking technologies) is not fully tested before the application traffic flows through it.

We outline some of the possible research topics below.

9.3.1   Scalable Common Control Plane for Optical Networks

Despite the recent upheaval of the global Telecom and IT sector, the arrival of optical technology is a reality, promising a bandwidth of 10 Terabits per second per fibre. Running broadband to every house will have a profound effect on society, in which the next several years would have as big an impact as computing did in the past several decades. A common and scalable control plane for optical networks is needed urgently to control next generation networks as well as to interoperate existing legacy networks. In addition to the next generation optical networks to be deployed, legacy Telecom networks already consist of tens of thousands of nodes. The scalability of a common control plane for such sized networks becomes a major research issue. The desired control plane hopes to provide, for network operators, fast provisioning and bandwidth trading in the near future. IP networks and technology have a well-established community such as IETF and IANA and IP control protocols are widely deployed and understood. However, IP protocols can only scale to hundreds of routers such as OSPF. IP has been designed for in-band, electrical, packet switched networks, but photonic packet processing has been proved to be a difficult task. Hence, the question is what should be the common control plane with scalability for optical networks.

9.3.2   Next Generation of TCP/IP

Optical networks have gone through a few generations in the past several years. In contrast, TCP/IP is a relatively old technology. IP was invented in the 1970s and TCP was created in the 1980s. Next generation of IP or IPv6 has started to address some of the issues in IPv4. The future network will be formed by a large number of mobile devices and Internet appliances. IP is too complex for such networks. The current TCP may not be an efficient solution for broadband networks. Next generation of TCP/IP is an interesting research area.

9.3.3   TCP/IP Performance Studies in Presence of a Number of Parallel Paths and Unidirectional LSPs

WDM-enabled optical networks possess a smaller operational footprint compared with the conventional single mode fibre network. This in fact indicates the existence of a large number of parallel paths (not just fibre lines but also wavelength channels), or equal cost multi-paths between a pair of nodes. For better utilization, optimal routing is introduced for load balancing among multi-paths (i.e. parallel transmission).

MPLS introduces an explicit routing mechanism through LSP setups. The unidirectional LSPs and optimal routing over equal cost multi-paths, to some extent, destroy the traffic symmetry, which causes the packet delivery and its acknowledgement to take different paths. For example, this could delay the TCP acknowledgement. The dynamic routing scheme may also cause TCP or application segments to arrive out-of-order. The impact due to the existence of a large number of parallel paths and unidirectional LSPs need further investigation.

9.3.4   Optical Packet Switching

It is unclear how all-optical packet switched networks perform in the absence of buffers. One may argue a large number of wavelength channels with wavelength conversion capability can provide an alternative solution to packet-level contention. However, further study is needed to investigate the performance. Optical memory itself is also a research issue.

9.3.5   Service Protection and Restoration

For carriers to sell services over optical networks, they have to implement service protection and/or restoration functions. From a network management point of view, optical signal QoS needs to be monitored, and fault management-related optical WDM intelligence needs to be provided. From a service management point of view, dynamic and real-time SLAs need to be supported and corresponding policies and constraints need to be enforced across the network. In the case of overlay networks, multi-layer protection or restoration schemes need to be correlated.

9.3.6   Optical Network Applications

To some extent, applications not only show the driving incentive of a (novel) technology but also give further directions on the technology development. Novel applications for IP/WDM optical networks and photonic technology are needed urgently.

9.3.7   Optical MIB Development

To allow intelligent, universal control across optical networks such as the universal control plane proposed in GMPLS, optical NE basic functionality and operations should be standardised and modelled into MIB. The MIB provides virtual NE resource abstraction for optical networks.

Bibliography

[Abou00]   O. S. Aboul-Magd, *et al.*, *IP over Optical Networks: A Framework*, http://www.ietf.org/internet-drafts/draft-many-ip-optical-framework-01.txt, work in progress, Aug 2000.

[Ande00]   W. Anderson, *et al.*, 'The MONET project a final report', *IEEE Journal of Lightwave Technologies*, Vol. 18, No. 12, pp. 1988–2009, Dec 2000.

[Ash98]   G. R. Ash, Dynamic Routing in Telecommunications Networks, McGraw-Hill, 1998.

[Awdu99]   D. Awduche, 'MPLS and traffic engineering in IP networks', *IEEE Communication Magazine*, Dec 1999.

[Awdu01]   D. Awduche and Y. Rekhter, 'Multiprotocol lambda switching: combining MPLS traffic engineering control with optical crosscon-nects', *IEEE Communication Magazine*, Mar 2001.

[Baro00]   S. Baroni, *et al.*, 'Analysis and design of backbone architecture alter-natives for IP optical networking', *IEEE Journal of Selected Areas in Communication*, Vol. 18, No. 10, pp. 1980–1994, Oct 2000.

[Bane97]   S. Banerjee, J. Yoo, and C. Chen, 'Design of wavelength-routed opti-cal networks for packet switched traffic', *IEEE Journal of Lightwave Technology*, Vol. 15, No. 9, pp. 1636–1646, Sep 1997.

[Bane99]   S. Banerjee, V. Jain, and S. Shah, 'Regular multihop logical topologies for lightwave networks', *IEEE Communications Surveys*, Vol. 2, No. 1, 1999.

[Bera95]   J. Beran, R. Sherman, M. Taqqu, and W. Willinger, 'Long-range dependence in variable-bit-rate video traffic', *IEEE Transactions on Communications*, Vol. 43, 1995.

[Blum00]   D. J. Blumenthal, 'All-optical label swapping networks and techni-ques', *IEEE Journal of Lightwave Technology*, Vol. 18, No. 12, pp. 2058–2075, Dec 2000.

[Brak95]   L. S. Brakmo and L. L. Peterson, 'TCP Vegas: End-to-end congestion avoidance on a global internet', *IEEE Journal of Selected Areas in Communication*, Vol. 13, No. 8, pp. 1465–1480, Oct 1995.

[Chan96]   G. K. Chang, G. Ellinas, K. Gamelin, M. Iqbal, and C. Brackett, 'Multi-wavelength reconfigurable WDM/ATM/SONET network testbed', *IEEE Journal of Lightwave Technology*, Vol. 14, No. 6, pp. 1320–1340, Jun 1996.

[Chan01]   G. K. Chang, *et al.*, 'A proof-of-concept, ultra-low latency optical label switching testbed demonstration for next-generation Internet networks', *OFC'2000*, Vol. 2, Section WD5, pp. 56–58, Baltimore, Maryland, Mar 2000.

[Chas01] J. Chase, A. Gallatin, and K. Yocum, 'End system optimizations for high-speed TCP', *IEEE Communication Magazine*, Apr 2001.

[Clar88] D. Clark. 'The design philosophy of the Internet protocols', *Proc. ACM SIGCOMM*, Sep 1988.

[Corm90] T. Cormen, C. Leiserson, and R. Rivest, *Introduction to Algorithms*, McGraw-Hill, 1990.

[Cott98] L. Cottrell, *PingER Tools*, http://www.slac.stanford.edu/xorg/icfa/ntf/tool.html, May 1998.

[Coul01] G. Coulouris, J. Dollimore, and T. Kindberg, *Distributed Systems: Concepts and Design*, Addison-Wesley, 3rd edition, 2001.

[Crov97] M. Crovella and A. Bestavros, 'Self-similarity in World Wide Web traffic: evidence and possible causes', *IEEE/ACM Transactions on Networking*, Vol. 5, No. 6, pp. 835–846, Dec 1997.

[Cunn99] D. G. Cunningham and W. G. Lane, *Gigabit Ethernet Networking*, MacMillan Technical Publishing, 1999.

[Davi00] B. Davie and Y. Rekhter, *MPLS: Technology and Applications*, Morgan Kaufmann, 2000.

[Deer90] S. Deering and D. Cheriton, 'Multicast routing in datagram internetworks and extended LANs', *ACM Transactions on Computer Systems*, Vol. 8, No. 2, pp. 85–110, May, 1990.

[Dijk59] E. Dijkstra, 'A note on two problems in connection of graphs', *Numerical Mathematics*, Vol. 1, pp. 269–271, 1959.

[Feld00] A. Feldmann, A. Greenberg, C. Lund, N. Reingold, J. Rexford, and F. True, 'Deriving traffic demands for operational IP networks: methodology and experience', *Proc. ACM SIGCOMM*, Aug 2000.

[Floy97] S. Floyd, V. Jacobson, C. Liu, S. McCanne, and L. Zhang, 'A reliable multicast framework for light-weight sessions and application level framing', *IEEE/ACM Transactions on Networking*, 1997.

[Fuma00] A. Fumagalli and L. Valcarenghi, 'IP Restoration vs. WDM protection: is there an optimal choice', *IEEE Network*, Nov 2000.

[Gers00] O. Gerstel and R. Ramaswami, 'Optical layer survivability an implementation perspective', *IEEE Journal on Selected Areas in Communications*, Vol. 18, No. 10, pp. 1885–1899, Oct 2000.

[Ghan00] N. Ghani, *et al.*, 'On IP-over-WDM integration,' *IEEE Communication Magazine*, Mar 2000.

[Gros98] D. Gross and C. Harris, *Fundamentals of Queuing Theory*, 3rd edition, John Wiley & Sons, 1998.

[GSMP98] IETF RFC 2297: *Ipsilon's General Switch Management Protocol Specification*, Mar 1998.

[Hala97] B. Halabi, *Internet Routing Architectures*, Cisco Press, New Riders Publishing, 1997.

[Hara98] H. Harai, M. Murata, and H. Miyahara, 'Performance analysis of wavelength assignment policies in all-optical networks with limited-range wavelength conversions', *IEEE Journal on Selected Areas in Communications*, Vol. 16, No. 7, pp. 1051–1060, Sep 1988.

[Harr93]   P. Harrison and N. Patel, *Performance Modeling of Communication Networks and Computer Architecture*, Addison Wesley, 1993.

[Henn96]   J. L. Hennessy and D. A. Patterson, *Computer Architecture: A Quantitative Approach*, 2nd edition, Morgan Kaufmann, 1996.

[Huit98]   C. Huitema, *IPv6 The New Internet Protocol*, 2nd edition, Prentice-Hall, 1998.

[Hunt00]   D. K. Hunter and I. Andonovic, 'Approaches to optical Internet packet switching', *IEEE Communications Magazine*, Vol. 38, pp. 116–122, Sep 2000.

[Jaco88]   V. Jacobson, 'Congestion Avoidance and Control,' *Proc. ACM SIGCOMM '88*, Aug. 1988.

[Jaco89]   V. Jacobson, *Traceroute*, ftp://ftp.ee.lbl.gov/traceroute.tar.Z, 1989.

[Kali99]   S. Kalidindi and M. Zekauskas, 'Surveyor: An Infrastructure for Internet Performance Measurements', *Proc. INET '99*, 1999.

[Klei75]   L. Kleinrock, *Queuing Systems Volume 1: Theory*, John Wiley & Sons, 1975.

[Klei76]   L. Kleinrock, *Queuing Systems Volume 2: Applications*, John Wiley & Sons, 1976.

[Labo91]   J. F. P. Labourdette and A. S. Acampora, 'Logically rearrangeable multihop lightwave networks', *IEEE Transactions on Communications*, Vol. 39, No. 8, pp. 1223–1230, Aug 1991.

[Labo94]   J. F. P. Labourdette, G. W. Hart, and A. S. Acampora, 'Branch-exchange sequences for reconfiguration of lightwave networks,' *IEEE Transactions on Communications*, Vol. 42, No. 10, pp. 2822–2832, Oct 1994.

[Lam97]   D. Lam, D. Cox, and J. Widom, 'Teletraffic modeling for personal communications services', *IEEE Communication Magazine*, Feb 1997.

[Labo00]   C. Labovitz, A. Ahuja, A. Bose and F. Jahanian, 'Delayed Internet Routing Convergence', *Proc. ACM SIGCOMM*, Stockholm, Sweden, Aug 2000.

[Lang01]   J. P. Lang, *et al.*, *Link Management Protocol*, http://www.ietf.org/internet-drafts/draft-ietf-ccamp-lmp-03.txt, 2001.

[Leib00]   B. T. Leib, 'The future of bandwidth trading', Derivates Strategy Magazine, Aug 2000.

[Lela94]   W. Leland, M. Taqqu, W. Willinger, and D. Wilson, 'On the self-similar nature of Ethernet traffic', *IEEE/ACM Transactions on Networking*, Feb 1994.

[Leon00]   E. Leonardi, M. Mellia, and M. Ajmone Marsan, 'Algorithms for the logical topology design in WDM all-optical networks', *Optical Network Magazine*, Vol. 1, No. 1, pp. 35–46, Jan 2000.

[Leun88]   C. H. C. Leung, *Quantitative Analysis of Computer Systems*, John Wiley & Sons, 1988.

[Li99]   T. Li, 'MPLS and the evolving Internet architecture', *IEEE Communication Magazine*, Dec 1999.

[Liu00a]   K. H. Liu, B. Wilson, and J. Wei, 'A wavelength scheduling application for WDM optical networks', *IEEE J. Selected Areas in Communications*, Vol. 18, No. 10, pp. 2041–2050, Oct 2000.

[Liu00b]   K. H. Liu, B. Wilson, and J. Wei, 'A management and visualization framework for reconfigurable WDM optical networks', *IEEE Network Magazine*, Vol. 14, No. 6, pp. 8–15, Nov 2000.

[Liu02]   K. H. Liu, C. Liu, J. Pastor, A. Roy, and J. Wei, 'Performance and testbed study of topology reconfiguration in IP over WDM networks', *IEEE Transactions on Communications*, 2002.

[Lync96]   N. A. Lynch, *Distributed Algorithms*, Morgan Kaufmann Publishers, 1996.

[Maed98]   M. W. Maeda, 'Management and control of transparent optical network', *IEEE Journal on Selected Areas in Communications*, Vol. 16, No. 7, pp. 1008–1023, Jul 1998.

[Mann01]   E. Mannie, editor, *Generalized Multi-Protocol Label Switching (GMPLS) Architecture*, http://www.ietf.org/internet-drafts/draft-ietf-ccamp-gmpls-architecture-00.txt, work in progress, Feb 2001.

[Matt00]   W. Matthews and L. Cottrell, 'The PingER project: active Internet performance monitoring for the HENP community', *IEEE Communication Magazine*, May 2000.

[McDy00]   D. McDysan, *QoS & Traffic Management for IP & ATM Networks*, McGraw-Hill, 2000.

[McGr00]   T. McGregor, H. Braun and J. Brown, 'The NLANR Network Analysis Infrastructure', *IEEE Communication Magazine*, May 2000.

[Mich97]   H. Michiel and K. Laevens, 'Teletraffic engineering in a Broad-Band era', *Proc. IEEE*, Dec 1997.

[Moy98a]   J. Moy, *OSPF Anatomy of an Internet Routing Protocol*, Addison-Wesley, 1998.

[Moy98b]   J. Moy, *OSPF Version 2*, IETF RFC 2328, Apr 1998.

[Mukh96]   B. Mukherjee, *et al.*, 'Some principles for designing a wide-area optical network', IEEE/ACM Transactions on Networking, Vol. 4, No. 5, pp. 684–696, 1996.

[Mukh97]   B. Mukherjee, *Optical Communication Networks*, McGraw-Hill, 1997.

[Mokh98]   A. Mokhtar and M. Azizoglu, 'Adaptive wavelength routing in all-optical networks', *IEEE/ACM Transactions on Networking*, Vol. 6, No. 2, pp. 197–206, Apr 1998.

[Mouf99]   H. Mouftah and J.Elmirghani (editors), *Photonic Switching Technology*, IEEE Press, 1999.

[Naga00]   N. Nagatsu, 'Photonic network design issues and applications to the IP backbone', *IEEE Journal of Lightwave Technology*, Vol. 18, No. 12, pp. 2010–2018, Dec 2000.

[Naru00]   A. Narula-Tam and E. Modiano, 'Dynamic load balancing in WDM packet networks with and without wavelength constraints', *IEEE Jour-*

	nal on *Selected Areas in Communications*, Vol. 18, No. 10, pp. 1972–1979, Oct 2000.
[Norr94]	I. Norros, 'A storage model with self-similar input', *Queueing Systems*, Vol. 16, pp. 387–396, 1994.
[Norr95]	I. Norros, 'On the use of Fractional Brownian Motion in the theory of Ethernet traffic', *IEEE J. Selected Areas in Communications*, Aug 1995.
[Nutt02]	G. Nutt, *Operating Systems: A Modern Perspective*, 2nd edition, Addison Wesley, 2002.
[Okam83]	H. Okamura, 'Multicommodity flows in graphs', *Discrete Applied Mathematics*, Vol. 6, pp. 55–62, 1983.
[Patt97]	D. Patterson and J. Hennessy, *Computer Organization and Design: The Hardware/Software Interface*, Morgan Kaufmann, 2nd edition, 1997.
[Paxs95]	V. Paxson and S. Floyd, 'Wide area traffic: the failure of Poisson modeling', *IEEE/ACM Transactions on Networking*, Jun 1995.
[Paxs98]	V. Paxson, *et al.*, *Framework for IP Performance Metrics*, IETF RFC 2330, May 1998.
[Pete00]	L. L. Peterson and B. S. Davie, *Computer Networks: A System Approach*, Morgan Kaufman, 2nd edition, 2000.
[Post80]	J. Postel, *User Datagram Protocol*, IETF RFC 768, Aug 1980.
[Post81a]	J. Postel, *Internet Protocol*, IETF RFC 791, Sep 1981.
[Post81b]	J. Postel, *Internet Control Message Protocol*, IETF RFC 792, Sep 1981.
[Post81c]	J. Postel, *Transmission Control Protocol*, IETF RFC 793, Sep 1981.
[Pumm95]	T. Pummill and B. Manning, *Variable Length Subnet Table for IPv4*, IETF RFC 1878, Dec 1995.
[Qiao00]	C. Qiao, 'Labeled optical burst switching for IP-over-WDM integration', *IEEE Communications Magazine*, Vol. 38, pp. 104–114, Sep 2000.
[Rama96]	R. Ramaswami and K. Sivarajan, 'Design of logical topologies for wavelength-routed optical networks', *IEEE J. Selected Areas in Communications*, Vol. 14, No. 5, pp. 858–867, Jun 1996.
[Rama98a]	R. Ramaswami and G. Sasaki, 'Multiwavelength optical networks with limited wavelength conversion', *IEEE/ACM Transactions on Networking*, Vol. 6, No. 6, pp. 744–754, 1998.
[Rama98b]	R. Ramaswami and K. Sivarajan, *Optical Networks: A Practical Perspective*, San Francisco: Morgan Kaufmann, 1998.
[Rama98c]	B. Ramamurthy and B. Mukherjee, "Wavelength Conversion in WDM Networking", *IEEE Journal on Selected Areas in Communications*, Vol. 16, No. 7, pp. 1061–1073, Sep 1998.
[Rama99]	L. G. Raman, *Fundamentals of Telecommunications Network Management*, IEEE Press, 1999.
[Rekh95]	Y. Rekhtar and T. Li, *A Border Gateway Protocol (BGP-4)*, IETF, RFC 1771, Mar 1995.
[Schw96]	M. Schwartz, *Broadband Integrated Networks*, Prentice-Hall, 1996.
[Siva00]	K. Sivalingam and S. Subramaniam (editors), *Optical WDM*

	Networks: Principles and Practice, Kluwer Academic Publishers, 2000.
[Stal97]	W. Stallings, *Data and Computer Communications*, 5th edition, Prentice-Hall, 1997.
[Stal98]	W. Stallings, *High Speed Networks TCP/IP and ATM Design Principles*, Prentice-Hall, 1998.
[Stev97]	W. R. Stevens, *TCP Slow Start, Congestion Avoidance, Fast Retransmit and Fast Recovery Algorithms*, IETF RFC 2001, Jan 1997.
[Stev98]	W. R. Stevens, *UNIX Network Programming, Networking APIs: Sockets and XTI*, Volume 1, 2nd edition, Prentice-Hall, 1998.
[Stra01]	J. Strand, A. L. Chiu, and R. Tkach, 'Issues for routing in the optical layer', *IEEE Communications Magazine*, Vol. 39, No. 2, pp. 81–87, Feb 2001.
[Ster99]	T. E. Stern and K. Bala, *Multiwavelength Optical Networks: A Layered Approach*, Reading, MA: Addison-Wesley, 1999.
[Suhi97]	E. Suhir, *Applied Probability for Engineers and Scientists*, McGraw-Hill, 1997.
[Supe01]	DARPA NGI SuperNet NC&M (Network Control and Management) Deliverable Reports 14, Bellcore/Telcordia Technologies, 2001.
[Tane95]	A. Tanenbaum, *Distributed Operating Systems*, Prentice-Hall, 1995.
[Tane96]	A. Tanenbaum, *Computer Networks*, 3rd edition, Prentice-Hall, 1996.
[Tane99]	A. Tanenbaum, *Structured Computer Organization*, 4th edition, Prentice-Hall, 1999.
[Thom97]	K. Thompson, G. Miller, and R. Wilder, 'Wide-area Internet traffic patterns and characteristics', *IEEE Network*, Nov/Dec, 1997.
[Vill]	C. Villamizar, *OSPF Optimized Multipath (OSPF-OMP)*, IETF Internet-Draft, work in progress.
[Wagn96]	R. E. Wagner, R. Alferness, A. Saleh, and M. Goodman, 'MONET: multiwavelength optical networking', *IEEE J. Lightwave Technology*, Vol. 14, No. 6, pp. 1349–1355, Jun 1996.
[Wase95]	O. J. Wasem, A. M. Gross, and G. A. Tiapa, 'Forecasting Broadband Demand Between Geographical Areas', *IEEE Communications Magazine*, Feb 1995.
[Waut96]	N. Wauters and P. Demeester, 'Design of the optical path layer in multiwavelength cross-connected networks', *IEEE J. Selected Areas in Communications*, Vol. 14, No. 5, pp. 881–892, Jun 1996.
[Wei00a]	J. Y. Wei, C. Liu, S. Park, K. H. Liu, S. Ramamurthy, H. Kim, and M. Maeda, 'Network control and management for the Next Generation Internet', *IEICE Transactions on Communications*, Vol. E83B, No. 10, pp. 2191–2209, Oct 2000.
[Wei00b]	J. Y. Wei and R. I. McFarland, 'Just-in-time signaling for WDM optical burst switching networks', *IEEE J. Lightwave Technology*, Vol. 18, No. 12, pp. 2019–2037, Dec 2000.
[Will95]	W. Willinger, M. Taqqu, R. Sherman, and D. Wilson, 'Self-similarity through high-variability: statistical analysis of Ethernet LAN traffic at the source level', *Proc. ACM SIGCOMM*, 1995.

[Wils00]   B. Wilson, N. Stofel, M. Post, J. Pastor, K. Liu, T. Li, K. Walsh, J Wei, and Y. Tsai, 'Network control and management for multiwavelength optical networks', *IEEE J. Lightwave Technology*, Vol. 18, No. 12, pp. 2038–2057, Dec 2000.

[Wu92]   T. Wu, *Fiber Network Service Survivability*, Artech House, 1992.

[Yoo00]   M. Yoo, C. Qiao and S. Dixit, "QoS performance of optical burst switching in IP-over-WDM networks", *IEEE Journal on Selected Areas in Communications*, Vol. 18, No. 10, pp. 2062–2071, Oct 2000.

[Yoo01]   M. Yoo, C. Qiao, and S. Dixit, 'Optical burst switching for service differentiation in the next-generation optical Internet', *IEEE Communications Magazine*, Vol. 39, No. 2, pp. 98–104, Feb 2001.

[Zhan93]   L. Zhang, *et al.*, 'RSVP: a new resource reservation protocol', *IEEE Network Magazine*, Sep 1993.

[Zhan00]   X. Zhang, J. Wei and C. Qiao, "Constrained multicast routing in WDM networks with sparse lighting", *IEEE INFOCOM 2000 – The Conference on Computer Communications*, No. 1, pp. 1781–1790, Mar 2000.

[Zhen99]   B. Zheng and M. Atiquzzaman, 'Traffic management of multimedia over ATM networks', *IEEE Communication Magazine*, Jan 1999.

Web Site List

java.sun.com	Sun Microsystems Java home page.
standards.ieee.org	Institute of Electrical and Electronics Engineers (IEEE) standard home page. IEEE maintains standards for LAN/MAN equipment.
watt.nlanr.net/AMP	Active Measurement Project (AMP) home page.
www.6bone.com	Internet Protocol version 6 network deployment.
www.abilene.iu.edu	Abilene Network Operations Center web site.
www.acm.org/sigcomm	ACM Special Interest Group (SIG) on Data Communication.
www.advanced.org/surveyor	The Surveyor Project Advanced Networks.
www.atmforum.org	The ATM forum
www.bluetooth.com	The official Bluetooth SIG web site.
www.comsoc.org	IEEE Communication Society web site.
www.darpa.mil	U.S. government DARPA web site.
www.etsi.org	European Telecommunications Standards Institute (ETSI). ETSI promotes standards in Europe for all areas in telecommunications, including fibre optics.
www.fibrechannel.com	Fibre Channel Industry Association (FCIA). FCIA is a professional organisation for Fibre Channel technology.
www.fols.org	Fibre Optics LAN section of the TIA. The section is a division of the TIA that focuses on fibre optic, local area network technologies and reports on the standards in this area.
www.gmpls.org	MPLS/GMPLS related standard documents, tutorials, and optical networking publications.
www.iana.org	Internet Assigned Numbers Authority home page.
www.icann.org	The Internet Corporation for Assigned Names and Numbers.
www.ietf.org	Internet Engineering Task Force (IETF) home page.
www.ifip.tu-graz.ac.at/TC6	International Federation for Information Processing, Technical Committee 6 on Networking.
www.ipv6forum.org	Internet Protocol version 6 forum.
www.isi.edu/div7/rsvp	USC RSVP protocol software release and related publications.
www.isoc.org/	Internet Society home page.

www.itu.int/ITU-T   International Telecommunication Union – Telecommunication Standardisation Sector home page.
ITU focuses on telecommunications standardisation on a worldwide basis.

www.lightreading.com   A global Web Site for optical networking.

www.mplsforum.org   The MPLS forum.

www.mplsrc.com   The MPLS resource center, which was founded in
January of 2000 to provide a clearinghouse for
information on the IETF's Multiprotocol Label
Switching Standard.

www.ngi-supernet.org   U.S. NGI SuperNet home page.

www.antd.nist.gov/itg/nistswitch The NIST Switch home page. NIST Switch is a label
switching development package that runs on Linux
and FreeBSD.

www.nlanr.net   National Laboratory for Applied Network Research
(NLANR). NLANR has as its primary goal to provide
technical, engineering, and traffic analysis support
of NSF High Performance Connections sites and
HPNSP (high-performance network service providers) such as the NSF/MCI very high performance
Backbone Network Service (vBNS).

www.odsi-coalition.com   Optical Domain Service Interconnect (ODSI). ODSI
is an open, informal coalition made up of more than
100 service providers and equipment vendors.

www.oiforum.com   Optical Internetworking Forum (OIF). OIF promotes
the development and deployment of interoperable
products and services for data switching and routing using optical networking technologies.

www.omg.org   Object Management Group web site.

www.opengroup.org   Open Group web site. Portal to the world of DCE.

www.optical-networks.com   The optical Network Magazine copublished by
SPIE - The International Society for Optical Engi-
neering and Kluwer Academic Publishers.

www.osa.org   Optical Society of America home page.

www.osa-jon.org   The Journal of Optical Networking published by
OSA via Internet.

www.scte.org   Society of Cable Telecommunications Engineers
(SCTE). SCTE is a non-profit professional association dedicated to advancing the careers of telecommunication professionals by providing
technical training, certification, and other pertinent
industry information.

www.sonet.com   An educational web site providing information on
SONET.

www.spie.org   International Society for Optical Engineering, the
web site for optics, photonics, and imaging.

www.t1.org	General information about Committee T1 and its Technical Subcommittees. Established in February 1984, Committee T1 develops technical standards and reports regarding interconnection and interoperability of telecommunications networks at interfaces with end-user systems, carriers, information and enhanced-service providers, and customer premises equipment.
www.tiaonline.org	Telecommunications Industry Association (TIA) home page. TIA promotes the interests of telecommunications manufacturers and service providers.
www.tmforum.org	The TeleManagement Forum (TMF) is a non-profit organisation providing network management standards to improve the management and operation of communications services.
www.vbns.net	The very high performance Backbone Network Service web site.
www.w3.org	World Wide Web Consortium.
www.wapforum.org	WAP forum, white papers and specifications.

Acronym List

A&A	Authorisation and Authentication
AAL	ATM Adaptation Layer
ABR	Available Bit Rate, Area Border Router
ACK	Acknowledge (e.g. TCP header)
ADM	Add/Drop Multiplexer
ADSL	Asymmetrical Digital Subscriber Line
AF	Assured Forwarding
AIN	Advanced Intelligent Network
ANSI	American National Standards Institute
API	Application Program Interface
APS	Automatic Protection Switching
ARP	Address Resolution Protocol
ARQ	Automatic Repeat Request
AS	Autonomous System
ASBR	AS Boundary Router
ASCII	American Standard Code for Information Interchange
ASIC	Application Specific Integrated Circuit
ASN	Abstract Syntax Notation
ASON	Automatic Switched Optical Network
ATM	Synchronous Transfer Mode
AWHG	Average Weighted Hop-distance Gain
BASE	Baseband
BE	Best Effort
Bellcore	Bell Communications Research (now Telcordia Technologies)
BER	Bit Error Rate (ratio)
BFR	Best-Fit Rate routing
B-ISDN	Broadband ISDN
BGMP	Border Gateway Multicast Protocol
BGP	Border Gateway Protocol
BLSR	Bi-directional Line Switched Ring
BML	Business Management Layer
bps	bits per second
Bps	Bytes per second
BPSR	Bi-directional Path Switched Ring
BROAD	Broadband
BSD	Berkeley Software Distribution
BT	British Telecom

CaSMIM	Connection and Service Management Information Modelling
CBT	Core-Based Tree
CBR	Constant Bit Rate
CCITT	International Telegraph and Telephone Consultative Committee, now International Telecommunications Union Telecommunication Standardisation Sector (see also ITU-T)
CDMA	Code Division Multiple Access
CE	Customer Edge
CI	Client Interface
CIDR	Classless Interdomain Routing
CLEC	Competitive Local Exchange Carrier
CLI	Command Line Interface
CLNP	Connectionless Network Protocol
CMIP	Common Management Information Protocol
COF	CORBA Common Object Facilities
COPS	Common Open Policy Service
CORBA	Common Object Request Broker Architecture
COS	CORBA Common Object Services
CPE	Customer Premises Equipment
CPU	Central Processing Unit
CRC	Cyclic Redundancy Check
CRM	Customer Relationship Management
CR-LDP	Constraint-based Routing Label Distribution Protocol
CSCW	Computer Supported Cooperative Work
CSMA-CD	Carrier sense Multiple Access with Collision Detection
CSPF	Constraint-based Shortest Path First routing
CTP	Connection Termination Point
CWDM	Coarse Wavelength Division Multiplexing
DARPA	Defense Advanced Research Projects Agency
DB	Decibel
DBR	Distributed Bragg Reflector
DCC	Data Communication Channel
DCN	Data Communication Network
DD	OSPF Database Description message
DEMUX	Demultiplexer
DFB	Distributed FeedBack laser
DHCP	Dynamic Host Configuration Protocol
diffserv	Differentiated Service
DLC	Digital Loop Carrier
DLCI	Data Link Connection Identifier
DM	Domain Manager
DNS	Domain Name System
DS0	Digital Signal Level 0 (64 Kbps)
DS1	Digital Signal Level 1 (1.544 Mbps)
DS1C	Digital Signal Level 1C (3.152 Mbps)

DS2	Digital Signal Level 2 (6.312 Mbps)
DS3	Digital Signal Level 3 (44.736 Mbps)
DS3C	Digital Signal Level 3C (91.053 Mbps)
DSCP	Differentiated Services Code Point
DSL	Digital Subscriber Line
DSLAM	Digital Subscriber Line Access Multiplexer
DSP	Digital Signal Processing
DSS	Digital Signal Standard
DWDM	Defense Wavelength Division Multiplexing
DOD	Department of Defense
DCX	Digital Cross Connect
DVMRP	Distance-Vector Multicast Routing Protocol
E1	ITU-T digital signal level 1 (2.048 Mbps)
E2	ITU-T digital signal level 1 (8.448 Mbps)
E3	ITU-T digital signal level 1 (34.368 Mbps)
E4	ITU-T digital signal level 1 (139.264 Mbps)
EBGP	Exterior Border Gateway Protocol
ECMP	Equal Cost Multiple Path
EDFA	Erbium-Doped Fibre Amplifier
EF	Expedited Forwarding
EGP	Exterior Gateway Protocol
ELAN	Emulated Local Area Network
ELED	Edge Light Emitting Diode
EML	Element Management Layer
EMS	Element Management System
EN	Event Notification message
ENDEC	Encoder/Decoder
ENNI	External Network-to-Network Interface
ERO	Explicit Route Object
ETReq	Explicit route Trail Request message
FBM	Fractional Brownian Motion
FCC	Federal Communications Commission
FDDI	Fibre Distributed Data Interface
FDM	Frequency Division Multiplexing
FEC	Forward Error Correction, Forwarding Equivalence Class
FIB	Forwarding Information Base
FIFC	First In First Out
FIN	Finish Flag (TCP header)
FQ	Fair Queuing
FlowSpec	Flow Specification
FSM	Finite State Machine
FTP	File Transfer Protocol
FTTC	Fibre to the Curb

FTTH	Fibre to the Home
FTTN	Fibre to the Neighbourhood
FWM	Four Wave Mixing

GbE	Gigabit Ethernet
Gbps	Gigabits per second
GMPLS	Generalised Multiprotocol Label Switching
GRE	Generic Route Encapsulation
GSMP	General Switch Control Protocol
GUI	Graphical User Interface

HDLC	High level Data Link Control
HDTV	High Definition Television
HTML	Hypertext Markup Language
HTTP	Hypertext Transfer Protocol

IAB	Internet Architecture Board
IANA	Internet Assigned Numbers Authority
IBGP	Interior Border Gateway Protocol
ICMP	Internet Control Message Protocol
ID	Identifier
IDL	Interface Definition Language
IDMR	Inter-Domain Multicast Routing
IDRP	Inter-Domain Routing Protocol
IEEE	Institute of Electrical and Electronics Engineers
IEEE 802.3	IEEE Ethernet Standard
IEEE 802.5	IEEE Token Ring Standard
IEEE 802.11	IEEE Wireless Network Standard
IETF	Internet Engineering Task Force
IGMP	Internet Group Management Protocol
IGP	Interior Gateway Protocol
ILD	Injection Laser Diode
ILP	Integer Linear Program
INNI	Internal Network-to-Network Interface
intserv	Integrated Service
IP	Internet Protocol
IPC	Interprocess Communication
IPng	IP next generation
IPSec	IP Security
IPv4	Internet Protocol, version 4
IPv6	Internet Protocol, version 6
IR	Interface Repository
IRTF	Internet Research Task Force
ISA	IETF Integrated Services Architecture
ISDN	Integrated Service Digital Network
IS-IS	Intermediate System to Intermediate System routing protocol

ISO	International Organisation for Standardisation
ISP	Internet Service Provider
ITU	International Telecommunication Union
ITU-T	International Telecommunications Union Telecommunication Standardisation Sector
JIT	Just In Time
Kbps	Kilobits per second
KSP	K-Shortest Path
LAN	Local Area Network
LANE	LAN Emulation
LATA	Local Access and Transport Area
LBS	Label-Based Switching
LC	Link Connection
LDAP	Lightweight Directory Access Protocol
LDP	Label Distribution Protocol
LEC	Local Exchange Carrier
LED	Light Emitting Diode
LIB	Label Information Base
LinkTP	Link Termination Point
LLR	Least Loaded Routing
LMP	Link Management Protocol
LOF	Loss of Frame
LOS	Loss of Signal
LSA	Link State Advertisement
LSP	Label Switched Path
LSR	Label Switch Router, Link State Request
LSAck	Link State Acknowledgement
LSU	Link State Update
LTE	Line Terminating Equipment
MAC	Media Access Control
MACA	Multiple Access with Collision Avoidance
MAN	Metropolitan Area Network
Mbone	the experimental Multicast Backbone over the Internet
Mbps	Megabits per second
MEMS	Micro Electro-Mechanical Systems
MIB	Management Information Base
MILP	Mixed Integer Linear Program
MIME	Multipurpose Internet Mail Extensions
MONET	Multiwavelength Optical Networking Consortium
MOSPF	Multicast Open Path Shortest First
MPλS	Multiprotocol Lambda Switching
MPLS	Multiprotocol Label Switching

MSN   Manhattan Street Network
MSS   Maximum Segment Size
MTNM   TMF Multi-Technology Network Management
MTU   Maximum Transmission Unit
MUX   Multiplexer

NAT   Network Address Translation
NBMA   NonBroadcast Multiple Access
NC&M   Network Control and Management
NDP   Neighbour Discovery Protocol
NE   Network Element
NEL   Network Element Layer
NetBIOS   Network Basic Input Output System
NFS   Network File System
NGI   Next Generation Internet
NIST   National Institute for Standards and Technology
NHRP   Next Hop Resolution Protocol
NIC   Network Interface Card
NMF   Network Management Forum
NML   Network Management Layer
NMS   Network Management System
NNI   Network to Network Interface
NPS   Network Process Status
NRZ   Non-Return to Zero
NRZI   Non-Return to Zero Inverted
NSAP   Network Service Access Point
NSF   National Science Foundation
NTG   Network Throughput Gain
NTON   National Transparent Optical Network
NTT   Nippon Telephone and Telegraph
NVS   Network Visualisation Service

OADM   Optical Add/Drop Multiplexer
OAM   Operations and Maintenance
OAM&P   Operations, Administration, Maintenance, and Provisioning
OBS   Optical Burst Switching
OC-1   Optical Carrier Level 1 (51.84 Mbps)
OC-3   Optical Carrier Level 3 (155.52 Mbps)
OC-12   Optical Carrier Level 12 (622.08 Mbps)
OC-48   Optical Carrier Level 48 (2488.32 Mbps)
OC-192   Optical Carrier Level 192 (9953.28 Mbps)
OCH   Optical Channel
ODSI   Optical Domain Service Interconnect (ODSI) industry coalition
O-E-O   Optical-Electrical-Optical
OIF   Optical Internetworking Forum
OLS   Optical Label Switching

OLSR	Optical Label Switching Router
OMG	Object Management Group
OMP	Optimised Multi Path
OMS	Optical Multiplex Section
OPR	Optical Packet Router
ORB	Object Request Broker
OSA	Optical Society of America
OSCP	Optical Switch Control Protocol
OSF	Open Software Foundation
OSI	Open Systems Interconnection
OSPF	Open Shortest Path First protocol
OSPF-OMP	OSPF Optimised Multi Path
OSS	Operations Support System
OTN	ITU-T Optical Transport Network
OTS	Optical Transmission Section
OVPN	Optical Virtual Private Network
OXC	Optical Cross Connect
PDFA	Praseodymium-Doped Fibre Amplifier
PDU	Protocol Data Unit
PE	Provider Edge
PHB	Per Hop Behaviour
PHY	Physical layer
PIM	Protocol Independent Multicast
PIM-DM	Protocol Independent Multicast Dense Mode
PIM-SM	Protocol Independent Multicast Sparse Mode
PNNI	Private Network-to-Network Interface
PON	Passive Optical Network
PoP	Point of Presence
POTS	Plain Old Telephone Service
POSIX	Portable Operating System Interface
PPP	Point to Point Protocol
PSTN	Public Switched Telephone Network
PTE	Path Terminating Equipment
PVC	Permanent Virtual Circuit
QA	Q-Adaptor
QoS	Quality of Service
RARP	Reverse Address Resolution Protocol
RBOCs	Regional Bell Operating Companies
RED	Random Early Detection
RF	Radio Frequency
RFC	Request For Comment
RI	Refractive Index
RIP	Routing Information Protocol

RMI	Remote Method Invocation
RMON	Remote Network Monitor
RMP	Reliable Multicast Protocol
RMTP	Reliable multicast Transfer Protocol
RPC	Remote Procedure Call
RPF	Reverse Path Forwarding
RSpec	Resource Specification
RSVP	Resource Reservation Protocol
RTCP	Real-Time Transport Control Protocol
RTO	Retransmission Timeout
RTP	Real-Time Transport Protocol
RTT	Round Trip Time

SAN	Storage Area Network
SAP	Service Access Point
SAR	Segmentation and Reassembly
SCSI	Small Computer Systems Interface
SDH	Synchronous Digital Hierarchy
SDU	Service Data Unit
SLED	Surface Light Emitting Diode
SLA	Service Level Agreement
SLIP	Serial Line Internet Protocol
SMDS	Switched Multimegabit Data Service
SMI	Structure of Management Information
SML	Service Management Layer
SMTP	Simple Mail Transfer Protocol
SNC	SubNetwork Connection
SNMP	Simple Network Management Protocol
SNR	Signal-to-Noise Ratio
SOA	Semiconductor Optical Amplifier
SONET	Synchronous Optical Network
SPE	Synchronous Payload Envelope
SPF	Shortest Path First
SRLG	Shared Risk Link Group
SS7	Signalling System Number 7
SSL	Secure Socket Layer
STE	Section Terminating Equipment
STM	Synchronous Transfer Mode
STP	Shielded Twisted Pair
STS-n	Synchronous Transport Signal Level n
SVC	Switched Virtual Circuit

T1	A 1.544 Mbps digital circuit (DS1)
T2	A 6.312 Mbps digital circuit (DS2)
T3	A 44.736 Mbps digital circuit (DS3)
TCP	Transmission Control Protocol

TDM	Time Division Multiplexing
TDMA	Time Division Multiplexing Access
Telnet	Remote Terminal Protocol
TE	Terminal Equipment, Traffic Engineering
TECP	Traffic Engineering to Control Protocol
TED	Traffic Engineering Database
TI	physical Transport Interface
TIA	Telecommunications Industry Association
TL1	Transaction Language 1
TLV	Type-Length-Value
TMN	Telecommunications Management Network
TOS	Type of Service
TReq	Trail Request message
TResp	Trail Response message
TSpec	Traffic Specification
TTL	Time To Live
TTP	Trail Termination Point

UBR	Unspecified Bit Rate
UDP	User Datagram Protocol
ULSR	Unidirectional Line Switched Ring
UML	Universal Modelling Language
UNI	User to Network Interface
UNI-C	User Network Interface - Client side (signaling functionality)
UNI-N	User Network Interface - Network side (signaling functionality)
UPSR	Unidirectional Path Switched Ring
URL	Universal Resource Locator
UTP	Unshielded Twisted Pair

VBR	Variable Bit Rate
VC	Virtual Channel
VCC	Virtual Channel Connection
VCI	Virtual Channel Identifier
VLAN	Virtual Local Area Network
VLSI	Very Large Scale Integration
VLSM	Variable Length Subnet Mask
VP	Virtual Path
VPC	Virtual Path Connection
VPI	Virtual Path Identifier
VPN	Virtual Private Network
VoIP	Voice over IP
VT	Virtual Tributary

WADM	Wavelength Add/Drop Multiplexer
WAMP	Wavelength Amplifier
WAN	Wide Area Network

WDM	Wavelength Division Multiplexing
WEFC	Wavelength Equivalent Forwarding Class
WFQ	Weighted Fair Queuing
WSXC	Wavelength Selective Cross Connect
WWW	World Wide Web
X.25	ITU-T packet switching protocol standard
X.400	ITU-T email standard
X.500	ITU-T directory service standard
XDM	Cross Domain Manager
XDR	External Data Representation

Index